Allele Mining for Genomic Designing of Grain Legume Crops

This book deliberates on the concept, strategies, tools, and techniques of allele mining in grain legume crops and its application potential in genome elucidation and improvement, including studying allele evolution, discovery of superior alleles, discerning new haplotypes, assessment of intra- and interspecific similarity, and also studies of gene expression and gene prediction. Available gene pools in global germplasm collections specifically consisting of wild allied species and local landraces for almost all major crops have facilitated allele mining. Development of advanced genomic techniques, including PCR-based allele priming and EcoTILLING-based allele mining, is being widely used now for mining superior alleles. Allele's discovery has become more relevant now for employing molecular breeding to develop designed crop varieties matching consumer needs, and also with genome plasticity to adapt to climate change scenarios. All these concepts and strategies, along with precise success stories, are presented in the chapters dedicated to the major grain legume crops.

1. The first book on the novel strategy of allele mining in grain legume crops for precise breeding.
2. Presents genomic strategies for mining superior alleles underlying agronomic traits from genomic resources.
3. Depicts case studies of PCR-based allele priming and EcoTILLING-based allele mining.
4. Elaborates on gene discovery and gene prediction in major grain legume crops.

This book will be useful to students and faculties in various plant science disciplines, including genetics, genomics, molecular breeding, agronomy, and bioinformatics; to scientists in seed industries; and also to policymakers and funding agencies interested in crop improvement.

Nextgen Agriculture: Novel Concepts and Innovative Strategies
Chittaranjan Kole
Series Editor

Allele Mining for Genomic Designing of Fruit Crops
Chittaranjan Kole, Kenta Shirasawa, and Anil Kumar Singh

Allele Mining for Genomic Designing of Cereal Crops
Chittaranjan Kole, Sharat Kumar Pradhan, and Vijay K Tiwari

Allele Mining for Genomic Designing of Vegetable Crops
Chittaranjan Kole, Tusar Kanti Behera, and Prashant Kaushik

Allele Mining for Genomic Designing of Oilseed Crops
Chittaranjan Kole, Manish Kumar Pandey, and Naveen Puppala

Sustainable Urban Agriculture: New Frontiers
Kheir Al-Kodmany, Madhav Govind, Sharmin Khan, and Chittaranjan Kole

Allele Mining for Genomic Designing of Grain Legume Crops
Chittaranjan Kole, C. Bharadwaj, and Abhimanyu Sarkar

For more information, please visit our website: Nextgen Agriculture: Novel Concepts and Innovative Strategies – Book Series – Routledge & CRC Press

Allele Mining for Genomic Designing of Grain Legume Crops

Edited by
Chittaranjan Kole, C. Bharadwaj, and
Abhimanyu Sarkar

CRC Press
Taylor & Francis Group
Boca Raton London New York

CRC Press is an imprint of the
Taylor & Francis Group, an **informa** business

First edition published 2025
by CRC Press
2385 NW Executive Center Drive, Suite 320, Boca Raton FL 33431

and by CRC Press
4 Park Square, Milton Park, Abingdon, Oxon, OX14 4RN

CRC Press is an imprint of Taylor & Francis Group, LLC

Library of Congress Cataloging-in-Publication Data
Names: Kole, Chittaranjan, editor. | Bharadwaj, C. (Chellapilla), editor. | Sarkar, Abhimanyu, editor.
Title: Allele mining for genomic designing of grain legume crops / edited by Chittaranjan Kole, C. Bharadwaj, and Abhimanyu Sarkar
Description: First edition | Boca Raton, FL : CRC Press, 2024 | Includes bibliographical references
Identifiers: LCCN 2024022038 (print) | LCCN 2024022039 (ebook) | ISBN 9781032471983 (hardback) | ISBN 9781032472058 (paperback) | ISBN 9781003385059 (ebook)
Subjects: LCSH: Legumes—Genetic engineering. | Plant genomes.
Classification: LCC SB317.L43 A45 2024 (print) | LCC SB317.L43 (ebook) | DDC 633.3—dc23/eng/20240617
LC record available at https://lccn.loc.gov/2024022038
LC ebook record available at https://lccn.loc.gov/2024022039

ISBN: 978-1-032-47198-3 (hbk)
ISBN: 978-1-032-47205-8 (pbk)
ISBN: 978-1-003-38505-9 (ebk)

DOI: 10.1201/9781003385059

Typeset in Times
by Apex CoVantage, LLC

Dedicated to

Dr. Norman Ernest Borlaug

The Father of Green Revolution, the Nobel Peace Prize Winner, and the Founder of the World Food Prize

Who inspired my research and academic works with his generous appreciation and advice.

Series Preface

Agriculture is going through a phase of transformation now. It has both challenges and opportunities. One of the major challenges it is facing is obviously food security. A population of about 9.3 billion projected by 2050 necessitates an increase of crop production by 60–70%. Another 10–15% increase would be needed to counter the reduction of crop production due to global warming and climate change. Nutrition security is another challenge with more than 800 million people suffering from malnutrition. Depletion of fossil fuel and increasing dependence on biofuel from crop residues is another challenge. Loss of biodiversity and land degradation also add to the list of challenges. The cultivable area is also decreasing fast due to housing, transport, and industrialization. Above all, climate change and global warming have a huge impact on agriculture, health, and environment.

Several novel concepts and strategies are emerging to address the above challenges. For example, the practice of precision farming will lead to precise and timely use of inputs resulting in higher yield and better quality of produce on the one hand and reduction of loss of inputs on the other. Similarly, urban agriculture and vertical agriculture will add a lot of cultivable areas in the future. Protected farming will facilitate growing crops under controlled conditions leading to reduction in biotic and abiotic stresses and precise utilization of inputs.

Agriculture requires maintenance of the philosophy farmers have been following since inception over thousands of years which are being sold today in new packages with sophisticatedly coined labels such as traditional agriculture, natural agriculture, sustainable agriculture, conservation agriculture, integrated agriculture, regenerative agriculture, and organic agriculture.

In the changing scenarios, next-generation agriculture will need designed crop varieties and breeds produced through designed practices. This book series aims at deliberations on several novel concepts and innovative strategies developed for next-generation agriculture and will hopefully benefit agricultural education, research, and outreach.

Chittaranjan Kole
Kolkata
January 10, 2024

Preface

The affordability, precision, and high-throughput nature of sequencing technologies in the last two decades have provided an unprecedented understanding of the genomes and target plant traits of economic importance. Further, the advances in third-generation sequencing technologies, including Hi-C, especially in the last decade, have made it possible for developing high-quality reference genomes, pangenomes, and gene annotations, including genome-wide structural variations. These developments have also increased the precision in gene and superior haplotype discovery, followed by mining allelic diversity among breeding lines and gene bank collections. High-throughput and cost-effective genotyping platforms have further boosted the adoption rate of genomic tools in the ongoing breeding programs across the world. Deployment of these advances in crops and making them well-equipped to scale up the integration of genomic interventions in the breeding programs will facilitate development of future-ready crops.

The aforementioned advances have made gene/allele mining more precise and cost-effective for the target traits preferred by farmers, consumers, and industry, as well as efforts toward mitigation of climate change impacts and achieving nutritional security prospective. It is important to accelerate the process of genomic designing to develop future-ready crops by bringing new allelic combination(s) at one or more loci underlying desired genes using modern breeding approaches, such as marker-assisted breeding, haplotype-based breeding, and gene/allele pyramiding, and also now through gene editing.

These new alleles originate through natural or induced mutations also at one or more genomic regions at a time. The traditional plant breeding methods practiced in the twentieth century depended solely on phenotypes for the development of new genotypes, primarily through selection and hybridization. The advent of DNA-based molecular markers in the 1980s facilitated indirect identification of the alleles by using linked molecular markers using genetic linkage mapping. Molecular breeding as a supplement to traditional breeding has successfully been employed in almost all major crops for their genetic improvement in a short time. It is worthwhile to mention that transgenic breeding has also contributed significantly to crop improvement at least for some major input target traits, such as herbicide tolerance, insect resistance, etc.; however, this technology is still facing challenges in societal adoption in different parts of the world, especially for food crops. Nevertheless, gene editing has been receiving positive vibes from the policymakers of different countries making a conducive environment for accelerating the use of this technology on a large scale.

Despite the huge loss of diversity during domestication, several national and international gene banks have millions of diverse germplasms to explore variability for the traits that are needed to breed climate-resilient and future-ready crops. Genetic diversity played a pivotal role during the entire period of conventional plant breeding, traditional and molecular-included. However, plant breeders have been using mainly the primary gene pool constituting the indigenous germplasm lines and local landraces as a genetic resource of new alleles. In the meantime, a drastic dearth of new alleles in the primary gene pool happened due to genetic erosion caused naturally or due to domestication over time, but aggravated because of the injudicious and irrational implementation of the so-called green revolution. Precisely, this highlighted the importance of allied wild crop relatives (WCRs), and thereby a mission of collection, conservation, characterization, and utilization of biodiversity. The emergence of advanced bioinformatics and the availability of cheaper and innovative genome sequencing technologies have empowered us to explore germplasm and identify a large number of donor alleles, and allele mining has become a reality!

The strategy of association mapping, the most popular trait mapping approach in animal and human sciences, has received huge adoption in plant genomics studies. The availability of large sequenced diverse panels with multilocation phenotyping data has provided much-needed insights into trait genomics through genome-wide association studies (GWAS), leading to the discovery of genes and superior haplotypes in many crops. This has led to the possibility of mining a large number of genes in the available accessions in crops. Mutations occur in the genome as single-nucleotide polymorphisms (SNPs) or insertion/deletions (InDels), resulting in the generation of new alleles or changes in existing alleles and allele combinations. Using exactly this principle, the approach of EcoTILLING has facilitated large-scale allele mining in several major crops. On the other hand, sequence-based allele mining has led to the discovery of allelic diversity both at the target gene and the whole genome. Recent advances in modern genomic tools and technologies, such as genome engineering, speed breeding, etc., facilitate the deployment of these new alleles in breeding programs in a relatively short time to accelerate the rate of genetic gains. Truly speaking, allele mining has revolutionized the strategy of utilizing genetic as well as genomic resources and become an important component of next-generation precision breeding and biotechnology!

Allele discovery has become more relevant now for employing molecular breeding to develop designed crop varieties matching consumer needs and with genome plasticity to adapt the climate change scenarios using targeted and precision breeding tools validated by both genomics and phenomics studies. All these concepts and strategies, along with precise success stories, are presented in over 50 chapters allocated under five volumes dedicated to the five crop groups, including cereals, oilseeds, grain legumes, fruits, and vegetables, under this book series.

This volume, titled *Allele Mining for Genomic Designing of Grain Legume Crops*, includes ten chapters dedicated to common bean, chickpea, cowpea, pea, lentil, mung bean, and lathyrus contributed by 64 scientists from 11 countries, including Australia, Bangladesh, China, Columbia, Ghana, India, Lebanon, Mexico, UAE, UK, and the USA. We express our thanks for their excellent contributions and timely cooperation.

This book will hopefully be useful to students, research scholars, and teaching faculties in the academia, to learn about fundamentals and applications of allele mining; to scientists practicing plant breeding in public and private institutes, in improvising their programs; and also, to policy-makers and funding agencies, in prioritizing research fields.

Contents

Abbreviations

AB	Ascochyta blight
AB-QTL	advanced backcross QTL
ADP	Andean diversity panel
AFLP	amplified fragment length polymorphism
AICRP	All India Coordinated Research Project
ANOVA	analysis of variance
ARC	Agricultural Research Council
AREC	agricultural research and extension centers
ASAP	allele-specific associated primers
AS-PCR	allele-specific PCR
AvgD	average root diameter
AVRDC	Asian Vegetable Research and Development Center
BC	backcross
BGM	Botrytis gray mold
BOAA	b-N-oxalyl-amino-L-alanine
BSA	bulked segregant analysis
BSH	broad-sense heritability
BSR-Seq	BSA RNA-Seq
BWA	Burrows–Wheeler aligner
CAP	coordinated agricultural program
CAZRI	Central Arid Zone Research Institute
CBB	common bacterial blight
CDDP	conserved DNA-derived polymorphism
CGKB	Cowpea Gene Space/Genomics Knowledge Base
CIAT	International Centre for Tropical Agriculture
CLS	*Cercospora* leaf spot
cM	centimorgan
CMS	cytoplasmic male sterility
CMV/CPMV	cowpea mosaic virus
COBB	cowpea bacterial blight
CR	charcoal rot
CTDB	Chickpea Transcriptome Database
CWRs	cultivated and wild relatives
CWRs	crop wild relatives
DArT	Diversity Arrays Technology
DAS	days after sowing
ddRAD	double-digest restriction enzyme–associated DNA
DdRADseq	double-digest restriction site–associated DNA sequencing
DEG	differentially expressed gene
DET	determinate
DH	doubled haploid
DHN	dehydrin
DRR	dry root rot
EC	electrical conductivity
EcoTILLING	Ecotype-TILLING
EMS	ethyl methanesulfonate
EST	expressed sequence tag

F2	second filial generation
F3	third filial generation
F4	fourth filial generation
F5	fifth filial generation
F6	sixth filial generation
FAO	Food and Agriculture Organization
FAOSTAT	FAO Corporate Statistical Database
FIGS	focused identification of germplasm strategy
FW	fusarium wilt
GA	gibberellic acid
GABC	genomic-assisted backcrossing
GBS	genotyping-by-sequencing
GE	genome editing
GE	genetic engineering
GEA	genome environment associations
GEBV	genomic estimated breeding value
GI	glycemic index
GMM	genic molecular markers
GMS	genic male sterility
GP	gene pool
GP	genomic prediction
GP1	primary gene pool
GP2	secondary gene pool
GP3	tertiary gene pool
GRC	genetic resources center
GS	genomic selection
GSS	gene space sequences
GWAS	genome-wide association study/studies
GWS	genome-wide selection
GWSS	genome-wide selection scans
h2	broad-sense heritability
HI	hydrolysis indices
HIS	heat stress index
HIT	heat index Thornthwaite
HPLC	high-performance liquid chromatography
HRM	high-resolution melting
HT	high-throughput
ICAR	Indian Council of Agricultural Research
ICARDA	International Center for Agricultural Research in the Dry Areas
ICIM	inclusive composite interval mapping
ICP	inductively coupled plasma
ICRISAT	International Crops Research Institute for the Semi-Arid Tropics
IDC	iron deficiency chlorosis
IITA	International Institute of Tropical Agriculture
InDels	insertions/deletions
INRA	Institut National de la Recherche Agronomique
IP	inorganic P
ISSR	inter simple sequence repeat
KASP	kompetitive allele-specific PCR
KTr	transpiration rate decline
LATH	lathyroides

LD	linkage disequilibrium
LEA	late embryogenesis abundant
LF	late flowering
LG	linkage group
Lip	light-independent photomorphogenesis
LIS	Legume Information System
MAB	marker-assisted breeding
MABB	marker-assisted backcross breeding
MABC	marker-assisted backcrossing
MAF	minor allele frequency
MAGIC	multiparent advanced generation intercross
MARS	marker-assisted recurrent selection
MAS	marker-assisted selection
MDP	Mesoamerican diversity panel
MLMx	mixed linear model
MLO	mildew locus O
MTAs	marker–trait association
MULLaRP	mung bean, urdbean, lentil, lathyrus, rajma, pea
MYMIV	mung bean yellow mosaic India virus
MYMV	mung bean yellow mosaic virus
NAM	nested association mapping
NARI	National Agricultural Research Institute
NBPGR	National Bureau of Plant Genetic Resources
NCGRB	National Center for Genetic Resources and Biotechnology
NGS	next-generation sequencing
NILs	near-isogenic lines
NORK	nodulation receptor kinase
NRC	number of root crosses
NRF	number of root forks
NRL	number of links
NRT	number of root tips
PA	phytic acid
PAGE	polyacrylamide gel electrophoresis
PAP	phytate P
PAV	presence absence variation
PCA	principal component analysis
PCR	polymerase chain reaction
PCS	principal component score
PGRFA	plant genetic resources for food and agriculture
PM	powdery mildew
PMD	Pea Marker Database
PPVFRA	Protection of Plant Varieties and Farmers' Rights Authority
PUFA	polyunsaturated fatty acid
PUpE	phosphorus uptake efficiency
PUtE	phosphorus utilization efficiency
QTL	quantitative trait locus
QTLs	quantitative trait loci
QTN	quantitative trait nucleotide
RAD	restriction site–associated DNA
RAPD	random amplified polymorphic DNA
Rchit	rice chitinase gene

RDA	Rural Development Administration
RDS	readily digestible starch
RFLP	restriction fragment length polymorphism
RIL	recombinant inbred line
RNAi	RNA interference
RS	resistant starch
RSA	root surface area
RV	root volume
SBE	starch-branching enzyme
SCADA	Supervisory Control and Data Acquisition
SCAR	sequence-characterized amplified region
SCN	soybean cyst nematode
SCoT	start codon–targeted polymorphism
SD	seed
SDP	Spanish diversity panel
SDS	slowly digestible starch
SIM	seed-induced mutant
SL	seedling
SNP	single-nucleotide polymorphism
SOLiD	supported oligonucleotide ligation and detection system
SRA	sequence read archives
SRAP	sequence-related amplification polymorphism
SSA	sub-Saharan Africa
SSAP	sequence-specific amplification polymorphism
SSD	single seed descent
SSH	subtractive suppressive hybridization
SSR	simple sequence repeat
STMS	sequence-tagged microsatellite sites
STS	sequence-tagged site
SVR	SNP for *V. radiata*
TALEN	transcription activator–like effector nuclease
TBS	TILLING-by-sequencing
TDF	transcript-derived fragment
TF	transcription factor
TFL	terminal flower
TILLING	targeting-induced local lesions in genomes
TP	total P
TRL	total root length
TSU	Tennessee State University
UCR	University of California–Riverside
UDS	University of Development Studies
UofA	University of Arkansas
USDA	United States Department of Agriculture
UTR	untranslated region
YMD	yellow mosaic disease
YSL	yellow stripe1-like
β-ODAP	beta-N-oxalyl-L-alpha, beta-diaminopropionic acid
Θcri	critical soil water content

Editor Biography

Prof. Chittaranjan Kole is an internationally reputed scientist with an illustrious professional career spanning over 40 years and globally recognized for his original contributions in the fields of plant genomics, biotechnology, and molecular breeding, leading to the publication of more than 150 quality research articles. He has edited over 170 books for the leading publishers of the world, including Springer, Wiley-Blackwell, and Taylor & Francis Group. His scientific contributions and editing acumen have been appreciated by seven Nobel laureates, including Profs. Norman Borlaug, Arthur Kornberg, Werner Arber, Phillip Sharp, Günter Blobel, Lee Hartwell, and Roger Kornberg. He has been honored with a number of fellowships, honorary fellowships, and national and international awards, including the Outstanding Crop Scientist Award conferred by the International Crop Science Society. Prof. Kole has served in all prestigious positions in academia, including as Vice-Chancellor of the BC Agricultural University and Project Coordinator of Indo-Russian Center of Biotechnology in India, and Director of Research of the Institute of Nutraceutical Research of Clemson University in the USA. He worked also in the Pennsylvania State University and Clemson University as Visiting Professor in the USA. He was awarded with the Raja Ramanna Fellowship by the Department of Energy, Government of India. He is also heading the International Climate-Resilient Crop Genomics Consortium, International Phytomedomics and Nutriomics Consortium, and Genome India International as Founding President. Prof. Kole has recently established the Prof. Chittaranjan Kole Foundation for Science and Society and is working for his founding chairman.

Dr. C. Bharadwaj is a principal scientist, Division of Genetics, ICAR-IARI, and chickpea breeder with 27 years of scientific experience in breeding legumes and pulses. He is also the honorary adjunct associate professor, the UWA Institute of Agriculture, University of Western Australia, Perth. Focusing on marginal and sub-marginal farmers, he has developed 18 high-yielding, climate-resilient chickpea varieties and is credited with developing the first marker-assisted bred chickpea variety for drought (Pusa Chickpea 10216) and for wilt (Pusa Chickpea Manav). He has published 180 papers in very high-impact journals, like *Nature* and *Nature Genetics*, apart from various book chapters, books, and technical reports. He was awarded with four international and five national awards, including the prestigious Springer Vegetos Dr. M. S. Swaminathan Award 2023, Best Scientist Award of ICAFS-2K23 in 2023, Agrivision Eminent Scientist Award 2022, and Agri Watch Research Recognition Award 2009. He has guided ten doctoral and nine postgraduate students as Chairperson. He is Fellow of the Indian Society of Pulses Research and Development, Kanpur, India; Fellow, Springer Nature–Vegetos; and Fellow of the Indian Society Genetics and Plant Breeding, New Delhi, India. He is a member of the Transgenic Evaluation Committee of DBT and an expert member in DBT for project evaluation. He is the core advisory group member for varietal descriptors development for chickpea, lentil, and cowpea for Bioversity International. He is a member of the editorial board for *Frontiers in Genetics*, *Frontiers in Plant Sciences*, *Indian Journal of Genetics and Plant Breeding*, *Legume Research*, and *Current Agriculture*.

Dr. Abhimanyu Sarkar is a molecular biologist with over 30 years of experience. He is Group Leader, Legume Genetics, at NIAB, Cambridge, UK. His areas of expertise include legume biology, speed breeding, gene discovery, and genomics. Before joining NIAB, he worked at the John Innes Centre, Norwich, UK, in Prof. Cathie Martin's group on the biology and genomics of the orphan legume grass pea (*Lathyrus sativus*). The group was responsible for sequencing the genome of grass pea, as well as developing various biological resources, such as mutagenized populations, recombinant inbred lines (RILs), and an EcoTILLING platform for grass pea. He had previously worked in New Zealand on white clover (*Trifolium repens*), a forage legume. He has also worked extensively on the molecular biology of various invertebrate organisms, including developing tools and methodologies for insect transgenesis using DNA transposable elements. He obtained his PhD in molecular and cell biology from the University of Maryland, College Park, USA; a master's degree in genetics from the Indian Agricultural Research Institute, Pusa, New Delhi, India; and a bachelor's degree in agriculture from the Govind Ballabh Pant University of Agriculture and Technology, Pantnagar, India.

Contributors

Daniel D. Ambachew
Department of Agricultural and Environmental Sciences
Tennessee State University
Nashville, TN, USA

Abhishek Anand
The Neotia University
West Bengal, India

Muralidhar Aski
Division of Genetics
Indian Institute of Agricultural Research
New Delhi, India

Ruchi Bansal
Division of Plant Physiology
Indian Institute of Agricultural Research
New Delhi, India

P. S. Basavaraj
School of Water Stress Management
ICAR-National Institute of Abiotic Stress Management
Baramati, Maharashtra, India

Chellapilla Bharadwaj
Principal Scientist
ICAR-Indian Agricultural Research Institute (IARI)
Pusa, Delhi, India

H. A. Bhargavi
Division of Crop Improvement
ICAR-Indian Grassland and Fodder Research Institute
Jhansi, Uttar Pradesh, India

Somnath Bhattacharyya
Professor and Officer in-Charge, Crop Research Unit, Genetics and Plant Breeding
Bidhan Chandra Krishi Viswavidyalaya
Mohanpur, Nadia, West Bengal, India

Matthew W. Blair
Department of Agricultural and Environmental Sciences
Tennessee State University
Nashville, TN, USA

Andrés J. Cortés
Corporación Colombiana de Investigación Agropecuaria – AGROSAVIA, CI La Selva
Las Palmas, Rionegro, Colombia
Departamento de Ciencias Forestales
Universidad Nacional de Colombia
Sede Medellín, Medellín, Colombia

Arpita Das
Department of Genetics and Plant Breeding
Bidhan Chandra Krishi Viswavidyalaya
Mohanpur, Nadia, WB, India

H. K. Dikshit
Division of Genetics
ICAR-Indian Agricultural Research Institute
New Delhi, India

G. P. Dixit
ICAR-Indian Institute of Pulses Research
Kanpur, Uttar Pradesh, India

Gayacharan
ICAR-National Bureau of Plant Genetic Resources
New Delhi, India

Padmavati G. Gore
Division of Germplasm Conservation
ICAR-National Bureau of Plant Genetic Resources
New Delhi, India

Debjyoti Sen Gupta
Division of Crop Improvement
ICAR-Indian Institute of Pulses Research
Kalyanpur, Kanpur, India

Alladin Hamwieh
International Center for Agricultural Research in the Dry Areas (ICARDA)
Beirut, Lebanon

B. S. Harshitha
Division of Genetics
ICAR-Indian Agricultural Research Institute
New Delhi, India

David Hickok
Department of Agricultural and Environmental Sciences
Tennessee State University
Nashville, TN, USA
Current Address: Cornell University, Ithaca, NY, USA

Arshad Husain
Division of Crop Improvement
ICAR-Indian Institute of Pulses Research
Kalyanpur, Kanpur, India

Sudhir Kumar Jha
ICAR-Indian Institute of Pulses Research
Kanpur, Uttar Pradesh, India

Uday Chand Jha
ICAR-Indian Institute of Pulse Research
Kanpur, UP, India

Nilesh Joshi
ICAR-Indian Agricultural Research Institute (IARI)
Pusa, Delhi, India

M. C. Keerthi
Division of Crop Protection
ICAR-Indian Institute of Horticultural Research
Bengaluru, Karnataka, India

Chittaranjan Kole
Chairman
Prof. Chittaranjan Kole Foundation for Science and Society
Kolkata, India
Chief Advisor
The Neotia University
West Bengal, India

Jitendra Kumar
Division of Crop Improvement
ICAR-Indian Institute of Pulses Research
Kalyanpur, Kanpur, India

Karthik Kumar M.
Division of Genetics
ICAR-Indian Agricultural Research Institute,
New Delhi, India

Kuldeep Kumar
ICAR-Indian Institute of Pulses Research
Kanpur, Uttar Pradesh, India

Neeraj Kumar
ICAR-Indian Agricultural Research Institute (IARI)
Pusa, Delhi, India

Sachin Kumar
Department of Genetics and Plant Breeding
Ch. Charan Singh University
Meerut, India

Yogesh Kumar
ICAR-Indian Institute of Pulse Research
Kanpur, UP, India

Neetu Singh Kushwah
Indian Institute of Pulses Research
Kanpur, India

Hui Li
Department of Agricultural and Environmental Sciences
Tennessee State University
Nashville, TN, USA

B. Manu
ICAR-Indian Institute of Pulses Research (IIPR)
Regional Research Center
Dharwad, Karnataka, India

Brijesh K. Mehta
Division of Crop Improvement
ICAR-Indian Grassland and Fodder Research Institute
Jhansi, Uttar Pradesh, India

Max C. Miller II
Department of Agricultural and Environmental Sciences
Tennessee State University
Nashville, TN, USA

Gyan Prakash Mishra
Division of Genetics
Indian Agricultural Research Institute
Delhi, India

Mousumi Murmu
Department of Genetics and Plant Breeding
Bidhan Chandra Krishi Viswavidyalaya
Mohanpur, Nadia, WB, India

Lahari Nakkalapudi
Department of Agricultural and Environmental Sciences
Tennessee State University
Nashville, TN, USA

Harsh Nayyar
Department of Botany
Panjab University
Chandigarh, India

Sarita Pandey
The Neotia University
West Bengal, India

Ashok Kumar Parihar
Crop Improvement Division
ICAR-Indian Institute of Pulses Research
Kanpur, Uttar Pradesh, India

Aditya Pratap
ICAR-Indian Institute of Pulses Research
Kanpur, Uttar Pradesh, India

Md. Mosiur Rahman
Pulses Research Center
Bangladesh Agricultural Research Institute
Gazipur, Bangladesh

K. R. Ramya
ICAR-National Bureau of Plant Genetic Resources
New Delhi, India

Abhishek Rathore
The International Maize and Wheat Improvement Center (CIMMYT)
Mexico

Sneha Priya Pappula Reddy
The UWA Institute of Agriculture
The University of Western Australia
Perth, Australia

Manish Roorkiwal
Khalifa Center for Genetic Engineering and Biotechnology (KCGEB)
UAE University
UAE

Soumyayan Roy
Department of Genetics and Plant Breeding
Bidhan Chandra Krishi Viswavidyalaya
Mohanpur, Nadia, WB, India

Ravi Prakash Saini
Seed Technology Division
ICAR-Indian Grassland and Fodder Research Institute
Jhansi, Uttar Pradesh, India

Abhimanyu Sarkar
National Institute of Agricultural Botany
Cambridge, UK

Sutanu Sarkar
Assistant Professor, Crop Research Unit, Genetics and Plant Breeding
Bidhan Chandra Krishi Viswavidyalaya
Mohanpur, Nadia, West Bengal, India

Kadambot H. M. Siddique
The UWA Institute of Agriculture
The University of Western Australia
Perth, Australia

Aditya Pratap Singh
Assistant Professor, Department of Plant Breeding and Genetics, School of Agriculture
GIET University
Gunupur, Rayagada, Odisha, India

Satya Taddi
Chinese University of Hong Kong
Shenzhen, China

Kuldeep Tripathi
Division of Germplasm Evaluation
ICAR-National Bureau of Plant Genetic Resources
New Delhi, India

Rajeev K. Varshney
State Agricultural Biotechnology Centre, Centre for Crop and Food Innovation
Murdoch University
Australia

Kiran Vishwakarma
Crop Improvement Division
ICAR-Indian Institute of Pulses Research
Kanpur, Uttar Pradesh, India

Xingbo Wu
Department of Agricultural and Environmental Sciences
Tennessee State University
Nashville, TN, USA
Current Address: University of Florida, Homestead, FL, USA

Sunaina Yadav
Division of Genetics
ICAR-Indian Agricultural Research Institute
New Delhi, India

Damba Yahaya
Department of Agricultural and Environmental Sciences
Tennessee State University
Nashville, TN, USA
Current Address: University of Development Studies, Tamale, Ghana

1 Allele Mining in Common Bean (*Phaseolus vulgaris* L)

Matthew W. Blair, Daniel D. Ambachew, Max C. Miller II, Hui Li, Lahari Nakkalapudi, and Andrés J. Cortés

1.1 INTRODUCTION

The genus *Phaseolus* is made up of 80 or so species, all from the New World (Gepts et al., 2008). Of these, five are domesticated crops: *Phaseolus vulgaris* (common bean), *Phaseolus acutifolius* (tepary bean), *Phaseolus lunatus* (lima or butter bean), *Phaseolus coccineous* (runner or scarlet bean), and *Phaseolus polyanthus* (year-long bean). Among these species, *P. vulgaris* is the most widely produced with large quantities worldwide, and then the more minor beans (Gepts, 2001, 2014). The wild relatives of these species are distributed entirely in the Americas, specifically along the Andes Mountains of South American, and with high diversity in Mexico and Central America. Common bean is considered the principal grain legume for direct human consumption in agri-food systems worldwide. The crop is cultivated throughout all major continental areas between 52°N and 32°S latitude and from sea level in temperate zones to elevations of more than 3,000 m in Andean South America (Omae et al., 2012). The crop is also cultivated for edible green leaves, green pods, or immature seeds (De Luque and Creamer, 2015), and in some regions, the straw is used as animal fodder (Ambachew et al., 2015). Figure 1.1 shows the five cultigens of *Phaseolus* and some variability in fresh versus dry consumption. The agronomic versus horticultural uses of these beans are possible in each species and lead to the terminologies of dry beans and snap or vegetable beans, which is the topic of the next two sections. However, we will concentrate on common bean, *P. vulgaris*, as the major species of international trade.

1.1.1 Dry Beans

The world production of common beans as a dry pulse crop (a.k.a. dry beans) is estimated to be 12 million metric tons annually, with 2.5 and 5.5 million metric tons alone in sub-Saharan Africa and Latin America plus the Caribbean region, respectively (Broughton et al., 2003).

The remaining 4 million metric tons are produced in countries like the United States, Turkey, Spain, Nepal, Italy, Iran, India, China, and Canada, where common bean is an important economic crop (Gepts et al., 2008). Export of the crop occurs from the former and latter countries with imports to all of Europe and to deficit countries in Africa, Asia, and especially Europe. The United Kingdom, for example, is a major importer of North American small white navy beans. Cuba and Mexico, despite being producer countries, are importers of black beans from North America as well. China exports a range of beans, including cream mottled and red kidney beans, to South America (Zhang et al., 2008), even if this region includes the largest-producing country in the world, Brazil (Blair et al., 2013a). Inter-regional trade of black and carioca beans from Argentina or Bolivia to Brazil is also important.

As far as the states in the USA producing dry beans, they are mainly in the northern sections of the country, from New York to Washington, passing through Michigan, Minnesota, and especially North Dakota (Figure 1.2). A parallel arc of production occurs in the lower tier of warmer Canadian

DOI: 10.1201/9781003385059-1

FIGURE 1.1 Five cultigens of *Phaseolus*.

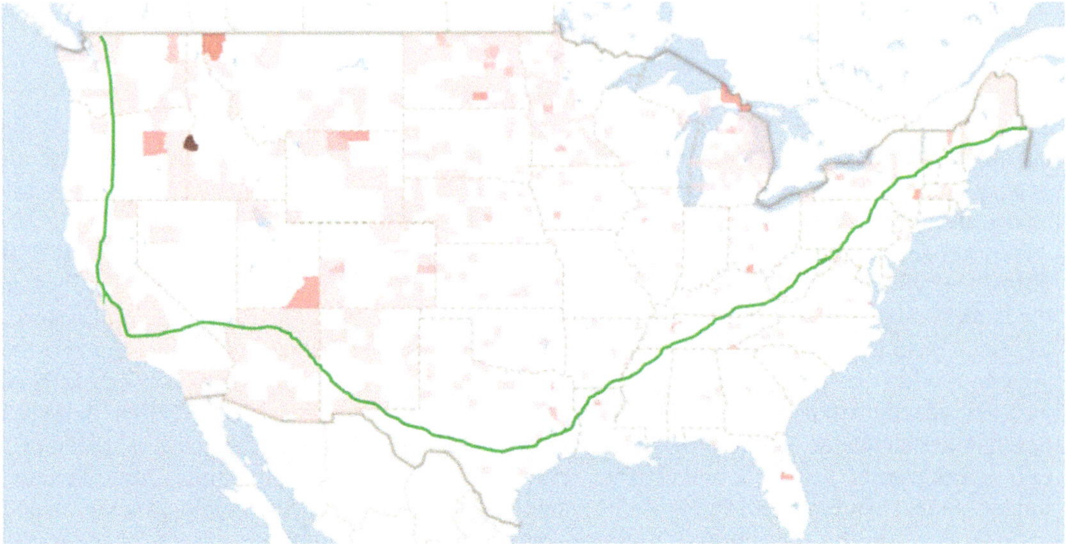

FIGURE 1.2 Regions of the United States involved in common bean production (by county), with line representing the approximate division between dry beans (north of line) and snap beans (south, east, and west of line). Some mixture of dry and snap beans occurs in the Midwest.

provinces form Ontario through Alberta to Saskatchewan, and then British Colombia. Production in Canada and the Northern states has increased, while it has decreased in Nebraska and Colorado, two areas that used to have large production. Seed production is higher in Idaho due to drier climate with fewer insect-borne viruses.

1.1.2 Snap Beans

As a horticultural product, the pods of common bean make the same species, *P. vulgaris*, a pre-eminent vegetable crop known as snap beans. The dollar value of snap beans is approximately half to equal to that of dry beans in the United States, despite lower acreage. The country grows almost a million acres of dry beans, but only a tenth of that to snap beans, but both are worth approximately 1 billion dollars a year. The location of snap bean production compared to dry bean production is quite divergent. Snap beans are a year-round crop produced during the winter in the state of Florida or imported from Mexico; it then starts to be produced in Southern Georgia in the springtime, Kentucky and Tennessee in the early summer, and Wisconsin in the late summer. This annual migration of planting acreages to snap bean is accompanied by the mechanical pickers that harvest the snap bean crop, keeping them busy throughout the year.

Other regions of snap bean production occur on the West Coast, moving from Southern California to Oregon's Willamette Valley. Given their economic importance, this chapter will emphasize snap beans as much as dry beans in the review of genome-wide association studies, even though a lot of snap bean research remains in the private sector domain, while dry bean research tends to be done by academic programs with the help of a few government-sponsored labs, namely, in the public sector.

Another focus of this review is on research that has helped make common beans more popular to consumers based on their nutritional qualities, while still being easy to produce for growers, despite the onset of climate changes, to which the crop is very sensitive. Common bean is a well-known candidate for providing easy-assimilated dietary proteins and essential micronutrients for undernourished communities, while offering secondary services as forages and soil fertilizers. However, major common bean cultivars typically lack adaptation to abiotic stresses (i.e., heat and drought), jeopardizing worldwide yields due to climate change (Cortés et al., 2020). Some biotic stresses remain difficult to control even in the developed world agriculture of the United States.

Here, we will emphasize two diseases (common bacterial blight and charcoal rot) that are difficult to control with pesticides and that are major problems associated with climate change, which creates conducive condition for the survival of the pathogen. In Tennessee, we have undertaken field screening (Figure 1.1) for both diseases using germplasm panels for allele mining.

1.2 GENOME-WIDE ASSOCIATION STUDIES (GWAS) IN COMMON BEAN: INTRODUCTORY EXAMPLES

Genome-wide association studies (GWAS) are the evaluation of marker–trait associations (MTAs) useful for trait discovery in structured or unstructured populations. Over the last decade and a half, GWAS research has proliferated for common beans, with a focus on abiotic stresses or quality traits as well as diseases, being major biotic stresses. Diseases evaluated have covered varietal reactions to fungal pathogens, such as those for angular leaf spot (Perseguini et al., 2016), anthracnose (Zuiderveen et al., 2016), and root rot caused by *Rhizoctonia solani* (Oladzad et al., 2019b). While common bacterial blight was well studied for major QTL, these have been confirmed by GWAS studies (Shi et al., 2011, Ambachew et al., 2021). Other biotic resistances have also been delved into, including virus resistance (Shi et al., 2012; Shi et al., 2021; Soler-Garzón et al., 2021) and resistance to nematode infection (Wen et al., 2019).

Among abiotic stresses, the most emphasized for GWAS evaluation have been heat and drought stresses (Blair et al., 2010a; Asfaw et al., 2017; Oladzad et al., 2019a), with some work on aluminum tolerance (Ambachew and Blair, 2021). The combination of effects from biotic and abiotic stresses, along with plant physiology (Moghaddam et al., 2016; Raggi et al., 2019), determined agronomic traits of interest, such as yield (Nemli et al., 2014; Moghaddam et al., 2016; Wu et al., 2020), root architecture (Kamfwa et al., 2015b, Ambachew and Blair, 2021), and nutrient content (Delfini et al., 2021a). Other characteristics have been cooking time and culinary quality traits (Cichy et al., 2015b) using diversity panels either from one or both Mesoamerican (small-seed) and Andean (large-seed) gene pool. Specialized traits for snap beans, such as herbicide resistance (Saballo et al., 2022) and stringiness of the pod suture (Liu et al., 2022), have been studied using GWAS. Current advancement in genetic markers for plants has reduced the cost of genotyping, allowing many marker–trait associations to be evaluated (Raggi et al., 2019).

In common bean, the most commonly used genotyping technology for GWAS has been single-nucleotide polymorphism (SNP) based on chip arrays such as Illumina or skim-type re-sequencing of partially reduced genomes, with the main example being genotyping-by-sequencing (GBS). Coupling genomic screening (Cortinovis et al., 2020) of common bean germplasm with phenotypic diversity via GWAS-type mixed linear models (MLMs) enables capturing predictive SNPs associated with target traits. Nowadays, there is an expanded assortment of genomic pipelines beyond classical GWAS models that appeal to these genomic associations to infer the genetic bases of complex traits in common bean and adaptation to abiotic stresses, like drought (Cortés and Blair, 2018) and heat (López-Hernández and Cortés, 2019; Elias et al., 2021). Some of these model alternatives, conceptually inspired in the GWAS technique, include genome–environment associations (GEA, Cortés et al., 2022) and genome-wide environmental scans (Rellstab et al., 2015), such as genome-wide selection scans (GWSS, Zahn and Purnell, 2016).

GWAS uses MLMs that explicitly incorporate random effects, such as kinship and population stratification (Kruglyak, 2008). GEA behaves just as a traditional GWAS, but instead of modelling a phenotypic vector, it decomposes an environmental-derived variable into its additive genetic factors. Meanwhile, GWSS relies on genomic background distributions surveyed with Bayesian tests for outliers (Antao et al., 2008). Any of these inferences is susceptible to be misleading (Maher, 2008; Pennisi, 2014) if demographic (Barton et al., 2019) and linkage disequilibrium (LD)–related genomic (Wray et al., 2013; Huber et al., 2016; Ellegren and Wolf, 2017) confounding factors (Lambert and Black, 2012; Wolf and Ellegren, 2017) are not properly handled. Even though MLM-based models are suitable to handle some of these spurious effects, the unveiled genetic architecture requires further validation (Abebe et al., 2015) via expression and transformation analyses. While GWAS has become popular to use in both types of common beans, GEA has only been possible in wild beans, where in situ data is collected from the small-seeded progenitor entries found in a few germplasm collections from countries where wild beans occur (Bolivia, Colombia, Mexico, USA). GEA has also been carried out for dry beans where long-lived *ex situ* gene bank collections have been established by the CGIAR system and by the USDA. For snap beans, the field of GEA analysis is entirely open, although some preliminary GWAS results have been published in the last three years (see Section 4.4).

Before entering into the details of GWAS/GEA results for each group of germplasm, we will give examples of the detailed phenotyping needed to conduct marker–trait association (MTA) analysis. These examples are from the context of Tennessee State University, which has research interests in both dry beans and snap beans with adaptation to the hot and humid conditions of the Southeastern United States. This region is made up of small- to medium-sized family farms in the case of specialty crop or vegetable production, with a growing movement to produce horticultural crops locally and organically. In the South, most farmers have the capacity to grow common bean at least twice a year based on spring and summer plantings. Dry beans used to be a common staple of the Southern diet together with corn but has waned, given urbanization. Mixed-crop vegetable farms, smaller organic farms, as well as backyard gardeners continue to produce both dry and snap beans in the region.

Early production, local consumption, and proximity to markets favor larger-scale snap bean production in a few regions, such as the Homestead area and Okeechobee basin of Florida, the lower tier of counties near the Florida/Georgia line, or on the Cumberland Plateau of Tennessee (Bost et al., 2013). Due to the hot climate, small-seeded Mesoamerican common beans do better than large-seeded Andean gene pool beans; however, Andean or Guatemala race snap bean varieties do well as well if planted in the cooler seasons of the year (winter in Florida, spring in Tennessee). Southeastern production of common bean in the summer has the most severe heat stress for the crop in the United States, given its timing of production and high night temperatures, with a range of warm season diseases as described later. In addition, soils of some farming areas are marginal and made up of Ultisols with low pH, low nutrient, and high aluminum constraints, such as abiotic stresses.

1.2.1 Example of Field Phenotyping: Disease Resistance Results

Ambachew et al. (2021) conducted a GWAS study with Mesoamerican beans to discover locally adapted germplasm with tolerance to two diseases common in Tennessee. The first was common bacterial blight (CBB), caused by *Xanthomonas axonopodis* pv. *phaseoli* (*Xap*), which is the most frequently encountered bacterial disease of common bean, causing 20–60% loss in yield of susceptible cultivars (Belete and Bastas, 2017; Zhu et al., 2016). CBB is regarded as a major common bean production constraint in warm parts of the United States (Akhavan et al., 2013). CBB is particularly severe in humid climates with significantly heavy rainfall. Characteristic symptoms are leaf spot and marginal necrosis with a narrow yellow halo around either. At first, the lesions on leaves are small, translucent, water-soaked spots which later develop dry brown centers and the halos. Lesions coalesce into irregularly shaped areas which may include the whole leaflet. Lesions on stems and pods tend to be more restricted and sunken. Those on the pods frequently turn from red to brown with age. With continued development, the vascular system may also turn brown, and surface cankers form on the stem (Akhavan et al., 2013). Management of CBB is challenging due to the seed-borne nature of the bacteria and its ability to produce large quantities of inoculum that increase disease spread (Belete and Bastas, 2017). No pesticides are available to control the disease, and it is found on both dry beans and snap beans. The only solution is the use of clean seed from dry land growing environments, which is not a solution for many organic producers.

The other important disease was charcoal rot, a soilborne root and stem disease caused by fungal pathogen *Macrophomina phaseolina* (Tassi) Goid. Charcoal rot is found throughout the tropics and subtropics and has a wide host range, consisting of over 500 plant species (Mengitsu et al., 2015). It affects some of the most economically significant crop plants, such as soybean, cotton, maize, sorghum (Adeyanju et al., 2015; Cloud and Rupe, 1991; Coser et al., 2017), and cowpea (Muchero et al., 2011), in North and South America, Asia and Africa, and some parts of Europe (Jana et al., 2003). Together with most of the legume crops, the common bean is a good host for the fungus, which causes a range of symptoms, depending on environmental conditions and age of the plant. In addition to charcoal rot, which is a stem or stalk rot disease, the pathogen also causes damping off and seedling blight in beans (Muchero et al., 2011). Charcoal rot is an important disease of common beans in the tropics, subtropics, and temperate regions with high temperatures and drought conditions.

Charcoal rot is widely distributed in the United States, and it is commonly found in the Southern states and occurs in the Midwest in seasons with hot, dry conditions (Hughes, 2013). The disease develops in the mid- to late summer, when plants are under stress, especially heat and drought stress. Infected plants may die prematurely and are often wilted and stunted and can cause significant yield losses.

Charcoal rot in the mature plant is associated with senescence, which is accelerated by water stress. In the United States, substantial losses have been associated with charcoal rot in major crops, like in soybean and sunflower, where the yield loss to charcoal rot is estimated to be millions of metric tons (Wrather and Koenning, 2006). It is reported that there is up to 65% yield loss in common beans (Miklas et al., 1998). Some common measures in controlling these diseases include the use of resistant varieties, chemical seed treatment, the use of disease-free seeds, and good cultural practices. The use of resistant dry bean cultivars is the most recommended means to ensure profitable and sustainable commercial bean production by farmers. Furthermore, this resistance must be functional at the field level, and therefore, the most appropriate phenotyping for the two diseases is through on-station experimentation, as shown in Figure 1.3 for the studies conducted over three years at Tennessee State University.

The results of the disease screening of the Mesoamerican gene pool common beans revealed the presence of variability among the 300 genotypes in the USDA-initiated diversity panel for the evaluated traits. The GWAS detected 14 significant SNP markers associated with CBB resistance distributed on five chromosomes, namely, Pv02, Pv04, Pv08, Pv10, and Pv11, but no loci for resistance to charcoal rot. A total of 89 candidate genes were identified in close vicinity (+100 kb) to the significant CBB markers, some of which could be directly or indirectly involved in plant defense to diseases. These results provide a basis to understand the complex inheritance of CBB resistance in Mesoamerican common beans and show that this biotic stress is unrelated to CR resistance, which was especially evident in one of the years when a late planting hit a drought period. Adapted common bean genotypes that could be used directly in dry bean production or which could be good parents in breeding programs for CBB and CR resistance were found with good yield potential for Southeastern US growing conditions despite the hot and humid conditions punctuated by intermittent drought found in this region.

FIGURE 1.3 Augmented design experiment for common bacterial blight and charcoal rot disease resistances of small- to medium-seeded Mesoamerican dry beans grown in Tennessee, in the Southeastern United States, for GWAS study. Total view shows ten incomplete blocks containing 30 genotypes and a set of controls per block: GNStar, GN#1Sel27, Coyne, ABC-Weighing from race Durango and IBC-3, OAC Rex and XAN176 from race Mesoamerica. Panel included pinto types (with 99 entries), small red (34), great northern (44), navy beans (54), pink (25), and black beans (44), for a total of 300 entries. The genotypes were arranged in plots consisting of two rows, with the same length and planting distances used previously (2 m long, 40 cm between rows, 10 cm within rows).

As seen in the preceding figure, short rows (2.5 m long) were used with 1 m alleys between blocks. Seeding rate was at approximately 12 seed/m. Inoculum was spread by repeated planting in the same overall field plot and by border rows of disease-susceptible snap beans, as shown in the photograph. In the case of data collection, CBB was evaluated in a 1-to-9 score, while charcoal rot was measured in percentage of plants infected per row, also known as incidence. Yield was measured in grams per plot and converted to kilograms per hectare based on plant numbers. An augmented design was used with ten incomplete blocks containing 30 genotypes and four controls (four disease-tolerant varieties). Even though the genotypes were not replicated, the controls were randomized in each incomplete block and replicated ten times to estimate the error variance. Two-way omnibus ANOVA was conducted on the weighted means of seed yield and CBB data from the field experiments using linear model (lm) function in R and two-stage nested procedure in SAS per year and three-stage for combined data across three years. The variance components were estimated by taking all the factors as random effects, including the genotype's race identification and commercial seed class as grouping variables. Data were collected on environmental conditions of precipitation events and minimum and maximum daily temperature across each of the three seasons. A greenhouse trial was conducted as well in this CBB study to use a single isolate of the disease and controlled inoculation but with GWAS is limited, given that disease screening must be done on adult plants, which take up a lot of bench space. Greenhouse trials for CBB or charcoal rot resistance are best done with a subset for confirmation of resistances identified in the field.

1.3 GREENHOUSE SCREENING OF TRAITS

Abiotic stresses can also be evaluated in greenhouse and rainout shelter screening as another method to apply more controlled conditions to the evaluation of common bean germplasm for salinity or aluminum stress tolerance (hydroponics) or heat and drought resistance (summer greenhouse with seedling assays or adult plant screening). Automated evaluation of physiological parameters is possible in greenhouse settings with specialized pot units grown on arrays of analytical balances which constantly monitor water loss and carbon/nitrogen fixation in the soil–plant system. In this section, we will highlight greenhouse hydroponic tank screening done for aluminum tolerance (Figure 1.4) performed in a greenhouse at Tennessee State University (Ambachew and Blair, 2021).

This study evaluated 267 genotypes of common bean that were part of the global Andean diversity panel (ADP) consisting mainly of genotypes assembled from public and private breeding programs in Africa and North America as well as elite lines and landraces from USDA.

Data were recorded on six plant variables: (1) total root length, (2) average root diameter, (3) number of root tips, (4) root volume, (5) number of forks, and (6) root surface area using WinRHIZO. Three environmental variables were considered during the 48-hour treatment period: (1) solution pH, (2) electrical conductivity (EC), and (3) temperature of the hydroponic system. These parameters around the roots were monitored in real time using supervisory control and data acquisition (SCADA). The recorded plant data on the roots were fit to linear mixed model per trial for the analysis of variance to test for genotype differences. Adjusted means considered replication and blocks within replication as random and genotypes as fixed effects. These were then used for Pearson correlation tests and for principal component analysis (PCA), where the first two vectors accounted for 94.5% and 93% of the explained variation under control and Al treatment conditions, respectively. Secondary data as percent change in the traits from control treatment and aluminum treatment were calculated from the adjusted means. Variance components and broad sense heritability (BSH) were calculated from the formula of Holland et al. (2003) and ranged from 57 to 95% without Al stress and 67 to 96% with Al stress, depending on the traits. Multivariate analysis included hierarchical clustering. Phenotypic data was compared to two sets of genotypic data: first, 13,906 SNPs from GBS data after filtering out markers with MAF <0.05, heterozygosity >0.02, and SNPs with more than two alleles, and then a set of 2,286 SNPs with high linkage disequilibrium LD (R2 >2)

GWAS with both datasets were performed with TASSEL software using mixed linear model (MLM), where population structure was controlled using the first three principal components calculated from LD-pruned SNP dataset. Kinship matrix was a covariate in the MLM model. Percent change datasets calculated from BLUPs of traits measured under control and aluminum treatment trials were used as a phenotype input for the association analysis. Manhattan plots and QQ plots were generated using the "CMPlot" R package. Significance levels were established using a Bonferroni correction at $p < 0.05$ based on the effective number of independent tests determined via SimpleM or LD pruning. The significant SNPs were positioned to the *Phaseolus vulgaris* V1 reference genome (G19833) using Jbrowse on Phtozome v13 candidate genes in +100k window positioning the significant SNP at the center. In total, 15 SNPs were significantly associated with root diameter, number of root forks, number of root links, root volume, root surface area, or total root length (Figure 1.5).

FIGURE 1.4 Greenhouse testing of common bean for aluminum toxicity stress tolerance with hydroponic tanks and WinRHIZO scanning of the root system traits.

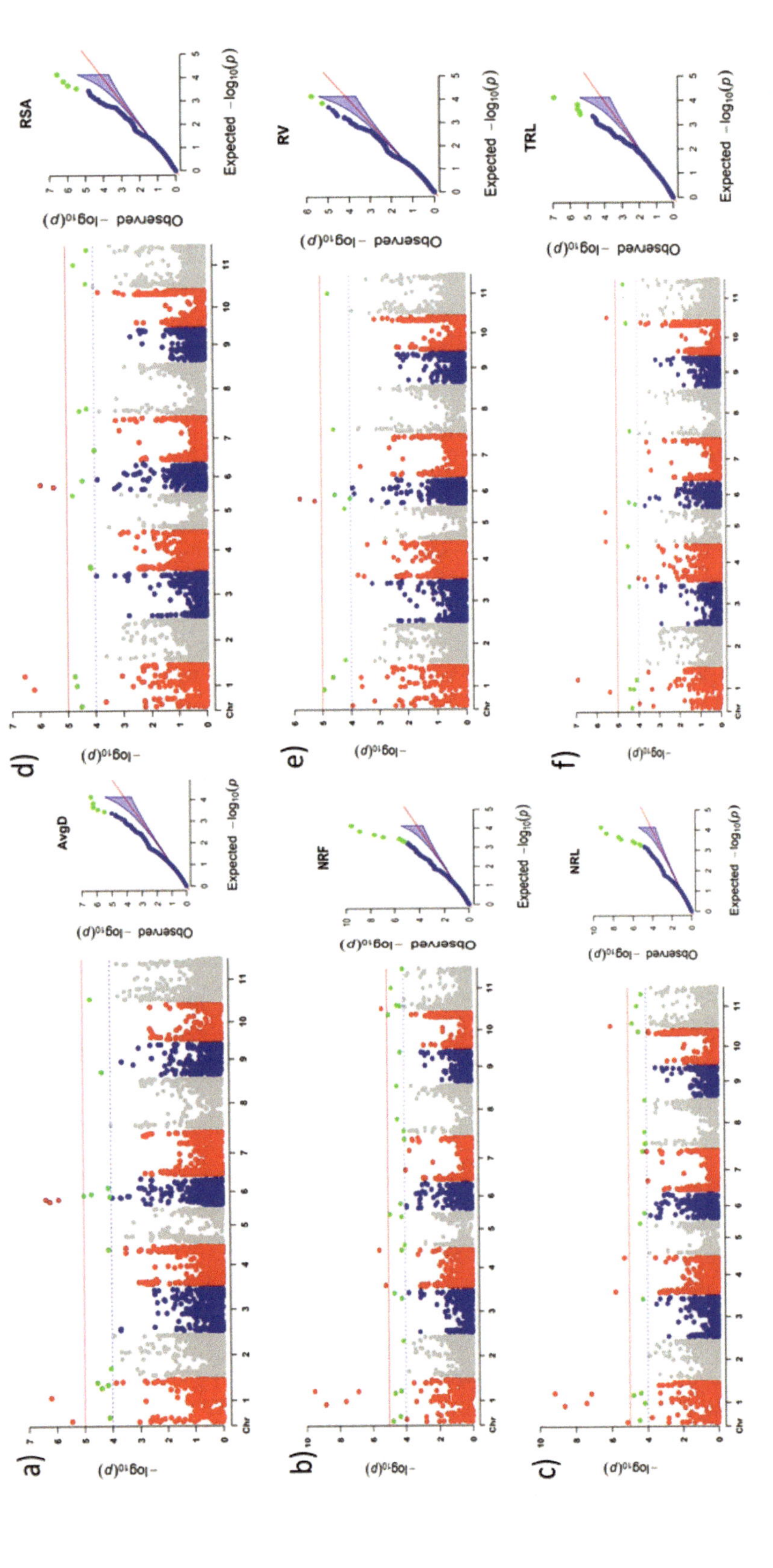

FIGURE 1.5 Manhattan and QQ plots of aluminum tolerance traits: *TRL*, total root length (cm); *RSA*, root surface area (cm^2); *RV*, root volume (mm^3); *AvgD*, average root diameter (mm); *NRT*, number of root tips; *NRF*, number of root forks; *NRC*, number of root crosses; *NRL*, number of links. SNPs are ordered by physical position and grouped by chromosome. The line indicates the genome-wide significance threshold. SNPs highlighted in red had p values lower than Bonferroni corrected significance cutoff value for a number of independent tests $p = 1 \times 10$.

1.4 GWAS IN COMMON BEAN: ESTIMATING THE GENOMIC BASES OF KEY AGRONOMIC TRAITS

Besides the examples given earlier, recent GWAS research has provided a basis for genetic dissection of some key additional traits, some being resistance to biotic stresses, other tolerances to abiotic stresses, and some that increase our understanding of common bean morphology. The technologies used have ranged from SNP detection on Illumina chip arrays detected with iScan machines to GBS with various platforms. Although the former is slightly more popular based on SNPs developed with a USDA project called CAP (coordinated agricultural program) and with Brazilian collaboration of EMBRAPA, the latter also has been used by service labs. One form of GBS is called ddRADseq and relies on random sequencing after digest by restriction enzymes. The ultimate SNP discovery technique is through re-sequencing, which has been less frequently done in common bean than in other crops due to the smaller research budget of this minor crop.

We summarize the GWAS studies done in common beans into two tables, the first of which describes studies in dry beans (Table 1.1), and the second in snap beans (Table 1.2), which have been the subject of even more recent results compared to dry beans, perhaps due to the smaller number of breeding programs working on beans as a vegetable than as a grain (pulse) crop.

A recent paper by Wu et al. (2021) is notable because of the large number of accessions used in the GWAS study (438) and the high investment made in re-sequencing. The detailed analysis uncovered 3.78 million SNP markers for complete genome saturation. The focus of the paper was on the genetic dissection of drought resistance based on root traits, including (1) surface area, (2) average diameter, (3) root volume, (4) total root length, (5) taproot length, (6) lateral root number, (7) root dry weight, (8) lateral root length, and (9) specific root weight/length ration. Not surprisingly, the results were extensive, with 2,542 and 408 significantly associated SNPs detected using a general linear model and a compressed mixed linear model, respectively. The overlap between models predicted a set of 389 SNPs of interest for marker-assisted breeding. Peak SNPs for loci were defined as candidate SNPs: a total of 196 loci containing 230 candidate SNPs were associated with nine traits: 104 for total root length, 47 for special root weight/length, 45 for root surface area, 23 for root volume, 19 for lateral root length, 4 for taproot length, 3 for root average diameter, and 3 for root dry weight. Around 41 of these SNPs were associated with more than one trait, and a total of 2,962 candidate genes were identified within the 196 loci.

In the category of GBS studies, rather than re-sequencing but instead with large SNP numbers detected is a study by Diaz et al., (2020). A total of 1,972,528 SNPs and indels were detected with *Ape*KI digestion for genome complexity reduction. The large number of loci is probably related to the inclusion of presence/absence variation (PAVs) and the number and source of genotypes included, which were approximately 600 recombinant inbred lines (RILs) from a multiparent advanced generation intercross (MAGIC) population developed using 8 Mesoamerican parents. The MAGIC population was tested under drought condition and evaluated for days to flowering, days to physiological maturity, yield, 100-seed weight, pod harvest index, zinc (Zn) concentration, and iron (Fe) concentration.

Despite the large investment in markers and RIL development, only 13 high-confidence QTL for seven traits were identified by Diaz et al. (2020): Pv01 contained one QTL explaining 35.8% of variance for days to flowering and one QTL explaining 5.5% for days to physiological maturity; the same region contained a yield QTL with phenotypic effects of up to 53 kg/ha; another QTL was also associated with seed zinc. Days to maturity hotspots were found on Pv06, containing QTL associated with days to physiological maturity and days to flowering; other alleles associated with late maturity were also present on Pv06. One QTL associated with pod harvest index was identified on Pv08, and another was found on Pv02. Two QTLs, one on Pv07 and one on Pv08, were associated with yield. Three QTLs were associated with micronutrient content, but none were consistently detected. One line, MGC583, had high yield under drought conditions as well as high accumulations of Fe and Zn. The results suggest that the MAGIC strategy is not very well adapted to common

TABLE 1.1
GWAS Studies of Common Bean (*Phaseolus vulgaris* L.) as a Dry Bean, Grain Legume

Title	Author(s), Year	Journal	Population	Genotyping/Enz./# of Markers	Traits
Genome-wide Association study Reveals Candidate Genes for Flowering Time Variation in Common Bean (*Phaseolus vulgaris* L.)	Raggi et al. (2019)	*Frontiers in Plant Science* 10:962	192 accessions primarily from Italy; some from Central and South America	Double-digest restriction site–associated DNA sequencing (ddRADseq)/*Sph*I and *Mbo*I 49,518 SNPs	Days to flowering
GWAS of pod morphological and color characters in common bean	García-Fernández et al. (2021)	*BMC Plant Biology* 21:184	301 accessions from the Spanish diversity panel	Genotyping-by-sequencing/*ApeK*I 32,812 SNPs	17 fresh pod traits grouped into 4 categories: (1) pod length, (2) pod cross section, (3) pod color, (4) number of seeds per pod
Genome-wide Association of Root Traits in the Andean Genepool of Common Bean (*Phaseolus vulgaris* L.) Grown With and Without Aluminum Toxicity	Ambachew and Blair (2021)	*Frontiers in Plant Science* 12. 628687	227 accessions from the Andean diversity panel	13,906 SNPs	Root diameter, root volume, number of root forks, number of root tips, number of root links, root surface area, total root length
Genome-wide association studies dissect the genetic architecture of seed shape and size in common bean	Giordani et al. (2022)	*G3 Genes|Genomes|Genetics* 12(4):jkac048	180 Brazilian accessions	Genotyping-by-sequencing/ *ApeKI*/ 10,362 SNPs	Length, width, perimeter, projected area, length-to-width ratio, circularity, 100-seed weight
Genetic dissection of drought resistance based on root traits at the bud stage in common bean	Wu et al. (2021)	*Theoretical and Applied Genetics* 134:1047–1061	438 Chinese accessions	Re-sequencing 3,783,033 SNPs and PAVs	Root surface area, root average diameter, root volume, total root length, taproot length, lateral root number, root dry weight, lateral root length, special root weight/length
Genetic mapping for agronomic traits in a MAGIC population of common bean (*Phaseolus vulgaris* L.) under drought conditions	Diaz et al. (2020)	*BMC Genomics* 21:799	636 and 599 RILS in years 1 and 2, from a MAGIC population developed using 8 Mesoamerican common bean elite lines	Genotyping-by-sequencing/ *ApeK*I 1,972,528 "variants," including SNPs and indels	Days to flowering, days to physiological maturity, yield, 100-seed weight, pod harvest index, zinc concentration, iron concentration

Genome-wide Association Study Identifies Genomic Regions for Important Morpho-Agronomic Traits in Mesoamerican Common Bean	Delfini et al. (2021b)	*Frontiers in Plant Science* 12:748829	178 Mesoamerican common bean accessions	Genotyping-by-sequencing/ *CviA*II 25,011 SNPs	Plant height, first pod insertion, total number of pods, pod length, number of nodules, number of locules per pod, number of seeds per pod, total seed weight per plant, 100-seed weight
Single and Multi-trait GWAS Identify Genetic Factors Associated with Production Traits in Common Bean Under Abiotic Stress Environments	Oladzad et al. (2019)	*G3 Genes|Genomes| Genetics* 5;9(6):1881–1892	155 Mesoamerican genotypes, 147 Andean genotypes, 5 tepary bean genotypes	Genotyping-by-sequencing/ *Mse*I and *Taqα*I/ 260,670 SNPs	Heat stress– and drought-related traits (yield, days to flowering, photosynthesis-related traits)
Genetic Factors Associated with Nodulation and Nitrogen Derived From Atmosphere in a Middle American Common Bean Panel	Oladzad et al. (2020)	*Frontiers in Plant Science* 11:576078	118 Mesoamerican common bean and 2 tepary bean genotypes	Genotyping-by-sequencing/ *ApeK*I 205,293 SNPs	Root nodule number, percent of N_2 derived from atmosphere
Genome-Wide Association and Regional Heritability Mapping of Plant Architecture, Lodging and Productivity in *Phaseolus vulgaris*	Resende et al. (2018)	*G3 Genes|Genomes| Genetics* 8(8):2841–2854	188 FAO accessions	DArTseq/*Pst*I and *Mse*I 11,564 SNPs	Plant architecture (branch insertion angles, length of guide, stem height, at which all pods are above the soil surface), lodging tolerance (degree of inclination in relation to vertical), yield (kg/ha)

TABLE 1.2
GWAS Studies of Snap Bean (*Phaseolus vulgaris* L.) as a Vegetable Crop

Title	Author, Year	Journal	Study Type/Population(s) Used	Marker Type and # of Markers	Traits Evaluated
Improving the Health Benefits of Snap Bean: Genome-Wide Association Studies of Total Phenolic Content	Myers et al. (2019)	*Nutrients* 11(10), 2509	Bean coordinated agriculture program (CAP) snap bean diversity panel of 149 accessions	Illumina Infinium Genechip BARCBEAN6K_3 single-nucleotide polymorphism (SNP) array (with 10,546 allele-specific probes). Genome-wide association studies (GWAS) analysis identified 11 loci for TPC and 13 loci for flower color.	Total phenolic content (TPC), flower, and pod color. Flowers were classified as white (vv), pink (vlaevlae), and purple (VV). The external color of fresh pods was quantified using a colorimeter. L^* represents brightness at 0 to 100, a^* *vector* represents colors from green to red, and b^* *vector* measures blue to yellow.
Genome-Wide Association Study Reveals That PvGUX1_1 Is Associated with Pod Strings in Snap Bean (*Phaseolus vulgaris* L.)	Liu et al. (2022)	*Biology*, 11(4), 611	138 accessions from the Institute of Vegetables and Flowers at the CAAS, including landraces, elite lines, and breeding lines	Whole-genome re-sequencing (WGS) 3,656,683 high-confidence SNPs.	ST1–2019 and ST1–2020 indicate the value of string length/pod length in each year. ST2–2019 and ST2–2020 indicate the presence of suture strings structure in each year.
Multiple Genomic Regions Govern Tolerance to Sulfentrazone in Snap Bean (*Phaseolus Vulgaris* L.)	Saballo et al. (2022)	*Frontiers in Agronomy*, 46	A subset of the snap bean association panel (SNAP) comprising 277 genotypes represents the diversity of snap beans grown in the United States	The original SNAP population was genotyped using GBS, and a total of 20,619 SNPs were selected. For PD, 150 marker–traits associations (MTAs); for BP, 167 MTAs were identified.	Tolerance to herbicide sulfentrazone reflecting plant population density (PD) and shoot biomass plant (BP), and snap bean total plant biomass (TPB).

bean, where useful variation is associated with certain seed classes that have poor combining ability, especially across gene pools Andean and Mesoamerican.

Oladzad et al. (2019) produced the next largest number of SNPs (260,670 loci) on a panel of 155 Mesoamerican, 147 Andean, and 5 tepary bean (out-group species: *P. acutifolius*) genotypes through double-digest GBS with *Mse*I and *Taq*I. Their goal was to conduct single- and multi-trait GWAS for production traits under heat and drought stress environments evaluating yield, days to flowering, and photosynthesis-related traits. One SNP located within a gene associated with heat stress response was associated with yield and was responsible for 14% phenotypic variance. Meanwhile, three other SNPs were associated with yield, each responsible for approximately 7% variance. Another SNP associated with days to flowering was responsible for 9.9% variance.

A subset of the same panel (only Middle American) was used for GWAS study of nodulation traits by Oladzad et al. (2020). In this case, 205,293 loci and only 118 common bean and 2 tepary genotypes were used for MTA research. The traits included root nodule number and percent of N_2 derived from the atmosphere. Interestingly, with so many markers and this low diversity panel, a single SNP block located on Pv06 was responsible for 11% to 13% of variation (0.01 or 0.05% cut-off thresholds, respectively), aligning with an ortholog of the *Arabidopsis thaliana* RAB GTPase, which was 20 Kb downstream of an auxin-responsive factor AUX/indole acetic acid (IAA)–related gene.

A modification of GBS called double-digest restriction site–associated DNA sequencing (ddRADseq) was used by Raggi et al. (2019) to evaluate candidate genes for flowering time variation in 192 accessions primarily from Italy, with some primary center of origin controls from Central and South America. The double digest with *Sph*I and *Mbo*I generated a large number of SNPs, in total 49,518, which were used in GWAS for days to flower. A total of eight significant SNPs were found on a total of four chromosomes, with seven of these correlating to possible candidate genes, such as homologs of the *Arabidopsis thaliana* genes: (1) photoperiod-independent early flowering 1 gene (PIE1) involved in multiple flowering pathways, (2) an NAD(P)-binding Rossmann-fold superfamily protein for oxidoreductase activity, (3) a meristem disorganization 1 (MD1) gene, (4) a protein phosphatase 2C 3-receptor involved in ABA signal transduction, (5) a sucrose transporter, and (6) a leucine-rich receptor-like protein.

Another modified GBS was conducted by Resende et al., (2018) using Diversity Arrays Technology (DArT) and double-digest restriction enzymes *Pst*I and *Mse*I to generate 11,564 SNP loci for GWAS evaluation of plant architecture (branch insertion angles, length of guide, stem height, at which all pods are above the soil surface), lodging tolerance (degree of inclination in relation to vertical), and yield (kg/ha). Their panel consisted of 188 Brazilian genotypes (91 landraces and 97 breeding lines) detecting eight QTL. The two most significant were associated to both traits on Pv01, explaining up to 28% variance for architecture and 30.4% for lodging. Two other QTLs were associated with plant productivity from one environment and explained 20% variance. Regional mapping detected 13 region-specific QTL: 6 QTLs on Pv01, Pv08, Pv09, and Pv11 for plant architecture; 3 QTLs for lodging; and one QTL associated with both.

A series of other GWAS studies using GBS has succeeded with fewer SNP loci detected, perhaps indicating the ideal number of markers for MTA scanning at between 10,000 and 20,000 for a small genome species like common bean. This number of SNPs is similar to the current number available for iSCAN technologies, although earlier studies were limited by having half this number, between 5,000 and 6,000. Examples of the lower-density GBS studies include three studies, highlighted next.

Giordani et al. (2022) evaluated 180 Brazilian accessions with GBS and single digest with *Ape*K1 restriction enzyme to uncover 10,362 SNPs useful in GWAS of seed shape and size traits in common beans. Specific measurements were taken on seed length, width, perimeter, projected area, length-to-width ratio, circularity, and 100-seed weight. A total of four genomic regions containing seven marker–trait associations were significant for length, width, area, perimeter, and circularity on Pv02, Pv08, and Pv11. A further two SNPs were significantly associated with length on Pv08 and Pv11, and one of these SNPs was also associated with perimeter and circularity. Two more

SNPs were associated with seed area on Pv02 and Pv08, and one of these was also associated with width. Variance explained for each MTA ranged from 5.18% to 23.7%. Surprisingly, no significant MTAs were detected for length-to-width ratio or 100-seed weight. This showed that most genotypes studied were Mesoamerican, without introgression of Andean alleles for larger seed size and weight.

GBS loci from a separate *Ape*K1 digest have been used for conducting GWAS for morphological and color characters in 301 accessions from a national germplasm collection from Spain by García-Fernández et al. (2021). In total, 32,812 SNPs were detected, and PAVs were not considered. This Spanish diversity panel was interesting because of the large amount of introgression between gene pools found in the genotypes, giving rise to new combinations of alleles. The analysis of 17 fresh and dry pod traits grouped into four categories – (1) pod length, (2) pod cross section, (3) pod color, and (4) number of seeds per pod – resulted in 63 significant single-locus quantitative train nucleotides (QTNs): 57 of these grouping into nine genomic regions (seven for pod length, one for cross section, and one for pod color).

Multi-locus GWAS in the same study identified 103 QTNs: the number of associations ranging from 21 to 37. In total, 14 QTNs were associated with pod cross-section traits, 52 with pod length, and 37 with pod color. A total of 166 QTNs comprising 62 QTLs were discovered: 23 for pod length, 18 for pod color, 6 for pod cross-section, 6 for number of seeds per pod. Most QTLs were located on Pv02, Pv04, Pv08, and Pv10, and 12 candidate genes within QTL were identified for pod size, 4 were found in QTL for seed size and color traits, and 11 candidate genes were identified within QTL associated with pod color. This GWAS study and those described previously are summarized in the table that follows, concentrating on dry grain phenotyping for common bean.

Fewer published GWAS studies have been carried out for snap bean, the vegetable common bean. This reflects the limited number of academic programs researching snap beans as compared to dry beans. While dry beans is researched in the 1862 land grand universities of the eight states (California, Colorado in the past, Idaho, Michigan, Minnesota, Nebraska, North Dakota, New York, and Washington), snap beans are only seriously studied in Oregon State. USDA centers in Lansing, MI, along with Pullman, WA, and Geneva, NY, support research through the ARS providing continuous stable funded projects in dry beans. A tropical research station of the USDA in Mayaguez, Puerto Rico, supports this effort, as do commodity groups called the US Dry Bean Council and the North American Pulse Association.

No such commodity group exists for snap bean, with production in Florida, Tennessee, and Wisconsin concentrated in the hands of specialized private farmers. While US snap beans are threatened by imports from Mexico, the reverse is true for dry beans, which are often exported from the United States to Mexico and beyond in the Caribbean to Cuba, Dominican Republic, and Haiti, and to Latin America, Europe, and the Middle East more generally. The snap bean studies conducted so far are summarized in the following table. It is notable that despite the economic importance of exported snap beans in Mexico, there are no studies from there. Similarly, Europe is a powerhouse in snap bean breeding but has not produced GWAS studies on the crop. Africa, Asia, Middle East, and Latin America are even further behind in snap bean research, despite export from each region to high-value markets. These include Kenya and Israel providing winter-produced snap beans to Europe. Meanwhile, China and India are huge national markets for the vegetable crop.

One exception to the lack of GWAS studies outside the USA was the breakthrough study in China of Liu et al. (2022), who found that the gene PvGUX1_1 was associated with lack of string fibers in snap bean, which gives rise to stringless pods, an important trait for this crop. The study was conducted at the Institute of Vegetables and Flowers at the Chinese Academy of Agriculture Sciences (CAAS), with 138 accessions, including landraces, elite lines, and breeding germplasm, using re-sequencing for a high degree of accuracy. The authors found an amazing total of 3,656,683 high-confidence SNPs and, of these, 205 loci associated with string presence. A major locus on Chr02 at 44.2 Mb was an MTA across two years and identified the *Phvu.002G270800* gene, a member of the glycosyltransferase family 8, which participates in synthesis of plant cell walls through xylan production and was the most probable determinant of pod suture formation.

Other studies of snap beans have included one by Myers et al. (2019), who used 149 accessions (a SNAP diversity panel) and approximately 5K SNPs from an Infinium chip made at the Beltsville Agricultural Research Center (BARC) of the USDA to evaluate the health benefits of snap beans, assuming these are mainly from phenolic compounds. GWAS identified 11 MTAs for phenolic content. As side projects, flower color genes were associated with 13 SNPs, and pod brightness was evaluated with a colorimeter in this same study.

Since flower and pod color can be visually selected for the utility of GWAS, these traits may be of less interest than the invisible trait of phenolics. The same panel of snap beans was used by Saballo et al. (2022) to discover tolerance to the herbicide sulfentrazone, but in this case using GBS genotyping, which produced 20,619 SNPs.

1.5 CROSS-CUTTING THEMES IN GWAS: BIOTIC STRESSES AND NUTRITIONAL QUALITY

Despite the agronomic differences in managing dry bean and snap beans being substantial, as subtypes of the same species, they suffer from mostly similar diseases and insect pests, although these vary a lot by region. Reviews of the biotic limitations of common bean have been done very frequently with mostly the same structure, so this reviewer will instead refer the reader to the American Phytophathological Society publication *Diseases of Common Bean*, the most comprehensive description of pathogens and their symptoms. Surprisingly, despite the evident possibility of GWAS for biotic stress trait analysis, very few early studies included this method.

1.5.1 Disease Resistance GWAS Studies in Dry Bean

Oladzad et al. (2019) was one of the first studying general agronomic traits to detect a single SNP associated with *Macrophomina phaseolina* responsible for 8.9% and 7.9% of variance under heat and drought stress conditions, respectively. In another, more recent study, Ambachew et al. (2021) did not find relevant SNPs for this important disease, perhaps because *M. phaseolina* can lead to total plant death by constricting the stem collar, and rating for percent damage is complicated by partial resistance to this disease and tolerance of drought and heat that increase the plant's susceptibility.

Oladzad et al. (2019) found in addition to the small effect QTL a total of three other SNPs that together explained 17% of variance in charcoal rot response. These authors suggested that a major cluster of malectin/receptor-like kinases could be involved.

Other diseases have also been studied using GWAS methodologies, with the majority of these studies concentrating on fungal pathogens, some of aerial plant parts and some affecting roots (Table 1.3). Since root disease resistance is often hard to assess visually without pulling up the plant, these studies, while often under controlled conditions, are among the most important for allele mining and molecular breeding goals.

As seen in the preceding table, several fungal root pathogens have been evaluated across different germplasm panels. Dramadri et al. (2020) evaluated 260 genotypes from the Andean diversity panel (ADP) for resistance to *Pythium ultimum* using genotypic information previously generated for the entire ADP based on the Illumina BARCBean6K-3 chip (5,389 SNPs) and on GBS-based markers (260,670 SNPs). They discovered eight significant SNPs using the larger GBS dataset, showing the value of more extensive genotyping when analyzing disease loci. The most significant SNP responsible for about 39% of variance was on Pv02, one other on Pv04 explained 36.2%, and two SNPs on Pv09 had similar significance. Using the bead chip data, eight significant SNPs were identified on Pv01 and Pv02, the most significant accounting for over 31% of variance, three others responsible for 29% altogether. The second chromosome had significant SNP responsible for 33%, with others on Pv02 responsible for 29–33%.

TABLE 1.3
GWAS Studies Carried Out for Biotic Stress Tolerance in Dry Beans and Snap Beans

Title	Author, Year	Journal	Population	Genotyping/Enz./# of Markers	Traits
Dry bean disease resistance					
Genome-Wide Association Study (GWAS) for Resistance to *Sclerotiniasclerotiorum* in Common Bean	Campa et al. (2020)	*Genes*, 11(12), 1496	294 lines from the Spanish diversity panel	Genotyping-by-sequencing/ *ApeK*I 9,070 SNPs	Resistance to white mold (*Sclerotiniasclerotiorum*)
Genome-wide association analysis of resistance to *Pythium ultimum* in common bean (*Phaseolus vulgaris*)	Dramadri et al. (2020)	*Plant Breeding*, 139(6), 1168–1180	260 Andean diversity panel genotypes	Marker datasets generated from ADP accessions genotyped by the Illumina BARCBean6K-3 Beadchip with 5398 SNPs and 260,670 SNPs generated for ADP using GBS	Pythium root rot (*Pythium ultimum*) resistance
Sources of Resistance to Common Bacterial Blight and Charcoal Rot Disease for the Production of Mesoamerican Common Beans in the Southern United States	Ambachew et al. (2021)	*Plants*, 10(5), 998	300 genotypes from the Mesoamerican diversity panel	Genotyping-by-sequencing 136,506 SNPs	Resistance to common bacterial blight (*Xanthomonas axonopodis*pv. *phaseoli*) and charcoal rot disease score 1–9, disease incidence, yield (kg ha-1)
Genome-Wide Association Study Reveals Genomic Regions Associated with Fusarium Wilt Resistance in Common Bean	Paulino et al. (2021)	*Genes*, 12(5), 765	205 Mesoamerican diversity panel	BeadChip BARCBean6K_3 5,298 SNPs	Disease severity rating of resistance to fusarium wilt (*Fusarium oxysporum* f. sp. *phaseoli*)
Genome-wide association and genomic prediction identifies soybean cyst nematode resistance in common bean including a syntenic region to soybean *Rhg1* locus	Wen et al. (2019)	*Hort. Research*, 6(1),9	363 accessions (Mesoamerican and Andean genotypes)	Genotyping-by-sequencing/ *Hind*III and *Bfa*I 84,416 SNPs	Resistance to soybean cyst nematode (*Heteroderaglycines*): number of cysts per plant and female index (mean number of females on tested accessions divided by mean number of females on soybean multiplied by 100)

Dry bean insect resistance					
Genome-wide association analysis of bean fly resistance traits in common bean	Nkhata et al. (2021)	*PLOS One*, 16(4)	99 genotypes, primarily from Africa	Diversity Arrays Technology sequencing (DArTSeq) 15,565 SNPs	Bean fly severity damage (scale of 1–9), plant mortality rate (number of plants dead per plot), pupa count
Snap bean biotic resistance					
Genome-Wide Association Study (GWAS) of White Mold Resistance in Snap Bean	Arkwaz et al. (2022)	*Genes*, 13(12), 2297	136 snap bean diversity panel, and 378 snap association panel	GBS analysis of the original SnAP population resulting in 34,442 high-quality SNPs	Field: disease severity index (DSI), including severity and incidence Greenhouse: seedling straw test; seedlings scored four days after inoculation at a 1–9 scale
Unravelling the Genetic Architecture of Rust Resistance in the Common Bean (Phaseolus vulgaris L.) by Combining QTL-Seq and GWAS Analysis	Wu et al. (2022)	*Plants*, 11(7), 953	Diversity panel, including 17 resistant and 71 susceptible snap bean accessions	For diversity panel, 603,910 high-quality SNPs identified from re-sequencing of 88 accessions	Rust resistance

Paulino et al. (2021), on the other hand, used the 205 genotypes from Mesoamerican diversity panel (MDP) along with 5,298 SNPs from the BARCBean6K chip to conduct GWAS research for fusarium wilt resistance in dry beans. This disease is caused by a fungus, *Fusarium oxysporum* f. sp. *phaseoli*, that enters from the soil into the plant and then becomes an internal pathogen, causing total collapse of susceptible genotypes. To date, GWAS has been done on severity rating data, which is a poor substitute for testing levels of oxidative stress and other parameters of cell death, especially in vascular tissues where the wilt occurs (Xue et al., 2015). Nonetheless, the GWAS study found 11 significant SNPs identified across seven genomic regions: 6 within genes, and 5 close to other candidate genes. Several clusters of candidate genes included transcription factors involved in protein regulation, genes involved in plant vascular diseases or drought resistance, and genes for precursors of flavonoid biosynthesis.

A final soilborne pathogen used for GWAS studies is the soybean cyst nematode (SCN: *Heterodera glycines*), where Wen et al. (2019) analyzed 363 common bean accessions of both the Mesoamerican and Andean gene pools, finding a significant cluster of MTAs in common bean that was syntenic to an SCN resistance locus of soybean (*Glycine max* L.). They used GBS with double digest by *Hind*III and *Bfa*I to discover 84,416 SNPs and counted the number of cysts per plant and the female index, which is the mean number of females on tested accessions divided by mean number of females on soybean check multiplied by 100. A total of six SNPs (five on Pv01 and one on Pv09) were significantly associated with soybean cyst nematode resistance. Four of the SNPs on chromosome Pv01 were syntenic with soybean chromosome 18, where SCN resistance has been mapped to the *Rhg1* gene in that species.

Moving to purely foliar diseases of common beans, Ambachew et al. (2021) were among the first to determine MTAs for common bacterial blight (CBB) resistance in both field and greenhouse environments. Due to the difficulty of quantifying the CBB causal pathogen *Xanthomonas axonopodis* pv. *phaseoli* (*xap*), these authors used a well-established disease resistance score from 1 to 9 that is visually correlated with the percentage of leaf affected by blight to associate the resistance in 300 genotypes from the MDP panel with 136,506 previously published SNPs to identify 14 significant loci on six chromosomes that provided resistance to the hot, humid production region of the Southeastern United States, where dry bean and snap bean production overlap.

In total, Ambachew et al. (2021) annotated 89 genes of interest around significant SNPs primarily on Pv10 (40) and Pv08 (35), although the effect of each SNP on variance explained the total was 41.7% and many of the genes near significant SNPs were associated with plant defense response. CBB is unlikely to be controlled by single gene resistance, and most studies have looked at quantitative trait loci (QTLs) in biparental mapping population using standard sources of resistance. The advantage of GWAS is not only in the allele mining of new genes to use in molecular breeding but also in the discovery of new sources of resistance from genotypes that were not previously considered to begin with as donor parents for use in classical or modern breeding. GWAS also gives the SNP marker that can be converted into an easy-to-use selectable KASP marker for rapid introgression into advanced lines by methods of speed breeding.

Studying another foliar pathogen, *Sclerotinia sclerotiorum*, the causal agent of white mold, Campa et al. (2020) evaluated 294 dry bean lines from the Spanish diversity panel for resistance to the disease. Using 9,070 SNPs generated by ApeK1 digest GBS analysis, the authors found 13 robust MTAs on six chromosomes with single-locus association and 22 on the same linkage groups with multi-locus association. These were grouped into 15 intervals, 8 overlapping with previously identified QTIs, while 7 as new R-loci. In total, 68 candidate genes were identified (22 related to pathogen recognition, 8 to signal transduction, and 31 to defense response).

1.5.2 Disease Resistance GWAS Studies in Snap Bean

Snap beans suffer from many of the diseases mentioned earlier but, in addition, must be carefully raised to avoid blemished pods. For this reason, GWAS is particularly important with snap bean. In a recent study for vegetable germplasm, Arkwazet al. (2022) studied white mold using 136 and 378 genotypes, respectively, from the previously discussed snap bean diversity and association panels and 34,442 high-quality SNPs to conduct GWAS analysis. They found 23 MTAs for field resistance based on a severity index and 11 MTAs for greenhouse resistance in a seedling test based on evaluation of recently germinated plants in infected straw at four days after inoculation with a 1–9 scale. In total, 11 of the MTAs were close to LRR-encoding genes and 15 were close to pathogenesis response protein genes.

Studying a further foliar disease of snap beans, Wu et al. (2022) used GWAS as a means of marker and germplasm discovery with Illumina re-sequencing to identify resistance to rust caused by *Uromyces appendiculatus*. They based the study on only 17 resistant and 71 susceptible snap bean accessions but used a high-density map of 603,910 SNPs. Despite there being more than a dozen well-studied, race-specific resistance genes in common bean (*Ur*-series), the authors found new loci among 64 non-redundant MTAs that were identified. These occurred in three genomic regions (Qur-1, Qur-2, and Qur-3) possessing 102, 231, and 100 predicted genes, respectively. Some of the resistance included immunity of no visible lesion or infected lesions without sporulation, which is usually not considered in the spectrum of resistance from lesion number, size, and halos. Notably, Qur-1 possessed four genes classified as leucine-rich repeat protein kinases. Qur-2 contained genes classified as NB-ARC domain-containing disease resistance proteins, disease-responsive dirigent-like proteins, and ABC transporter family proteins Qur-3 had 6 genes for NB-ARC domain disease resistance proteins.

1.5.3 Insect Resistance GWAS Studies

To our knowledge, there has been only one insect resistance study using GWAS so far, and it was conducted in sub-Saharan Africa, where insect pests tend to be much more serious on common bean than in the rest of the world. Nkhata et al. (2021) analyzed bean fly resistance using 99 genotypes primarily from his region and a service technology mentioned before called DArTSeq to generate 15,565 SNPs. Bean fly severity damage was evaluated on a 1–9 scale backed up by data on plant mortality rate (number of plants dead per plot). In addition, the authors conducted pupae counts, which get to the main symptom of insect damage, which is lodging of the plants when the stem is devastatingly infected.

As results, bean fly severity damage, plant mortality rate, and pupal counts had 10, 4, and 12 significant MTAs. These were located on 5, 3, and 8 chromosomes, respectively. A total of 11 candidate genes for pupae and 1 for plant mortality rate were identified. The number of pupae at the base of the common bean stem determines the likelihood of plant death caused by the insect. Some plants may be more attractive to the bean fly if they have not been bred in Africa, for example, most exotic snap bean germplasm, which must be protected with strong neonicatanoid systemic insecticides like Imidacloprid, which is considered an environmental hazard.

1.5.4 Nutritional GWAS Studies

Another cross-cutting theme for dry beans and snap beans, and for both the agronomic and horticultural worlds they represent, is the nutritional quality of their seed or other consumed plant parts (leaves, pods, etc.) for humans or, in some cases, livestock (for example, stems are used as fodder for goats and sheep in Ethiopia). Although the protein and mineral accumulation of different plant parts is of interest for all these sub-products, the majority of GWAS studies other than the one on phenolics of snap bean pods by Myers et al. (2019) have been conducted on dry grain (Table 1.4). These include a recent study by Caproni et al. (2020) of 179 European landraces and 13 cultivars finding a common area for Fe and Zn with 12 MTAs on Pv01 after screening with 49,518 SNPs. Erdogmus et al. (2020) evaluated 173 diverse genotypes with 19,204 SNP data from DArTseq in three sites in Turkey (Bornova and Menemen, Izmir) to evaluate calcium (Ca) and manganese (Mn)

TABLE 1.4
Nutritional Studies of Common Beans (*Phaseolus vulgaris* L.)

Title	Authors	Journal (Year)	Study Type/Population	Marker Type and Number	Traits Evaluated
Genome-Wide Association Studies of Mineral Content in Common Bean	Ochatt et al. (2021)	*Frontiers in Plant Science* 12:636484.	174 Croatian landraces of common beans	17,514 polymorphic markers; 8,092 SNP from DArTseq.	Seed mineral content: (1) iron, (2) calcium, (3) nitrogen, (4) phosphorous, (5) magnesium, (6) zinc
European landrace diversity for common bean biofortification: a genome-wide association study	Caproni et al. (2020)	*Scientific Reports* 10:1	179 landraces and 13 cultivars of 5 generations of single seed descent	49,518 SNPs	(1) Iron, (2) zinc, and (3) phytate concentrations
Portuguese Common Bean Natural Variation Helps to Clarify the Genetic Architecture of the Legume's Nutritional Composition and Protein Quality	Mendes et al. (2021)	*Plants* 11:1–26	106 accessions from Portuguese plant germplasm	16,689 SNPs of illumine Infinium BARCBean6k_3 Bead Chip assay and DArTseq	Macronutrients: proteins, carbohydrates, fat, moisture resistance, starch, fiber; amino acids and trypsin inhibitor activity
Unlocking the genomic regions associated with seed protein contents in Turkish common bean germplasm through genome-wide association study	Baloch and Nadeem (2022)	*Turkish Journal of Agriculture and Forestry* 11:46	182 accessions from 19 provinces of Turkey	7,900 DArTseq markers	11 DArTseq markers associated with protein content
Genome-wide association studies of Ca and Mn in the seeds of the common bean (*Phaseolus vulgaris* L.)	Erdogmus et al. (2020)	*Genomics* 112:6–11	173 diverse genotypes; experiment carried out in Turkey (Bornova and Menemen, Izmir, Turkey)	19,204 SNPs of DArTseq	Calcium: 6 SNPs from Bornova and 6 SNPs from Menemen Manganese: 8 (2015) and 4 (2016) from Bornova, 8 (2015) and 11 (2016) from Menemen
Genome-wide association study for grain mineral content in a Brazilian common bean diversity panel.	Delfini et al. (2021c)	*Theoretical and Applied Genetics* 134:9	178 Mesoamerican accessions from Brazilian diversity panel grown at IDR-Paraná Research Stations in 3 sites (Londrina, Ponta Grossa, and Guarapuva)	707 SNPs used after removing high-LD SNPs	7 P, 4 K, 4 Ca, 11 Mg, 2 Cu, 6 Mn, 5 S, 7 Zn, and 2 Fe QTNs detected
Genome-Wide Association Study of Seed Folate Content in Common Bean	Martin et al. (2021)	*Frontiers in Plant Science* 12	96 accessions	5,361 molecular markers with Illumina Infinium platform	6 QTLs associated with seed foliate content

concentrations. Genotype–environment interactions were strong for the uptake of the elements in that study, as evidenced by six different MTAs each found for Ca in Bornova and Menemen, and different years having varying MTAs for Mn (8 in 2015 and 4 in 2016) from Bornova, and 8 in 2015 and 11 in 2016 for Menemen, with none in Izmir.

A year later, Ochatt et al. (2021) conducted a GWAS study with 174 Croatian landraces and 17,514 polymorphic markers to evaluate MTAs for various macro- and micronutrients. Their study involved measuring the following seed minerals: (1) magnesium (Mg), (2) nitrogen (N), and (3) phosphorous (P), as well as (4) Ca, (5) Fe, and (6) Zn. Single QTL was found for Ca on Pv08, and Mg on Pv09. Five QTLs were found for P, four on Pv07 and one on Pv08. Two for Zn on Pv06. No QTL for K, Fe, Mn.

Delfini et al. (2021c) used a Brazilian common bean diversity panel of 178 Mesoamerican accessions grown at IDR-Paraná Research Stations in the cities of Londrina, Ponta Grossa, and Guarapuava to discover MTAs for some of the same minerals described previously. After removing high-LD SNPs, only 707 loci were used, but 11 still uncovered for Mg, 7 QTL each for P and Zn, 6 for Mn, 4 each for Ca and K, 2 each for Cu and Fe. Other studies with few markers included one by Rasool et al. (2022) for P, K, and S varying among 96 Himalayan genotypes.

Another micronutrient, folate, being a type of vitamin B precursor, was studied by Martin et al. (2021) using 96 accessions and 5,361 Illumina Infinium markers, finding six QTLs associated with seed folate content. Other types of vitamin B would be of interest to evaluate in common bean along with sources of its other valuable nutraceutical properties. Arguably, the most important macronutrient from common bean is nitrogen, as this is reflected in protein content. This nutrient, plus starch, sugars, and fiber, have recently been considered by several authors.

Protein GWAS studies in common bean have been undertaken so far in Eurasia, despite the trait's importance in fighting malnutrition in the Global South, especially in Africa and Latin America, where meat is rarely available or getting more and more expensive. Baloch and Nadeem (2022) looked at 182 accessions from 19 provinces of Turkey to determine seed protein content. With 7,900 DArTseq markers, they found 11 MTAs for this trait on chromosomes Pv02, Pv04, Pv05, Pv07, and Pv09. Mendes et al. (2021) evaluated 106 accessions from Portugal using a total of 16,689 SNPs combined from Infinium BARCBean6k_3 Chip assay and DArTseq evaluations to find MTAs for all the macronutrients, including proteins (MTAs on six chromosomes), as well resistance starches (on 4 chromosomes). Fiber with MTA on Pv07 and the amino acid profile (Pv02 and Pv09), were also evaluated.

1.6 NEXT FRONTIERS

A major question to leverage allele mining and pre-breeding of common bean is whether there is sufficient heritable variation in traits associated with yield, tolerance to abiotic and biotic stresses, and adaptation, and whether the nature of such variation, typically polygenic, is a realistic selection target given epistatic interactions and expression effects.

In this context, genotyping of germplasm and reference collections comprising crop wild relatives (CWR) and landraces that span heterogeneous habitats offer an invaluable source of natural standing variation. Allelic clines that correlate with habitat gradients are signatures of the natural selection imprint (Forester et al., 2016), shaped through thousands of generations by the natural equilibrium between genotypes and their environment that favors survival (Hancock et al., 2011), often compromising productivity. Therefore, pre-breeding schemes must foresee utilizing these novel sources of variation without scarifying sustainable yields.

1.6.1 GWAS with Crop Wild Relatives

Crop wild relatives for common bean are mainly in the same species of *P. vulgaris* (Cortes et al., 2012a; Cortés et al., 2012b; Blair et al., 2016). However, these can be considered difficult to use for breeding, although recent pioneering studies have found them useful for GWAS studies and gene discovery (López-Hernández and Cortés, 2019). Other related species like tepary bean – *Phaseolus acutifolius* (Mejía-Jiménez et al., 1994; Muñoz et al., 2003; Buitrago-Bitar et al., 2021; Burbano-Erazo et al., 2021) – constitute gene pools of adaptive diversity for abiotic stress tolerance. Yet despite their tolerance to drought and heat stress, their utilization is limited partly because of poor characterization of their genetic background. Therefore, a key research area is to expand modern molecular pre-breeding tools (Ahmad et al., 2020), such as genome re-sequencing (Fuentes-Pardo and Ruzzante, 2017; Wu et al., 2020) and genomic prediction (Desta and Ortiz, 2014), and couple them with introgression and hybrid breeding schemes to select well-adapted common bean genotypes for production in environments with lack of rainfall.

1.6.2 Specialized Screening of Traits

Laboratory or growth chamber analysis of traits is another way to mine for germplasm that contains novel phenotypic traits and which might therefore have novel alleles for a trait of interest in common beans. Examples of specialized screening of traits usually fall in the area of seed quality, such as (1) use of chemical digest and ICP evaluation of minerals (reviewed in Blair 2013c), (2) HPLC evaluation of seed coat extracts of tannins (Caldas and Blair, 2009) or seed extractions of phytates (Diaz et al., 2010), and (3) Mattson cooker to evaluate the cooking time of dry grain (Cichy et al., 2015).

1.6.3 Introgression Breeding to Boost Adaptation and Resistance

The genus *Phaseolus* comprises five cultivated species with contrasting geographies and adaptations (Debouck, 2000; Bitocchi et al., 2017). Common and lima beans (*P. lunatus* L.) both exhibit parallel domestications in the Andean and Mesoamerican gene pools (Blair et al., 2012b), with comparable genomic selection signatures (Cortés et al., 2018), despite the latter being better adapted to warm and humid environments (Chacon-Sanchez and Martinez-Castillo, 2017). Meanwhile, tepary bean (*P acutifolius* A. Gray) is the most heat- and drought-tolerant (Parsons and Howe, 1984), as it is an annual autogamous species from northwest Mexico near the arid border with the USA (Blair et al., 2012a; Gujaria-Verma et al., 2016; Mhlaba et al., 2018a; Mhlaba et al., 2018b). Hybrid breeding has played a key role in bidirectional introgression of novel adaptive and resistant alleles (Burgarella et al., 2019; Kumar et al., 2020). Leverage hybrid pre-breeding in common bean is an innovation found in the literature for a couple of decades (Muñoz et al., 2003) due to heterogeneous interspecific adaptation and trait segregation.

Even more heat and drought tolerance may be found within wild relatives of tepary bean. For instance, *P. parvifolius* (Freytag) is an ephemeral bean geo-referenced to desert localities that appears at the borders of dry upland forests after desert rains (Muñoz et al., 2003; Cruz Jimenez-Galindo, 2018). Year and runner beans (*P. dumosus* and *P. coccineus*, respectively) also harbor resistance allelic diversity to biotic stresses, such as fungi, due to their adaptation to humid and cold environments (Spataro et al., 2011; Guerra-Garcia et al., 2017).

All these related species and wild gene pool have been utilized as donors of allelic diversity for common bean in terms of adaptation (Porch et al., 2013), abiotic stress tolerance (Mejía-Jiménez et al., 1994; Muñoz et al., 2003; Souter et al., 2017; Buitrago-Bitar et al., 2021; Burbano-Erazo et al., 2021), biotic stress resistance (Garzon et al., 2008), and even nutrient accumulation (Blair et al., 2010b; Blair and Izquierdo, 2012). These examples illustrate that despite intrinsic phylogenetic distance (Crespo et al., 2018), interspecific introgression pre-breeding offers an efficient alternative to unlock diversity and pyramid target alleles into common bean cultivars.

A common denominator of the introgression breeding approaches described earlier is that they rely on classical backcross (BC) populations (Blair et al., 2006). In its recurrent version, the offspring of the wild donor and the cultivar is recurrently crossed with the latter in order to enrich the stability of elite genomic features that confer agronomic advantage, while retaining the few regions

TABLE 1.5
GWAS or GEA Studies of Wild Bean (*Phaseolus vulgaris* L.)

Title	Authors	Journal (Year)	Study Type/Population	Marker Type and Number	Traits Evaluated
Signatures of Environmental Adaptation During Range Expansion of Wild Common Bean (*Phaseolus vulgaris*)	Ariani et al. (2019)	*bioRxiv*, 571042	A panel of 246 wild *P. vulgaris* accessions, including 157 Mesoamerican, 77 southern Andean, and 12 Northern Peru–Ecuador	19,126 SNPs from GBS using the *Cvi*AII restriction enzyme; in total, 49 genes identified associated with the bio-climatic variables selected	Seven bio-climatic variables, including isothermality, max temperature of warmest month, minimum temperature of coldest month, temperature annual range, annual precipitation, precipitation of driest month, and precipitation of warmest quarter
Last-Generation Genome–Environment Associations Reveal the Genetic Basis of Heat Tolerance in Common Bean (*Phaseolus vulgaris* L.)	López-Hernández et al. (2019)	*Frontiers in Genetics*, (10) 954	78 geo-referenced wild accessions spanning the two gene pools	GBS identified 23,373 SNP markers, from which 120 loci associated with the 3 bioclimatic indices, and 22 genes revealed related to heat tolerance in plants	Three different bioclimatic indices (HIT, HSI, and PCA1) incorporating six bioclimatic variables from WorldClim (annual mean temp., maximum temp. of warmer month, mean temp. of the wettest quarter, mean temp. of the driest quarter, mean temp. of the warmest 4-month period, and average of absolute maximum temp. during the reproductive phase)
Genome wide association study discovers genomic regions involved in resistance to soybean cyst nematode (*Heterodera glycines*) in common bean	Jian et al. (2019)	*PLOS One*, 14(2), e0212140	USDA core collection consisting of 423 PIs, including 10 wild accessions	5,398 SNPs from the Illumina BARCBean6K_3 Infinium SNP array, from which 14 and 23 significant marker–trait associations found for the Middle American and Andean subpopulations, respectively	Resistance to SCN HG type 0 indicated by a female index (FI) for each accession grown in the greenhouse

that converse exotic trait variation. However, alternating backcrosses between the exotic donors and the elite cultivars as part of congruity BC schemes serve to overcome interspecific incompatibilities and embryo abortion that are otherwise reinforced in the endogamous advanced recurrent generations of simpler BC populations (Muñoz et al., 2003; Burbano-Erazo et al., 2021).

GWAS-enabled markers-assisted selection (MAS) is also a key technology to speed up backcrossing and improve selection accuracies at each generation. MAS-coupled BC has successfully been implemented in common bean for the improvement of yield traits (Blair et al., 2006), resistance to biotic stresses (Garzon et al., 2008), tolerance to abiotic pressures (Muñoz et al., 2003; Buitrago-Bitar et al., 2021; Burbano-Erazo et al., 2021), and nutrient accumulation (Blair et al., 2010b; Blair and Izquierdo, 2012).

More recently, Conejo-Rodriguez et al. (2022) have managed to recover parental phenomic proportions in some of the hybrids between common bean and tepary bean and its wild relative *P. parvifolius*. These BC populations constitute a pre-breeding baseline to design customized cultivars (Varshney et al., 2021a) and further define downstream goals for target traits. Some of the parental species may themselves be re-adopted as novel food sources in certain regions (von Wettberg et al., 2020). However, classical BC schemes, or even more modern MAS-BC, may perform poorly when trying to filter for the multiple genomic regions with modest effects that determine complex polygenic traits (Boyle et al., 2017), such as yield, adaptation (Barghi et al., 2020), and abiotic stress tolerance (Westbrook et al., 2020). In order to improve the retention of the exotic polygenic bean phenotype together with its underlying genetic basis, BC may be coupled with genomic prediction (GP) models, enabling genomic-assisted backcrossing (GABC, Cortés and López-Hernández, 2021).

1.7 GENOMIC-ENABLED PREDICTIVE BREEDING FOR BEAN IMPROVEMENT

Common bean offers a key source of high dietary protein and micronutrient contents for the poorest. Yet common bean is generally considered susceptible to drought and heat, stresses that are expected to become more frequent due to climate change. Many of the valuable traits of common bean, such as its nutritional quality, are polygenic, affected by genotype–environment interactions and of low heritability, making them difficult to select for.

Therefore, leveraging molecular markers to speed up the pre-breeding discovery and mobilization of natural variation into common bean cultivars is a main avenue of research. Ultimately, envisioning avenues to better couple phenotypic screening and multi-dimensional section indices with genomic innovations will unlock hidden allelic variation to be utilized within pre-breeding cycles targeting common bean sustainable yield, adaptation, and biotic stress resistance under changing climate scenarios and given the intense food security situation of many developing countries that rely on food imports and aid but which may be able to produce higher amounts of common bean if the appropriate varieties are bred for their situations.

Polygenetic variance often hampers molecular breeding due to the difficulty to simultaneously select for the multiple alleles that shape the complex phenotype (Boyle et al., 2017). Overcoming this bottleneck in common bean genetic improvement requires updating gene-based GWAS and MAS approaches by polygenic additive models in which Fisherian infinitesimal allelic contributions are permissible (Fisher, 1930).

GP-derived genomic estimated breeding values (GEBVs), in turn, allow ranking genotypes according to their genetic value (de los Campos et al., 2015). The most prominent genomic-enabled predictive breeding strategy is genomic selection (Migicovsky and Myles, 2017), built upon genomic prediction models *sensu strictu* (Desta and Ortiz, 2014; Crossa et al., 2017), and modern machine learning algorithms (Crossa et al., 2019; Montesinos-López et al., 2021a, 2021b; Tong and Nikoloski, 2021).

Predictive breeding has already started permeating common bean improvement for traits such as yield (Lobaton et al., 2018; Keller et al., 2022), seed morphology (Sadohara et al., 2021), phenology (Rosado et al., 2020), cooking time (Diaz et al., 2021b), biotic stress resistance (Diaz et al., 2021a; Shi et al., 2021), and abiotic stress tolerance (Keller et al., 2020).

Predictions may improve if inputted with *de novo* GWAS-retrieved markers, yet results from allele mining with GWAS are already encouraging by themselves as to improve the timing and the accuracy of the selection cycles. However, further improvement of complex polygenic traits in common bean would definitively benefit by coupling predictive breeding not only with introgression breeding (i.e., GABC) initiatives (Burgarella et al., 2019; Kumar et al., 2020), as discussed in the previous section, but also with the diversifying selection (McCouch, 2004) and speed breeding (Watson et al., 2018; Alves et al., 2020; Varshney et al., 2021b) paradigms.

Even technologies traditionally developed for simpler monogenic Mendelian traits, like *de novo* domestication (Fernie and Yan, 2019) and genome editing (Ahmar et al., 2021), are starting to be implemented with success in polygenic genomic architectures of non-model plants (Lemmon et al., 2018). The extensions remain to be seen in bean improvement.

1.8 PERSPECTIVES

Common bean molecular breeding has entered the SNP marker (Hyten et al., 2008; Hyten et al., 2010; Souza et al., 2010; Cortés et al., 2011; Galeano et al., 2012; Blair et al., 2013b) and genomic era in the last decade (Schmutz et al., 2014; Blair et al., 2018). However, the common bean phenotyping urgently demands equivalent developments, comparable to those from the phenomic discipline, so that genome–phenotype inferences are not decoupled. Similarly, common bean breeding must start estimating genetic parameters and testing populations across the environmental continuum as part of the nascent enviromics approach (Cooper and Messina, 2021; Costa-Neto and Fritsche-Neto, 2021; Resende et al., 2021). Only then can common bean improvement may fulfil the plant breeding triangle (Crossa et al., 2021).

New paradigms demand novel platforms. For instance, modernized computational analytical tools are required to condense and interpret intricate trans-disciplinary multi-dimensional data (Tirnaz et al., 2022). Nonetheless, for such computational platforms to succeed, open networks (McCouch et al., 2016) of crop mobilization (McCouch et al., 2020) are equally paramount. After all, unlocking genetic and phenotypic novelty not only requires widespread genomic and phenotypic characterizations but also demands targeting isolated pockets of cryptic diversity (Ulian et al., 2020), usually conserved by national gene banks and breeding programs (Cerón-Souza et al., 2021).

Therefore, the common bean research and farmers community would benefit from open-access (Spindel and McCouch, 2016) platforms bridging germplasm conservation, parental screening (Blair et al., 2013a), pre- and breeding efforts, seed multiplication, and seed deployment initiatives (Peláez et al., 2022). Allele mining is the keystone of all these projects and is an opportunity to re-prioritize diversity in crop breeding.

1.9 AUTHOR CONTRIBUTIONS

MWB conceived and wrote this chapter with insights and contributions from AJC and DDA. The following are the contributed tables and research: HL (snap beans and wild relatives), MCMII (abiotic and biotic stresses), and LN (nutritional traits). All authors made substantial contributions in preparing and editing this manuscript. Figures were prepared by DDA and MWB.

1.10 FUNDING

The United States Department of Agriculture (USDA), through the National Institute of Food and Agriculture (NIFA) Evans Allen Program (TENX-07), funded the authors from Tennessee State University, along with a grant from Agriculture and Food Research Initiative (AFRI) for plant breeding program A1141 grant number AFRI-2021–67014–33935, for a PhD assistantship to MCMII. Agrosavia funded AJC through the Vetenskapsrådet (grants 2016–00418 and 2022–04411) and Kungliga Vetenskapsakademien (grant no. BS20170036), and the British Council (grant 527023146) funded AJC as PI during the drafting of this chapter. The Graduate Research School in

Genomic Ecology (GENECO) from Lund University funded AJC mobility to participate in synergistic meetings with MWB. Fulbright's US Specialist Program supported the visit of MWB to AJC during the summer of 2019 in Rionegro (Antioquia, Colombia).

ACKNOWLEDGMENTS

MWB thanks discussions with students from Tennessee State University: D. Diani, A. Hines, C. Randle, J. Vickers; from Emory University: M. Jenkins and H. Pachon; along with researchers at Agrosavia–La Selva station: L. F. López-Hernández, I. Cerón-Souza, C. I. Medina, D. Peláez, and P. H. Reyes-Herrera. MWB thanks C. H. Galeano for Agrosavia Organization of the Fulbright's US Specialist Program. AJC thanks M. J. Torres for support while drafting part of the chapter.

REFERENCES

Abebe TD, Naz AA, Leon J (2015) Landscape genomics reveal signatures of local adaptation in Barley (*Hordeum vulgare* L.). *Front Plant Sci* 6:813.

Adeyanju A, Little C, Yu J,Tesso T (2015) Genome-Wide Association Study on Resistance to Stalk Rot Diseases in Grain Sorghum. *G3* 5(6):1165–1175.

Ahmad R, Akbar AM, Mukhtar Balal R (2020) From markers to genome based breeding in horticultural crops: An overview. *Phyton* 89:183–204.

Ahmar S, Ballesta P, Ali M, Mora-Poblete F (2021) Achievements and challenges of genomics-assisted breeding in forest trees: From marker-assisted selection to genome editing. *Int J Mol Sci* 22:10583.

Akhavan A, Bahar M, Askarian H, Lak MR, Nazemi A et al. (2013) Bean common bacterial blight: pathogen epiphytic life and effect of irrigation practices. *SpringerPlus* 2:41.

Alves FC, Balmant KM, Resende MFR, Kirst M, Los Campos G (2020) Accelerating forest tree breeding by integrating genomic selection and greenhouse phenotyping. *Plant Genome* 13(3):e20048.

Ambachew D, Blair MW (2021) Genome wide association mapping of root traits in the andean genepool of common bean (*Phaseolus vulgaris* L.) grown with and without aluminum toxicity. *Front Plant Sci* 12:628687.

Ambachew D, Joshua J, Mmbaga MT, Blair MW (2021) Sources of resistance to common bacterial blight and charcoal rot disease for the production of Mesoamerican common beans in the southern United States. *Plants* 10(5):998.

Ambachew D, Mekbib F, Asfaw A, Beebe SE, Blair MW (2015) Trait associations in common bean genotypes grown under drought stress and field infestation by BSM bean fly *Crop J* 3(4):305–316.

Antao T, Lopes A, Lopes RJ, Beja-Pereira A, Luikart G (2008) Lositan: A workbench to detect molecular adaptation based on a fst-outlier method. *BMC Bioinf* 9(1):1–5.

Ariani A, Gepts P (2019) Signatures of environmental adaptation during range expansion of wild common bean (*Phaseolus vulgaris*). *BioRxiv*:571042.

Arkwazee HA, Wallace LT, Hart JP, Griffiths PD, Myers JR (2022) Genome-wide association study (GWAS) of white mold resistance in snap bean. *Genes* 13(12):2297.

Asfaw A, Ambachew D, Shah T, Blair MW (2017) Trait associations in diversity panels of the two common bean (*Phaseolus vulgaris* L.) gene pools grown under well-watered and water-stress conditions. *Front Plant Sci* 8:733.

Baloch FS, Nadeem MA (2022) Unlocking the genomic regions associated with seed protein contents in Turkishcommon bean germplasm through genome-wide association study. *Turk J Agric For* 46(1):113–128.

Barghi N, Hermisson J, SchlöTterer C (2020) Polygenic sdaptation: A unifying framework to understand positive selection. *Nat Rev Genet* 21:769–781.

Barton N, Hermisson J, Nordborg M (2019) Why structure matters. *Elife* 8:e45380.

Belete T, Bastas KK (2017) Common Bacterial Blight (*Xanthomonas axonopodis* pv. phaseoli) of Beans with Special Focus on Ethiopian Condition. *J Plant Pathol Microbiol* 8:403.

Bitocchi E, Rau D, Bellucci E, Rodriguez M, Murgia M et al. (2017) Beans (*Phaseolus* Ssp.) as a model for understanding crop evolution. *Front Plant Sci* 722.

Blair MW, Brondani RVP, Díaz LM, Del Peloso MJ (2013a) Diversity and population structure of common beans from Brazil. *Crop Sci* 53:1983–1993.

Blair MW, Cortes AJ, Farmer AD, Huang W, Ambachew D et al. (2018) Uneven recombination rate and linkage disequilibrium across a reference SNP map for common bean (*Phaseolus vulgaris* L.). *PLoS One* 13:e0189597.

Blair MW, Cortés AJ, Penmetsa RV, Farmer A, Carrasquilla-Garcia N et al. (2013b) A high-throughput SNP marker system for parental polymorphism screening, and diversity analysis in common bean (*Phaseolus vulgaris* L.). *Theor Appl Genet* 126:535–548.

Blair MW, Cortés AJ, This D (2016) Identification of an erecta gene and its drought adaptation associations with wild and cultivated common bean. *Plant Sci* 242:250–259.

Blair MW, Galeano CH, Tovar E, Muñoz MC, Castrillón AV et al. (2010a) Development of a Mesoamerican intra-genepool genetic map for quantitative trait loci detection in a drought tolerant × susceptible common bean (*Phaseolus vulgaris* L.) cross. *Mol Breed* 29:71–88.

Blair MW, Izquierdo P (2006) QTL analysis of yield traits in an advanced backcross population derived from a cultivated Andean x wild common bean (*Phaseolus vulgaris* L.) cross. *Theorand Appl Genet* 112:1149–1163.

Blair MW, Izquierdo P (2012) Use of the advanced backcross-QTL method to transfer seed mineral accumulation nutrition traits from wild to andean cultivated common beans. *Theor Appl Genet* 125:1015–1031.

Blair MW, Izquierdo P, Astudillo C, Grusak MA (2013c). A legume biofortification quandary: variability and genetic control of seed coat micronutrient accumulation in common beans. *Front Plant Sci* 4:275.

Blair MW, Monserrate F, Beebe SE, Restrepo J, Flores JO (2010b) Registration of high mineral common bean germplasm lines NUA35 and NUA56 from the red-mottled seed class. *J Plant Registr* 4:55–59.

Blair MW, Pantoja W, Munoz CL (2012a) First use of microsatellite markers in a large collection of cultivated and wild accessions of tepary bean (*Phaseolus acutifolius* A. Gray). *Theor Appl Genet* 125:1137–1147.

Blair MW, Soler A, Cortés AJ (2012b) Diversification and population structure in common beans (*Phaseolus vulgaris* L.). *PLos One* 7:e49488.

Bost S, Canaday C, Hale F, Hensley D, Wszelaki A (2013) In: *A Crop Profile for Beans (Snap) in Tennessee.* University of Tennessee, Knoxville, TN, USA.

Boyle EA, Li YI, Pritchard JK (2017) An expanded view of complex traits: From polygenic to omnigenic. *Cell* 169:1177–1186.

Broughton WJ, Hernandez G, Blair MW, Beebe SE, Gepts P et al. (2003) Beans (*Phaseolus* spp.) – model food legumes. *Plant Soil* 252:55–128.

Buitrago-Bitar MA, Cortés AJ, López-Hernández F, Londoño-Caicedo JM, Muñoz-Florez JE et al. (2021) Allelic diversity at abiotic stress responsive genes in relationship to ecological drought indices for cultivated tepary bean, *Phaseolus acutifolius* A. Gray, and its wild relatives. *Genes* 12:556.

Burbano-Erazo E, León-Pacheco R, Cordero-Cordero C, López-Hernández F, Cortés A et al. (2021) Multi-environment yield components in advanced common bean (*Phaseolus vulgaris* L.) × tepary bean (*P. acutifolius* A. Gray) interspecific lines for heat and drought tolerance. *Agronomy* 11:1978.

Burgarella C, Barnaud A, Kane NA, Jankowski F, Scarcelli N et al. (2019) Adaptive introgression: An untapped evolutionary mechanism for crop adaptation. *Front Plant Sci* 10:4.

Caldas GV, Blair MW (2009) Inheritance of seed condensed tannins and their relationship with seed-coat color and pattern genes in common bean (*Phaseolus vulgaris* L.). *Theor Appl Genet* 119(1):131–42.

Campa A, García-Fernández C, Ferreira JJ (2020) Genome-wide association study (GWAS) for resistance to *Sclerotinia sclerotiorum* in common bean. *Genes* 11(12):1496.

Caproni L, Raggi L, Talsma EF, Wenzl P, Negri V (2020) European landrace diversity for common bean biofortification: A genome-wide association study. *Sci Rep* 10(1):1–13.

Cerón-Souza I, Galeano CH, Tehelen K, Jiménez HR, González C (2021) Opportunities and challenges to improve a public research program in plant breeding and enhance underutilized plant genetic resources in the tropics. *Genes* 12:1584.

Chacon-Sanchez MI, Martinez-Castillo J (2017) Testing domestication scenarios of lima bean (*Phaseolus lunatus* L.) in Mesoamerica: Insights from genome-wide genetic markers. *Front Plant Sci* 8:1551.

Cichy KA, Porch TG, Beaver JS, Cregan P, Fourie D et al. (2015) A *Phaseolus vulgaris* Diversity Panel for Andean Bean Improvement. *Crop Sci* 55:2149–2160.

Cichy KA, Wiesinger JA, Mendoza FA (2015) Genetic diversity and genome-wide association analysis of cooking time in dry bean (*Phaseolus vulgaris* L.). *Theor Appl Genet* 128(8):1555–67.

Cloud GL, Rupe JC (1991) Morphological instability on a chlorate medium of isolates of *Macrophomina phaseolina* from soybean and sorghum. *Phytopathology* 81:892–895.

Conejo-Rodriguez DF, Urban MO, Santaella M, Gereda JM, Contreras AD et al. (2022) Using phenomics to identify and integrate traits of interest for better-performing common beans: A validation study on an interspecific hybrid and its acutifolii parents. *Front Plant Sci* 13:1008666.

Cooper M, Messina CD (2021) Can we harness "enviromics" to accelerate crop improvement by integrating breeding and agronomy? *Front Plant Sci* 12:735143.

Cortés AJ, Blair MW (2018) Genotyping by sequencing and genome – environment associations in wild common bean predict widespread divergent adaptation to drought. *Front Plant Sci* 9:128.

Cortés AJ, Chavarro MC, Blair MW (2011) SNP marker diversity in common bean (*Phaseolus vulgaris* L.). *Theor Appl Genet* 123:827–845.

Cortés AJ, Chavarro MC, Madriñán S, This D, Blair MW (2012a) Molecular ecology and selection in the drought-related Asr gene polymorphisms in wild and cultivated common bean (*Phaseolus vulgaris* L.). *BMC Genet* 13(1):1–14.

Cortés AJ, López-Hernández F (2021) Harnessing crop wild diversity for climate change adaptation. *Genes* 12(5):783.

Cortés AJ, López-Hernández F, Blair MW (2022) Genome–environment associations, an innovative tool for studying heritable evolutionary adaptation in orphan crops and wild relatives. *Front Genet* 13:910386.

Cortés AJ, López-Hernández F, Osorio-Rodriguez D (2020) Predicting thermal adaptation by looking into populations' genomic past. *Front Genet* 11:564515.

Cortés AJ, Skeen P, Blair MW, Chacón-Sánchez MI (2018) Does the genomic landscape of species divergence in *Phaseolus* beans coerce parallel signatures of adaptation and domestication? *Front Plant Sci* 9:1816.

Cortés AJ, This D, Chavarro C, Madriñán S, Blair MW (2012b) Nucleotide diversity patterns at the drought-related Dreb2 encoding genes in wild and cultivated common bean (*Phaseolus vulgaris* L.). *Theor Appl Genet* 125:1069–1085.

Cortinovis G, Frascarelli G, Di Vittori V, Papa R (2020) Current state and perspectives in population genomics of the common bean. *Plants* 9(3):330.

Coser SM, Chowda Reddy RV, Zhang J, Mueller DS, Mengistu A et al. (2017) Genetic Architecture of Charcoal Rot (Macrophomina phaseolina) Resistance in Soybean Revealed Using a Diverse Panel. *Front Plant Sci* 8:1626.

Costa-Neto G, Fritsche-Neto R (2021) Enviromics: Bridging different sources of data, building one framework. *Crop Breed Appl Biotechnol* 21:e393521S393512.

Crespo S, Rivera-Peña M, Rosero-Alpala DA, Muñoz-Florez JE, Rao IM et al. (2018) Pollen viability of tepary bean (*Phaseolus acutifolius* A. Gray.) mutant lines under water stress conditions and inoculation with rhizobia. *Acta Agron* 67:319–325.

Crossa J, Fritsche-Neto R, Montesinos-Lopez OA, Costa-Neto G, Dreisigacker S et al. (2021) The modern plant breeding triangle: Optimizing the use of genomics, phenomics, and enviromics data. *Front Plant Sci* 12:651480.

Crossa J, Martini JWR, Gianola D, Perez-Rodriguez P, Jarquin D et al. (2019) Deep kernel and deep learning for genome-based prediction of single traits in multienvironment breeding trials. *Front Genet* 10:1168.

Crossa J, Perez-Rodriguez P, Cuevas J, Montesinos-Lopez O, Jarquin D et al. (2017) Genomic selection in plant breeding: Methods, models, and perspectives. *Trends Plant Sci* 22:961–975.

Cruz Jimenez-Galindo J (2018) Screening for drought tolerance in tepary and common bean based on osmotic potential assays. *Plant* 6:24.

De Los Campos G, Sorensen D, Gianola D (2015) Genomic heritability: What is it? *PLoS Genet* 11:e1005048.

De Luque JJR, Creamer B (2015) Major constraints and trends for common bean production and commercialization, establishing priorities for future research. *Agron Colomb* 32:423–431.

Debouck DG (2000) Biodiversity, ecology and genetic resources of *Phaseolus* beans-seven answered and unanswered questions. In *The Seventh Ministry of Agriculture, Forestry and Fisheries (MAFF), Japan, International Workshop on Genetic Resources, Ibaraki, Japan, 13–15 October, 1999: Part 1, wild legumes*. National Institute of Agrobiological Resources (NIAR), pp. 95–123.

Delfini J, Moda-Cirino V, dos Santos Neto J, Zeffa DM, Nogueira AF et al. (2021a) Genome-wide association study for grain mineral content in a brazilian common bean diversity panel. *Theor Appl Genet* 134:2795–2811.

Delfini J, Moda-Cirino V, dos Santos Neto J, Zeffa DM, Nogueira AF et al. (2021b) Genome-wide association study identifies genomic regions for important morpho-agronomic traits in Mesoamerican common bean. *Front Plant Sci* 12:748829.

Delfini J, Moda-Cirino V, Dos Santos Neto J, Ruas PM, Sant'Ana GC et al. (2021c) Population structure, genetic diversity and genomic selection signatures among a Brazilian common bean germplasm. *Sci Rep* 11(1):2964.

Desta ZA, Ortiz R (2014) Genomic selection: Genome-wide prediction in plant improvement. *Trends Plant Sci* 19:592–601.

Dia LM, Arredondo V, Ariza-Suarez D, Aparicio J, Buendia HF et al. (2021a) Genetic analyses and genomic predictions of root rot resistance in common bean across trials and populations. *Front Plant Sci* 12:629221.

Diaz S, Ariza-Suarez D, Izquierdo P, Lobaton JD, de la Hoz JF et al. (2020) Genetic mapping for agronomic traits in a MAGIC population of common bean (*Phaseolus vulgaris* L.) under drought conditions. *BMC Genom* 21:799.

Diaz S, Ariza-Suarez D, Ramdeen R, Aparicio J, Arunachalam N et al. (2021b) Genetic architecture and genomic prediction of cooking time in common bean (*Phaseolus vulgaris* L.). *Front Plant Sci* 11:622213.

Diaz LM, Arredondo V, Ariza-Suarez D, Aparicio J, Buendia HF et al. (2021a) Genetic Analyses and Genomic Predictions of Root Rot Resistance in Common Bean Across Trials and Populations. *Front Plant Sci* 12:629221.

Díaz AM, Caldas GV, Blair MW (2010) Concentrations of condensed tannins and anthocyanins in common bean seed coats. *Food Res Int* 43(2):595–601.

Dramadri IO, Amongi W, Kelly JD, Mugisha Mukankusi C (2020) Genome-wide association analysis of resistance to *Pythium ultimum* in common bean (*Phaseolus vulgaris*). *Plant Breed* 139(6):1168–1180.

Elias JCF, Gonçalves-Vidigal MC, Ariani A, Valentini G, Martiniano-Souza MDC et al. (2021) Genome-environment association analysis for bio-climatic variables in common bean (*Phaseolus vulgaris* L.) from Brazil. *Plants* 10:1572.

Ellegren H, Wolf JBW (2017) Parallelism in genomic landscapes of differentiation, conserved genomic features and the role of linked selection. *J Evol Biol* 30:1516–1518.

Erdogmus S, Ates D, Nemli S, Yagmur B, Asciogul TK et al. (2020) Genome-wide association studies of Ca and Mn in the seeds of the common bean (*Phaseolus vulgaris* L.). *Genomics* 112(6):4536–4546.

Fernie AR, Yan J (2019) De novo domestication: An alternative route toward new crops for the future. *Mol Plant* 12:615–631.

Fisher RA (1930) *The Genetical Theory of Natural Selection*. Oxford University Press, Oxford.

Forester BR, Jones MR, Joost S, Landguth EL, Lasky JR (2016) Detecting spatial genetic signatures of local adaptation in heterogeneous landscapes. *Mol Ecol* 25:104–120.

Fuentes-Pardo AP, Ruzzante DE (2017) Whole-genome sequencing approaches for conservation biology: Advantages, limitations and practical recommendations. *Mol Ecol* 26:5369–5406.

Galeano CH, Cortés AJ, Fernandez AC, Soler A, Franco-Herrera N et al. (2012) Gene-based single nucleotide polymorphism markers for genetic and association mapping in common bean. *BMC Genet* 13:48.

García-Fernández C, Campa A, Garzón AS, Miklas P, Ferreira JJ (2021) GWAS of pod morphological and color characters in common bean. *BMC Plant Biol* 21:184.

Garzon LN, Ligarreto GA, Blair MW (2008) Molecular marker-assisted backcrossing of anthracnose resistance into andean climbing beans (*Phaseolus vulgaris* L.). *Crop Sci* 48:562–570.

Gepts P. (2014) The contribution of genetic and genomic approaches to plant domestication studies. *Curr Opin Plant Biol* 18(5):1–9.

Gepts P, Aragao F, Barros E, Blair MW, Brondani R, Broughton W, Hernández G, Kami J, Lariguet P, McClean P, Melotto M, Miklas P, Pedrosa-Harand A, Porch T, Sánchez F (2008) Genomics of Phaseolus beans, a major source of dietary protein and micronutrients in the tropics. In: Moore PH, Ming R (eds) *Genomics of Tropical Crops*. Springer Publ., Chp 5, pp. 113–143.

Giordani W, Gama HC, Chiorato AF, Garcia AAF, Vieira MLC (2022) Genome-wide association studies dissect the genetic architecture of seed shape and size in common bean. *G3 Genes Genomes Genet* 12(4):jkac048.

Guerra-Garcia A, Suarez-Atilano M, Mastretta-Yanes A, Delgado-Salinas A, Pinero D (2017) Domestication genomics of the open-pollinated scarlet runner bean (*Phaseolus coccineus* L.). *Front Plant Sci* 8:1891.

Gujaria-Verma N, Ramsay L, Sharpe AG, Sanderson L-A, Debouck DG et al. (2016) Gene-based SNP discovery in tepary bean (*Phaseolus acutifolius*) and common bean (P. Vulgaris) for diversity analysis and comparative mapping. *BMC Genom* 17:239.

Hancock A., Brachi B, Faure N, Horton MW, Jarymowycz LB et al. (2011) Adaptation to climate across the *Arabidopsis Thaliana* genome. *Science* 334:83–86.

Holland JB, Nyquist WE, Cervantes-Martínez CT (2003) Estimating and interpreting heritability for plant breeding: An update. *Plant Breed Rev* 2003:9–112.

Huber CD, Degiorgio M, Hellmann I, Nielsen R (2016) Detecting recent selective sweeps while controlling for mutation rate and background selection. *Mol Ecol* 25:142–156.

Hughes T (2013). Disease Profile: Charcoal Rot of Soybean. 608, 3938.

Hyten DL, Song Q, Choi IY, Yoon MS, Specht JE et al. (2008) High-throughput genotyping with the goldengate assay in the complex genome of soybean. *Theor Appl Genet* 116:945–952.

Hyten DL, Song Q, Fickus EW, Quigley CV, Lim J-S et al. (2010) High-throughput SNP discovery and assay development in common bean. *BMC Genom* 11:475.

Jain S, Poromarto S, Osorno JM, McClean PE, Nelson BD (2019) Genome wide association study discovers genomic regions involved in resistance to soybean cyst nematode (*Heterodera glycines*) in common bean. *PLoS One* 14(2):e0212140.

Jana T, Sharma T, Prasad RD, Arora DK (2003) Molecular characterization of *Macrophomina phaseolina* and *Fusarium* species by a single primer RAPD technique. *Microbiol Res* 158: 249–257.

Kamfwa K, Cichy KA, Kelly JD (2015b) Genome-wide association analysis of symbiotic nitrogen fixation in common bean. *Theor Appl Genet* 128(10):1999–2017.

Keller B, Ariza-Suarez D, De La Hoz J, Aparicio JS, Portilla-Benavides AE et al. (2020) Genomic prediction of agronomic traits in common bean (*Phaseolus vulgaris* L.) under environmental stress. *Front Plant Sci* 11:1001.

Keller B, Ariza-Suarez D, Portilla-Benavides AE, Buendia HF, Aparicio JS et al. (2022) Improving association studies and genomic predictions for climbing beans with data from bush bean populations. *Front Plant Sci* 13:830896.

Kruglyak L (2008) The road to genome-wide association studies. *Nat Rev Genet* 9:314–318.

Kumar S, Hilario E, Deng CH, Molloy C (2020) Turbocharging introgression breeding of perennial fruit crops: A case study on apple. *Hortic Res* 7:47.

Lambert CG, Black, LJ (2012) Learning from our gwas mistakes: From experimental design to scientific method. *Biostatistics* 13:195–203.

Lemmon ZH, Reem NT, Dalrymple J, Soyk S, Swartwood KE et al. (2018) Rapid improvement of domestication traits in an orphan crop by genome editing. *Nat Plants* 4:766–770.

Liu Z, Gao S, Zhang H, Xu Z, Qian W (2022) Genome-wide association study reveals that PvGUX1_1 is associated with pod stringlessness in snap bean (*Phaseolus vulgaris* L.). *Biology* 11(4):611.

Lobaton JD, Miller T, Gil J, Ariza D, De La Hoz JF et al. (2018) Resequencing of common bean identifies regions of inter–gene pool introgression and provides comprehensive resources for molecular breeding. *Plant Genome* 11:170068.

López-Hernández F, Cortés AJ (2019) Last-generation genome–environment associations reveal the genetic basis of heat tolerance in common bean (*Phaseolus vulgaris* L.). *Front Genet* 10:22.

Maher B (2008) Personal genomes: The case of the missing heritability. *Nature* 456(7218):18–21.

Martin CJ, Torkamaneh D, Arif M, Pauls K P (2021) Genome-wide association study of seed folate content in common bean. *Front Plant Sci* 12:696423.

Mccouch S (2004) Diversifying selection in plant breeding. *PLoS Biol* 2:1507–1512.

Mccouch S, Navabi K, Abberton M, Anglin NL, Barbieri RL et al. (2020) Mobilizing crop biodiversity. *Mol Plant* 13:1341–1344.

Mccouch SR, Wright MH, Tung CW, Maron LG, Mcnally KL et al. (2016) Open access resources for genome-wide association mapping in rice. *Nat Commun* 7:10532.

Mejía-Jiménez A, Muñoz C, Jacobsen HJ, Roca WM, Singh SP (1994) Interspecific hybridization between common and tepary beans: Increased hybrid embryo growth, fertility, and efficiency of hybridization through recurrent and congruity backcrossing. *Theor Appl Genet* 88:324–331.

Mendes FA, Leitão ST, Correia V, Mecha E, Rubiales D et al. (2021) Portuguese common bean natural variation helps to clarify the genetic architecture of the legume's nutritional composition and protein quality. *MDPI Plants* 11(1):26.

Mengistu A et al. (2015) In: Hartman GL, Rupe JC, Sikora EJ, Domier LL, Davis JA, Steffey KL (eds) *Compendium of Soybean Diseases and Pests*. American Phytopathological Society, St. Paul, MN, pp. 67.

Mhlaba ZB, Amelework B, Shimelis H, Modi AT, Mashilo J (2018a) Genetic differentiation among selected tepary bean collections revealed by morphological traits and simple sequence repeat markers. *Acta Agric Scand Sect B Soil Plant Sci* 68:608–618.

Mhlaba ZB, Mashilo J, Shimelis H, Assefa AB, Modi AT (2018b) Progress in genetic analysis and breeding of tepary bean (*Phaseolus acutifolius* A. Gray): A review. *Sci Hortic* 237:112–119.

Migicovsky Z, Myles S (2017) Exploiting wild relatives for genomics-assisted breeding of perennial crops. *Front Plant Sci* 8:460.

Miklas PN, Stone V, Urrea CA, Johnson E, Beaver JS (1998) Inheritance and QTL Analysis of Field Resistance to Ashy Stem Blight in Common Bean. *Crop Sci* 38: 916–921.

Moghaddam SM, Mamidi S, Osorno JM, Lee R, Brick M et al. (2016) Genome-wide association study identifies candidate loci underlying agronomic traits in a middle American diversity panel of common bean. *Plant Genome* 9(3) https://doi.org/10.3835/plantgenome2016.02.0012.

Montesinos-López OA, Montesinos-López A, Hernandez-Suarez CM, Barrón-López JA, Crossa J (2021a) Deep-learning power and perspectives for genomic selection. *Plant Genome* 14:e20122.

Montesinos-López OA, Montesinos-LóPez A, PéRez-RodríGuez P, BarróN-LóPez JA, Martini JWR et al. (2021b) A review of deep learning applications for genomic selection. *BMC Genom* 22:19.

Muchero W, Ehlers JD, Close TJ, Roberts PA (2011) Genic SNP markers and legume synteny reveal candidate genes underlying QTL for *Macrophomina phaseolina* resistance and maturity in cowpea [*Vigna unguiculata* (L) Walp.]. *BMC Genom* 12:8.

Muñoz LC, Blair MW, Duque MC, Tohme J, Roca W (2003) Introgression in common bean X tepary bean interspecific congruity-backcross lines as measured by AFLP markers. *Crop Sci* 44:637–645.

Myers JR, Wallace LT, Moghaddam SM, Kleintop AE, Echeverria D et al. (2019) Improving the health benefits of snap bean: Genome-wide association studies of total phenolic content. *MDPI Nutrients* 11(10):2509.

Nemli S, Asciogul TK, Kaya HB, Kahraman A, Esiyok D et al. (2014) Association mapping for five agronomic traits in the common bean (*Phaseolus vulgaris* L.). *J Sci Food Agric* 94:3141–3151.

Nkhata W, Shimelis H, Melis R, Chirwa R, Mzengeza T et al. (2021) Genome-wide association analysis of bean fly resistance and agro-morphological traits in common bean. *PLoS One*, 16(4):e0250729.

Ochatt SJ, Kamfwa K, Kantar MB, Carovíc-Stanko K, Gunjača J (2021) Genome-wide association studies of mineral content in common bean. *Front Plant Sci* 12:636484.

Oladzad A, González A, Macchiavelli R, de Jensen CE, Beaver J et al. (2020) Genetic factors associated with nodulation and nitrogen derived from atmosphere in a middle American common bean panel. *Front Plant Sci* 11:576078.

Oladzad A, Porch T, Rosas JC, Moghaddam SM, Beaver J et al. (2019a) Single and multi-trait GWAS identify genetic factors associated with production traits in common bean under abiotic stress environments. *Gen Genom Genet* 9(6):1881–1892.

Oladzad A, Zitnick-Anderson K, Jain S, Simons K, Osorno JM et al. (2019b) Genotypes and Genomic Regions Associated With *Rhizoctonia solani* Resistance in Common Bean. *Front Plant Sci* 10:956.

Omae H, Kumar A, Shono M (2012) Adaptation to High Temperature and Water Deficit in the Common Bean (*Phaseolus vulgaris* L.) during the Reproductive Period. *J Bot* 2012:803413.

Parsons LR, Howe TK (1984) Effects of water stress on the water relations of *Phaseolus vulgaris* and the drought resistant *Phaseolus acutifolius. Physiol Planta* 60:197–202.

Paulino JF de C, Almeida CP de, Bueno CJ, Song Q, Fritsche-Neto R et al. (2021) Genome-wide association study reveals genomic regions associated with fusarium wilt resistance in common bean. *MDPI Genes* 12(5):765.

Peláez D, Aguilar PA, Mercado M, LóPez-Hernández F, Guzmán M et al. (2022) Genotype selection, and seed uniformity and multiplication to ensure common bean (*Phaseolus vulgaris* L.) var. liborino. *MDPI Agronomy* 12(10):2285.

Pennisi E (2014) Disputed islands. *Science* 345:611–613.

Perseguini JM, Oblessuc PR, Rosa JR, Gomes KA, Chiorato AF et al. (2016) Genome-Wide Association Studies of Anthracnose and Angular Leaf Spot Resistance in Common Bean (*Phaseolus vulgaris* L.). *PLoS One* 11(3):e0150506.

Porch T, Beaver J, Debouck D, Jackson S, Kelly J et al. (2013) Use of wild relatives and closely related species to adapt common bean to climate change. *MDPI Agronomy* 3:433–461.

Raggi L, Caproni L, Carboni A, Negri V (2019) Genome-wide association study reveals candidate genes for flowering time variation in common bean (*Phaseolus vulgaris* L.). *Front Plant Sci* 10:962.

Rasool S, Mahajan R, Nazir M, Bhat KA, Shikari AB et al. (2022) SSR and GBS based GWAS study for identification of QTLs associated with nutritional elemental in common bean (*Phaseolus vulgaris* L.). *Sci Hortic* 306:111470.

Rellstab C, Gugerli F, Eckert AJ, Hancock AM, Holderegger R (2015) A practical guide to environmental association analysis in landscape genomics. *Mol Ecol* 24:4348–4370.

Resende RT, Piepho HP, Rosa GJM, Silva-Junior OB, Silva FFE et al. (2021) Enviromics in breeding: Applications and perspectives on envirotypic-assisted selection. *Theor Appl Genet* 134:95–112.

Resende RT, de Resende MDV, Azevedo CF, Silva FF, Melo LC et al. (2018) Genome-wide association and regional heritability mapping of plant architecture, lodging and productivity in *Phaseolus vulgaris. G3 Genes Genomes Genet* 8(8):2841–2854.

Rosado RDS, Cruz CD, Barili LD, De Souza Carneiro JE, Carneiro PCS et al. (2020) Artificial neural networks in the prediction of genetic merit to flowering traits in bean cultivars. *Agriculture* 10:638.

Saballos AI, Soler-Garzón A, Brooks M, Hart JP, Lipka AE et al. (2022) Multiple genomic regions govern tolerance to sulfentrazone in snap bean (*Phaseolus vulgaris* L.). *Fron Agron* 4:869770.

Sadohara R, Long Y, Izquierdo P, Urrea CA, Morris D et al. (2021) Seed coat color genetics and genotype × environment effects in yellow beans via machine-learning and genome-wide association. *Plant Genome* 15(1):e20173.

Schmutz J, Mcclean PE, Mamidi S, Wu GA, Cannon SB et al. (2014) A reference genome for common bean and genome-wide analysis of dual domestications. *Nat Genet* 46:707–713.

Shi A, Gepts P, Song Q, Xiong H, Michaels TE et al. (2021) Genome-wide association study and genomic prediction for soybean cyst nematode resistance in USDA common bean (*Phaseolus vulgaris*) core collection. *Front Plant Sci* 12:624156.

Shi C, Chaudhary S, Yu K, Park SJ, Navabi A et al. (2011) Identification of candidate genes associated with CBB resistance in common bean HR45 (*Phaseolus vulgaris* L.) using cDNA-AFLP. *Mol Biol Rep* 38(1):75–81.

Shi C, Yu K, Xie W, Perry G, Navabi A et al. (2012) Development of candidate gene markers associated to common bacterial blight resistance in common bean. *Theor Appl Genet* 125:1525–1537.

Soler-Garzón A, Oladzad A, Beaver J, Beebe S, Lee R et al. (2021) NAC candidate gene marker for *Bgm*-1 and interaction with Qtl for resistance to bean golden yellow mosaic virus in common bean. *Front Plant Sci* 12:628443.

Souter JR, Gurusamy V, Porch TG, Bett KE (2017) Successful introgression of abiotic stress tolerance from wild tepary bean to common bean. *Crop Sci* 57:1160–1171.

Souza TLPOD, Barros EGD, Bellato CM, Hwang E-Y, Cregan PB (2010) Single nucleotide polymorphisms (SNPs) in common bean. *Bean Improv Coop* 53:108–109.

Spataro G, Tiranti B, Arcaleni P, Bellucci E, Attene G et al. (2011) Genetic diversity and structure of a worldwide collection of *Phaseolus coccineus* L. *Theor Appl Genet* 122:1281–1291.

Spindel JE, Mccouch SR (2016) When more is better: How data sharing would accelerate genomic selection of crop plants. *New Phytol* 212:814–826.

Tirnaz S, Zandberg J, Thomas W, Marsh J, Edwards D et al. (2022) Application of crop wild relatives in modern breeding: An overview of resources, experimental computational methodologies. *Front Plant Sci* 13:1008904.

Tong H, Nikoloski Z (2021) Machine learning approaches for crop improvement: Leveraging phenotypic and genotypic big data. *J Plant Physiol* 257:153354.

Ulian T, Diazgranados M, Pironon S, Padulosi S, Liu U et al. (2020) Unlocking plant resources to support food security and promote sustainable agriculture. *Plants People Planet* 2:421–445.

Varshney RK, Barmukh R, Roorkiwal M, Qi Y, Kholova J et al. (2021a) Breeding custom-designed crops for improved drought adaptation. *Adv Genet* 2(3):e202100017.

Varshney RK, Bohra A, Roorkiwal M, Barmukh, R., Cowling WA et al. (2021b) Fast-forward breeding for a food-secure world. *Trends Genet* 37:1124–1136.

Von Wettberg E, Davis TM, Smýkal P (2020) Wild plants as source of new crops. *Front Plant Sci* 11:591554.

Watson A, Ghosh S, Williams MJ, Cuddy WS, Simmonds et al. (2018) Speed breeding is a powerful tool to accelerate crop research breeding. *Nat Plants* 4:23–29.

Wen L, Chang H-X, Brown PJ, Domier LL, Hartman GL (2019) Genome-wide association and genomic prediction identifies soybean cyst nematode resistance in common bean including a syntenic region to soybean *Rhg1* locus. *Hortic Res* 6(1):9.

Westbrook JW, Zhang Q, Mandal MK, Jenkins EV, Barth LE et al. (2020) Optimizing genomic selection for blight resistance in American chestnut backcross populations: A trade-off with American chestnut ancestry implies resistance is polygenic. *Evol Appl* 13:31–47.

Wolf JB, Ellegren H (2017) Making sense of genomic islands of differentiation in light of speciation. *Nat Rev Genet* 18(2):87–100.

Wrather JA, Koenning SR (2006) Estimates of disease effects on soybean yields in the United States 2003 to 2005. *J Nematol* 38(2):173–80.

Wray NR, Yang J, Hayes BJ, Price AL, Goddard ME et al. (2013) Pitfalls of predicting complex traits from SNPs. *Nat Rev Genet* 14:507–515.

Wu J, Wang L, Fu J, Chen J, Wei S et al. (2020) Resequencing of 683 common bean genotypes identifies yield component trait associations across a north-south cline. *Nat Genet* 52:118–125.

Wu L, Chang Y, Wang L, Wu J, Wang S (2021) Genetic dissection of drought resistance based on root traits at the bud stage in common bean. *Theor Appl Genet* 134:1047–1061.

Wu X, Wang B, Xi Y, Wang Y, Tian S et al. (2022) Unravelling the genetic architecture of rust resistance in the common bean (*Phaseolus vulgaris* L.) by combining QTL-Seq and GWAS analysis. *MDPI Plants* 11(7):953.

Xue R, Wu J, Zhu Z, Wang L, Wang X et al. (2015) Differentially Expressed Genes in Resistant and Susceptible Common Bean (*Phaseolus vulgaris* L.) Genotypes in Response to *Fusarium oxysporum* f. sp. phaseoli. *PLoS One* 10(6):e0127698.

Zahn LM, Purnell BA (2016) Genes under pressure. *Science* 354:52.

Zhang X, Blair MW, Wang S (2008) Genetic diversity of Chinese common bean (*Phaseolus vulgaris* L.) landraces assessed with simple sequence repeat (SSR) markers. *Theor Appl Genet* 117:629–640.

Zhu J, Wu J, Wang, L, Blair MW, Zhu Z et al. (2016) QTL and candidate genes associated with common bacterial blight resistance in the common bean cultivar Longyundou 5 from China. *Crop J* 4(5):344–352.

Zuiderveen GH, Padder BA, Kamfwa K, Song Q, Kelly JD (2016) Genome-Wide Association Study of Anthracnose Resistance in Andean Beans (*Phaseolus vulgaris*). *PLoS One* 11(6):e0156391.

2 In Quest of Novel Alleles for Stress Tolerance in Chickpea

C. Bharadwaj, Neeraj Kumar, Uday Chand Jha, Manish Roorkiwal, Alladin Hamwieh, Rajeev K. Varshney, Kadambot H. M. Siddique, Yogesh Kumar, Sneha Priya Pappula Reddy, Nilesh Joshi, Chittaranjan Kole, and Harsh Nayyar

2.1 INTRODUCTION

Chickpea (*Cicer arietinum* L.) is one of the major grain legume crops grown primarily in dryland during the cool seasons. It is a self-pollinated, annual diploid (2n = 2x = 16) species with a genome size of 738 Mb (Nene et al., 2012; Varshney et al., 2013). Other names for it include garbanzo bean or Bengal gram, and it belongs to the genus *Cicer*. It contains 44 species overall; among these, 43 are wild (34 perennial wild species and 9 annual), and only 1 is cultivated or farmed. It serves as a major source of dietary protein for humans at the lowest cost when compared to other protein sources. Its consumption is widespread in many parts of the world, owing to its high nutritional value. The chickpea seed matrix is perfect for a healthy diet because it provides carbohydrates (50–58%), protein (15–22%), fat (3.8–10.20%), moisture (7–8%), and micronutrients (1%) (Jukanti et al., 2012; USDA, 2021). Chickpea is the world's second most important pulse crop, accounting for approximately 14.24 million tons of total production, of which 80% is *desi*, and the remaining 20% is *kabuli* type (FAOSTAT, 2020; Merga and Haji, 2019). India is a key consumer and importer of chickpeas, although in recent years, the country's output of chickpeas has greatly expanded, rising from 9.38 million tons in 2016–2017 to 13.98 million tons in 2021–2022 (MoA & FW, GOI, 2022), making it the world's largest producer. Due to its symbiotic nitrogen fixation abilities, it is known to have an efficient N_2-fixing system and thus performs well with crop rotation programs.

Pathogenic fungi, bacteria, viruses, and nematodes cause diseases such as fusarium wilt, dry root rot, Ascochyta blight, collar rot, bacterial blight, filiform virus, and root nematode in the Indian subcontinent (Nene and Sheila, 1996). On the other hand, chickpea is grown mainly in arid and semi-arid environments on soils of poor agricultural quality. In these regions, the abiotic stresses, such as extreme temperatures, drought, and salinity at various growth stages during the productive season, are the major adverse environmental factors that limit its production. Chickpea wild relatives and landraces are known to thrive in harsh terrain and environmental conditions, indicating a high tolerance to biotic and abiotic stresses. Exploring and comprehending the extent of natural variation for various stress tolerance traits among wild and cultivated landraces, as well as understanding the molecular mechanics underlying the trait, would be extremely beneficial in developing tolerant cultivars (Jaganathan et al., 2015; Kumar et al., 2020).

Up to the last decade, chickpea was regarded as an "orphan crop" due to the restricted genomic resources available, and thus, the breeders relied much upon traditional breeding approaches (Varshney et al., 2012). A lot of progress has been achieved recently in the efforts to create genetic resources and use them to create superior chickpea varieties utilizing contemporary breeding techniques (Soren et al., 2020; Kumar et al., 2021; Bharadwaj et al., 2021). Next-generation sequencing technologies have been used extensively in the development of molecular markers. The various

DOI: 10.1201/9781003385059-2

markers linked with desirable traits and dense genetic maps have been developed by utilizing these resources.

MAJOR BIOTIC CONSTRAINTS OF CHICKPEA

Several biotic and abiotic stresses limit chickpea production. A total of 172 pathogens have been identified as causing diseases in chickpeas. Ascochyta blight (AB) and Botrytis gray mold (BGM) are the severe diseases that could result in up to 100% yield losses in favorable conditions (Nene and Sheila, 1996). The major biotic constraints influencing chickpea production are depicted in Figure 2.1 and are briefly described in what follows.

2.1.1 Fusarium Wilt

Fusarium oxysporum f. sp. *ciceris*, which belongs to *Gibberella* clade, causes chickpea wilt. Eight races were identified based on disease severity in a set of differential cultivars. A total of eight pathogenic *Foc* races (races 0, 1A, 1B/C, 2, 3, 4, 5, and 6) have been reported worldwide (Jendoubi et al., 2017). Racial distribution in India was studied by Dubey et al. (2011a). There were three races from each of the following states. There are eight races: race 1 from Andhra Pradesh and Karnataka; race 2 and race 4 from Uttar Pradesh; race 2 from Bihar; race 3 and race 8 from Punjab; races 3, 6, and 7 from Madhya Pradesh; race 3 and race 8 from Rajasthan; race 4 from Delhi and Haryana; race 4 from Punjab; race 4 and 7 from Maharashtra; race 4 and 8 from Punjab; races 5 and 6 from Chhattisgarh and Jharkhand; and races 7 and 8 from Gujarat. The pathogen infects at every stage of development, from seedling to bearing stage, that is, blooming and pod-forming stage. When the sensitive stage occurs at the same time as a rapid rise in temperature and water stress during the blooming and podding stages, this is the essential time for high incidence. In extremely

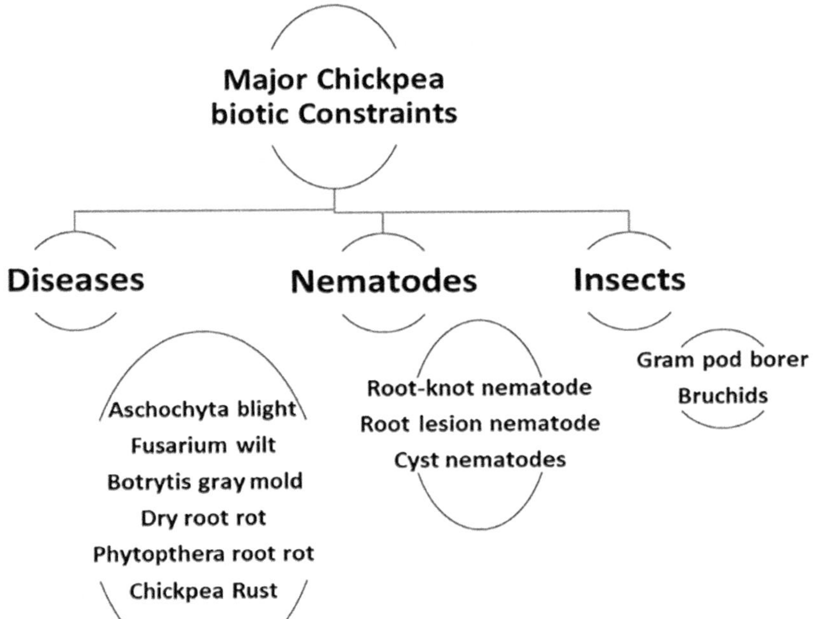

FIGURE 2.1 Schematic depiction of major biotic constraints impacting chickpea production and productivity in field.

sensitive cultivars, wilt signs emerge 25 days after seeding (e.g., JG-62). The symptoms of disease are more prominent between 6 and 8 weeks after the beginning of flowering and up to the pod stage. Thus, early and late stages of wilt may be separated. The yield is lost more due to early wilt than late wilt (Chen et al., 2011). Common wilt symptoms are comprised of trickling petioles and rachis, uphill yellowing and withering of leaves from the bottom of the plant, drying, and ultimately, death of plants. Affected stele and root tissues can be seen to have interior browning of vascular tissues (Westerlund et al., 1974).

2.1.2 Dry Root Rot

DRR is the second major disease of chickpea, caused by *Rhizoctonia bataticola* (Taub.). There have not been many reports on *R. bataticola* of chickpea pathotypes. By utilizing a range of differentials from chickpeas on blotter paper and sick soil, six pathotypes have been found (Dubey et al., 2011b). The predication has been made that hyphal fusion, mutation, and mitotic crossing are the major factors determining pathogenic diversity. But *R. bataticola* isolates from different agroecological zones in India showed cultural, morphological, and molecular variation, according to Sharma et al. (2012). The seedling stage and the reproductive phase are both impacted by the DRR pathogen. But older plants are significantly more prone to disease than younger ones (Sharma and Pande, 2013). Flowering and podding, like those of the wilt, are the most acute stages of infection. If the temperature drops below 30°C while the plants are in their sensitive stage, they change color to a unique straw hue with brown stems and lower leaves. Black-color taproots with diminished lateral and fine roots were visible when the infected plants were removed. Later, the roots become weak, torn, and eventually die. When the stem is split apart at an advanced stage, the distinctive signs are dark-colored small *sclerotial* bodies (Nene and Sheila, 1996).

2.1.3 Collar Rot Disease

Sclerotium rolfsii is responsible for the deadly soilborne disease known as collar rot. It is a rapidly spreading fungal infection that significantly reduces seedling viability, with early seedling mortality ranging from 54.70 to 95.00% and yield losses of between 22 and 50% (Shrivastava et al., 1984). All stages of the crop are impacted by this disease. The first observable sign in the field is stem discoloration. The collar area, which is close to the soil level, is occupied by a cottony white mold. Later, the plant's whole body dies, with the aboveground portion of the stem being girdled. The emergence of chlorotic or drying branches or whole plants that are spread across the landscape is the disease's defining feature. A symptom of both this disease and dry root rot is the development of *sclerotia* in the stems of the infected plants when they are cut open. Four states in India that grow chickpeas – Andhra Pradesh, Karnataka, Chhattisgarh, and Madhya Pradesh – have seen yield losses varying from 7.10 to 10.50% (Ghosh et al., 2013).

2.1.4 Ascochyta *Blight* (AB)

The fungus *Ascochyta rabiei*, which grows on plant waste remains in the soil, is mostly responsible for chickpea AB. This is the main foliar disease affecting Southern Europe, Northern Africa, and West Asia. It has been known to occur in colder parts of the globe (Chongo et al., 2003). Disease symptoms are evident in each and every part of the plant that is located aboveground, significantly reducing both quality and quantity. The fungus is a hemi-biotroph, switching from a biotrophic to a necrotrophic phase. As a result, the disease's symptoms could manifest three to four days following the incubation period. On the leaf surface, a great deal of tiny dark pycnidia are produced, although they are scattered at random along the stems and pods of diseased tissues. Stele lesions cause girdling and weakening, resulting in breakage. The disease mostly spreads through contaminated seeds generated by infected crops. Fungus develops on the surface of the seed coat of infected seeds, cotyledons, and developing seedlings (Kaiser and Hannan, 1988). The transmission of the disease mainly takes place through seed and soil. Primary inoculum is obtained from diseased host plant waste remains in the soil. The disease thrives in moist, cloudy environments in the cool night with temperature of 10°C and daytime temperatures of 15–23°C. In Punjab and parts of Himachal Pradesh, particularly, winter rains cause epidemics. After winter rains, the sickness intensifies in February and March (Pande et al., 2005).

2.1.5 Botrytis Gray Mold (BGM)

Carranza (1965) reported an outbreak of this disease caused by *B. cinerea* in Argentina, resulting in crop loss of almost 95%. In general, infection arises in areas where the weather is cool, overcast, and humid. BGM has been reported on chickpea all over the world. Several examples from Argentina (Haware and McDonald, 1992; Davidson et al., 2007) demonstrate nearly full yield loss (about 100%) under favorable environmental circumstances. The disease became prevalent during 1978–1979 in India and was responsible for the destruction of approximately 20,000 ha of chickpea cultivation (Grewal et al., 1992; Haware, 1998). But in Nepal, this disease affects nearly all years of crop production, resulting in an average yield deprivation of 15% (Joshi, 1992). Yield is affected by the start of infection, its progression, the severity of the disease, meteorological conditions, and pathogen inoculum levels.

2.1.6 Rust Disease

Rust disease in chickpea is caused by the basidiomycete *Uromyces ciceris-arietini*, which is a hemiform fungus with unknown pycnial and aecial phases. It has been reported in several Indian states, including Maharashtra and Tamil Nadu, and had been widespread in Bihar, Punjab, and Uttar Pradesh. The study on outbreaks of rust chickpea was carried out by Asthana (1957) Seoni-Malwa in India on local cultivars, with 100% infection in a few genotypes within eight hours. Patil et al. (2016) found an 80–100% yield reduction owing to rust in chickpea. The disease is spread predominantly by airborne *uredospores*. Rust pustules and *urediniospores* on leaves cause the rusty look along with minute circular or oval brick-colored red or orange powdered pustules. Plants in advanced stages of infection may dry up early, resulting in a significant yield reduction.

2.1.7 Chickpea Stunt

In all regions of India where chickpeas are grown, the viral stunt disease is widespread and tends to strike in January and February each year. If it happens before blooming, the yield is lost completely, and if it happens after flowering, the yield is lost between 75% and 90% (Horn et al., 1995). The chickpea chlorotic dwarf virus (CpCDV), which is transmitted by leaf hopper; the most frequently associated virus chickpea luteo virus (CpLV); and the aphid-vectored bean leaf roll virus (BLRV) have all been identified as present in chickpea stunt disease (Kumari et al., 2006; Horn et al., 1993). Infected plants exhibit leaf chlorosis or leaf reddening (depending on the chickpea variety). The field can quickly identify infected plants by their stunted growth and yellow, orange, or brown coloring. The visual indicators of stunt disease include redness of leaf in desi chickpea and yellowing in *kabuli* chickpea genotypes.

2.1.8 Cyst Nematodes

Heterodera ciceri is a chickpea parasite, also called cyst nematodes, causing stunting and chlorosis, with condensed blooms and pods, as well as poorly formed roots and fewer *Rhizobium* nodules, resulting in yield penalty (Greco et al., 1988). The severity of disease incidence and yield loss due to cyst nematode can be minimized by the development of resistant varieties. Nonetheless, the cultivated chickpea *C. arietinum* showed no resistance to *H. ciceri* (Singh and Ocampo, 1997). Interestingly, the Ladiz type of *C. reticulatum* displayed superior degree of resistance, and cultivars of chickpea derived from this line with much-enhanced and high-yielding kabuli cultivar have been made available for cultivation (Malhotra et al., 2002)

2.1.9 Root-Knot Nematodes

Root-knot nematodes (RKN) are major families of plant parasitic nematodes. There are three species of RKN which have been found to pervade chickpea in diverse conditions. Two species, namely, *Meloidogyne incognita* and *M. javanica*, are known to occur in the subtropics of Asia, Africa, and South America, respectively. *Meloidogyne artiella* is spotted in the Mediterranean region (Thompson et al., 2000). RKN-infected chickpea fields show discontinuous regions of plant growth because of high infection, which also results in growth retardation, minimal branching, and greenish yellow to yellow leaves. Large distinctive galls are frequently produced on the roots of diseased plants as a result of tissues surrounding RKN feeding sites proliferating frequently. RKN infestation entails the establishment of lasting feeding sites in plant roots, where the nematodes encourage the proliferation of large cells that serve as sinks for the plant photosynthates that the worm selectively accesses. Nematode feeding can result in xylem vessels clogging, which restricts the movement of nutrients and water, inhibits plant growth, and reduces yield. Plants that are affected regularly exhibit stunted growth and greenish yellow to yellow leaves (Castillo et al., 2008). The wilt/root rot complex are exacerbated by RKN destruction; infection by nematodes can weaken plant defenses, causing wilt, and even resistant genotypes can become sensitive. In chickpea crops, Rao and Krishnappa (1995) reported a 28% prevalence of the wilt disease related to the combination of *M. incognita* and *F. oxysporum f.* sp. *Ciceris.*

2.1.10 Root Lesion Nematodes

Lesion nematodes (*Pratylenchus* spp.) are a serious stumbling block in chickpea production. They produce root lesions, resulting in decreased crop growth and output. They can also become more severe with wilt infection caused by *Fusarium oxysporum* (Castillo et al., 2008).

2.1.11 Gram Pod Borer

Helicoverpa armigera, the gram pod borer, is without a doubt the most destructive pest of chickpea across Asia's tropics and subtropics, inflicting devastating crop losses. The moths deposit their eggs on leaves, flowers, flower buds, and pods. Eggs are originally pale yellow, lustrous, and chiseled, and as they age, they turn brown. The larvae feed on crops from seedling through maturity. The larva, in its early stages, feeds by rubbing against the fragile areas of leaves or stems. Larvae in their second and following instars consume flowers, buds, and foliage. Unlike third-instar larvae, which feed on the grains through concentric holes in the pod walls, fully matured larvae feed on the seeds with their heads within the pod and the remaining bodies outside. A fully matured larva can destroy 7–40 pods. The crop suffers little harm during the early vegetative stage since the plant is tolerant and can withstand as much as 60% damage. Attack of insect pest during the formation of pod and seed development stages causes significant harm to the plant.

2.1.12 Bruchids

Bruchids are the most common storage pests of legumes; *Callosobruchus chinensis* (L.) and *Callosobruchus maculatus* (Fab.) are extensive, although *C. analis* is restricted to Southeast Asia (Rao et al., 2013). Despite their widespread distribution, they are thought to be particularly essential in tropical and subtropical environments. After the egg phase, which lasts for about seven days, a chalky-colored grub holes inside the seed from the bottom of the egg, consumes it, and transforms into a full-grown pupa within the seed, where the eggs are laid on the seed's surface. The small cylindrical cut hole from where adults come out lives for an average period of 12 to 14 days. On the basis of relative humidity and temperature, one generation could take four to five weeks. Bruchids result in both quantitative (loss of marketable weight and loss of edible grain) and qualitative (presence of insects and other pollutants; preference by consumer and market price) losses from storing. The damaged grains are fully empty, have low germination capacity, and are therefore unsuitable for food.

2.2 CHICKPEA CWRs AND ALLELES THEY HARBOR

The cultivation of chickpea was started a long way back, around 10,000–12,000 years ago, in the Fertile Crescent of Iran, Turkey, and Israel/Jordan (Redden and Berger, 2007). The secondary centers of chickpea variety are expected to be Ethiopia (including the Mediterranean west of Greece/Crete), Central Asia to the Indian subcontinent, Asia Minor to Iran, and the Caucasus. Chickpea CWRs are found all over the world, from sea level (*C. judiacum*) to 5,000 m altitude (*C. microphyllum*) (Abbo et al., 2007). Three gene pools have been postulated in chickpea based on their crossability. The primary gene pool consists of *C. arietinum*, *C. echinospermum*, and *C. reticulatum*. They are also included in the core gene pool. Utilizing crossability, karyotype, isozyme allele frequencies, grain protein, and random amplified polymorphic DNA (RAPD) analysis to assess genetic diversity, scientists have found a close genetic relationship between *C. arietinum*, *C. reticulatum*, and *C. echinospermum*, forming a significant crossability group (Ahmad, 1999). Annual species *C. judaicum* and *C. pinnatifidum* make up the secondary gene pool; crossing *C. bijugum*, *C. judaicum*, and *C. pinnatifidum* with species in the major crossability group has had minimal success (ICARDA, 1998). Moreover, *C. bijugum*, *C. judaicum*, and especially, *C. pinnatifidum* have been identified as sources of tolerance to biotic and abiotic stresses because of their extremely high genetic diversity and high levels of genetic variation (Singh et al., 1998) and thus have great potential as sources of resistance for future chickpea breeding. The three annual species *C. yamashitae*, *C. chorassanicum*, and *C. cuneatum*, as well as 34 perennial species, make up the tertiary gene pool, which is the most distantly related grouping (Van der Maesen et al., 2007). Despite the fact that tertiary gene pools of perennial and annual species are genetically comparably isolated from the domestic gene pool, annual wild relatives have garnered more attention than perennial wild relatives as potential sources of disease resistance genes (Singh et al., 2007).

Only *C. reticulatum*, one of the eight annuals of wild *Cicer* species, is easily crossable with cultivated chickpea, producing a sterile hybrid. However, *C. judaicum*, *C. bijugum*, and especially, *C. pinnatifidum* are good sources of resistance to chickpea stresses (Singh et al., 1998). Tissue culture and embryo rescue procedures have been utilized to break down crossability barriers and create interspecific hybridizations (Badami et al., 1997). Various studies have found the causes of resistance in wild relatives (Tables 2.1). Researchers are keen to discover novel sources of resistance genes and to incorporate these genes into breeding programs. They are searching for resistant donors from primitive cultivars (landraces), wild species and their relatives, screened germplasm, and cultivated varieties. However, introducing resistance genes through conventional breeding programs into elite genotypes of chickpeas is a time- and resource-consuming task. Even if they have no significant influence on crop potential, wild species and landraces must be evaluated. In order to enhance the chickpea genetics, the basic gene pool has widely been utilized.

2.3 ADVANCE GENOMIC RESOURCES FOR BIOTIC STRESS RESILIENCE IN CHICKPEA

Modern plant breeding techniques can employ molecular technologies to aid in the discovery and selection of desired traits (Moose and Mumm, 2008). Crop biotic stress molecular research has

TABLE 2.1
Chickpea Wild Relatives and Resistant Sources for Diseases and Insect Tolerance

Sr. No.	Biotic Stress	Species	References
1.	Fusarium wilt	*C. arietinum, C. reticulatum, C. bijugam, C. judaicum, C. pinnnatifidum, C. echinospermum, C. cuneatum*	Nguyen et al. (2004); Ahmad et al. (2013)
2.	Ascochyta blight	*C. arietinum, C. judaicum, C. reticulatum, C. montbretii, C. bijugam, C. pinnnatifidum, C. cuneatum, C. echinospermum*	Singh et al. (2014); Li et al. (2017)
3.	Botrytis gray mold	*C. judaicum, C. bijugam, C. pinnnatifidum, C. reticulatum*	Singh et al. (2014); Hegde et al. (2018)
4.	Root-knot nematode	*C. bijugum, C. judaicum, C. pinnnatifidum, C. reticulatum, C. echinospermum*	Singh et al. (2014)
5.	Root lesion nematode	*C. echinospermum, C. reticulatum*	Thompson et al. (2011)
6.	Cyst nematode	*C. bijugam, C. pinnnatifidum, C. reticulatum*	Ahmad et al. (2013); Singh et al. (2010)
7.	*Phytophthora* root rot resistance	*C. reticulatum, C. bijugum, C. pinnnatifidum, C. Echinospermum*	Knights et al. (2008)
8.	Rust	*C. bijugam, C. reticulatum, C. echinospermum*	Sillero et al. (2012)
9.	Bruchids	*C. reticulatum*	Eker et al. (2018)
10.	Pod borer	*C. bijugum, C. reticulatum, C. echinospermum, C. cuneatum, C. pinnatifidum, C. Microphyllum*	Sharma et al. (2006); Golla et al. (2018)

concentrated on finding and tagging genomic locations for resistance genes (Table 2). Disease resistance markers are more advanced; the development of improved cultivars by incorporating disease resistant genes from donors into superior cultivars via marker-assisted selection (MAS) is the most effective use of MAS and marker-assisted backcross breeding (MABB) (Varshney et al., 2014). The linked markers are used to map genes or discover quantitative trait loci (QTLs) for biotic stress resistance (FW, AB, *Phytophthora* root rot, and BGM). DNA markers have been utilized extensively to develop inter- and intraspecific linkage maps for chickpea (Gaur et al., 2012; Stephens et al., 2014). Chickpea major disease AB and its resistance are governed by quantitative traits, with multiple QTLs being found (Santra et al., 2000; Taran et al., 2010). For the development of long-lasting resistance against diseases, advanced molecular genetic tools have been developed to enable successful pyramiding of resistance genes (Millan et al., 2006). A recent study identified a 100 kb area (AB4.1) on chromosome 4 was shown to be strongly related with AB resistance in genome-wide association studies (GWAS). The AB4.1 region co-located with a wide QTL interval of 7 Mb, 30 Mb was previously found in three distinct mapping populations genotyped at low density using SSR or SNP markers. These technologies have greatly improved the ability to breed chickpeas for AB resistance (Li et al., 2017).

Also, three QTLs that have been reported for Botrytis gray mold are also identified for resistance in chickpea, and the linked markers allow for the pyramiding of various genes or QTLs to increase resistance levels (Anuradha et al., 2011). *C. reticulatum* acc. ILWC 292 was shown to be resistant to BGM, whereas GPF2 (*Cicer arietinum* L.) is resistant to AB. To discover QTLs responsible for resistance to AB and BGM, 187 F_8 recombinant inbred lines (RILs) generated from an interspecific cross of GPF2 *C. reticulatum* acc. ILWC 292 were employed. Three QTLs for AB and four QTLs for BGM resistance were discovered using composite interval mapping. Two QTLs for AB and

TABLE 2.2
List of QTLs or Genes Along with Their Location and Markers Associated Described for Biotic Stress Resistance in Chickpea

Trait	Name of Population	Locations	Markers Associated with Genes or QTL(s)	Genetic Effects	References
Ascochyta blight	FLIP84–92C 9 PI 599072	LG1, LG6	UBC733b, UBC181a, Dia4	50.3 and 45%	Santra et al. (2000)
	ICCV96029 9 CDC Frontier	LG3, LG4, LG6	TA64, TS54, TA176	56%	Taran et al. (2007)
	ICCV 96029/CDC Luna,	LG2, LG4	TR19, TS54	48%	Anbessa et al. (2009)
	Lasseter 9 ICC3996	LG4	SNP_40000185	45%	Stephens et al. (2014)
	S95362 9 Howzat	LG4	TA146, TA72	59%	Stephens et al. (2014)
Fusarium wilt	WR-315 9 C-104	N/A	CS-27, UBC170	Foc-1	Mayer et al. (1997)
	ICC-4958 9 PI 498777	N/A	CS-27, UBC-855	Foc-4	Ratnaparkhe et al. (1998)
	JG-62 9 Surutato-77	N/A	CS-27	Foc-1, Foc-4	Tullu et al. (1999)
	ICC-4958 9 PI 498777	LG2, LG3	CS-27, UBC-170	Foc-0, Foc-4, Foc-5	Tekeoglu et al. (2000)
	ICC-4958 9 PI 498777	LG2	CS27, TA96, TA27	Foc-1, Foc-4, Foc-5	Winter et al. (2000)
	CA2139 9 JG62	LG3	OPJ20(600), TR59	37.8%	Cobos et al. (2005)
	JG62 9 Vijay	LG2	TA110, TA96, H1B06y	Foc-1, Foc-2, Foc-3	Gowda et al. (2009)
	C 214 9 WR 315	LG6	CaM1402, CaM1125, TA22	29.2%	Sabbavarapu et al. (2013)
Botrytis gray mold	JG62 9 ICCV2	LG3, LG6	SA14, TA25, TA159	43.6%	Anuradha et al. (2011)
Rust	ILC-72 9 Cr5–10	LG7	TA18, TA180	31%	Madrid et al. (2008)

three QTLs for BGM were constant throughout both years. These QTLs can be fine-mapped for the deployment of resistance to AB and BGM in superior chickpea cultivars utilizing marker-assisted selection (Kushwah et al., 2021). Chickpeas are susceptible to the deadly ailment fusarium wilt, and genetic mapping studies have found that the pathogen resistance genes for races 0, 1, 2, 3, and 5 are present on the chickpea map on loci 2 (LG2) (Castro et al., 2010). LG/2 is a hotspot for fusarium wilt resistance due to the clustering of six resistance genes (Millan et al., 2006; Sharma and Muehlbauer, 2007). Near-isogenic lines for the race 5 resistance gene have been generated, and map-based cloning and fine mapping are the potential approach and can also be deployed for this gene (Castro et al., 2010). Resistance to fusarium wilt and Ascochyta blight was mapped via five QTLs, out of which three QTLs were discovered for resistance to race 1, including a significant QTL on CaLG02 and a minor QTL each on CaLG04 and CaLG06. Major QTLs on CaLG02 and CaLG04 were discovered for race 3 (Garg et al., 2018). FW resistance is successfully pyramided for various racial groups (foc 1, 2, 3, 4, and 5). The introgression lines created in the Pusa 391 genetic background were then subjected to foreground selection using three SSR markers (GA16, TA 27, and TA 96), while a recent example of the use of genomic resources in product delivery used 48 SSR markers that were evenly distributed on all chromosomes (Bharadwaj et al., 2022). Transcriptomic profiling report with gene regulatory network of chickpea reveals the contrasting responses and candidate gene signature of wilt disease. The reported DEGs, which are controlled by miRNA, can be a promising genomic resource for exploration of silencing technology in fusarium wilt management (Taxak et al., 2017).

With respect to chickpea rust, there is very limited information available to date. Recently, in a RIL population generated from an interspecific cross between *C. arietinum* (ILC72) and *C. reticulatum* (Cr5–10), the Cr5–10 line was used as a donor and known to be a single dominant gene governing rust resistance in chickpea. This gene, Uca1/uca1, was situated on LG7 at a distance of 3.9 cM.

2.4 ABIOTIC STRESS

Apart from the previously discussed biotic constraints, chickpea production is also severely affected by several abiotic factors, limiting crop yields, depending on the agroecological zone and cropping region. Among abiotic stresses, drought, salinity, heat, and cold stresses are the most important yield-limiting factors (Tapan et al., 2015; Kumar et al., 2017; Gunes et al., 2008). Droughts, heat waves, and cold spells are more common in dry and semi-arid regions where chickpea is traditionally grown, and climate change events increase the frequency of these stresses (Solomon et al., 2007). The progress for breeding tolerant chickpea cultivars to these stresses has been slowed due to their complex nature of screening, phenotyping, and genetics governing stress responses (Soren et al., 2020; Shimray et al., 2022). Understanding these complex traits, it is essential to integrate physiological, biochemical, and molecular understanding of responses for crop improvement (Kumar et al., 2021a, 2021b). Minor genes or quantitative genes are majorly responsible for tolerance to abiotic stresses. Many factors and how they interact during abiotic stress conditions have an impact on crop growth and production. The current section attempts to comprehend and describe the history of significant abiotic stresses, such as drought, salinity, heat, and cold, as well as their breeding strategies, to improve resistance in chickpea, as well as contemporary molecular breeding techniques.

2.4.1 Drought Stress

Drought stress is one of the major abiotic stresses limiting chickpea yield globally. Aiming at developing drought-tolerant chickpea genotypes, substantial research have been conducted at the molecular level, and substantial genomics resources have been developed. Many important QTLs and causal candidate gene(s) conferring drought tolerance have been reported based on biparental QTL mapping and genome-wide association mapping approach (Hamwieh et al., 2013; Varshney et al., 2014; Jaganathan et al., 2015; Kale et al., 2015). Additionally, the ICAR-AICRP has released BGM 10216, which confers drought resistance, for commercial planting in India's central and southern zones (Roorkiwal et al., 2020; Bharadwaj et al., 2021).

Using traditional biparental-based QTL mapping and GBS technology and comprehensive phenotyping of many physiological and yield-contributing traits across the multiple environments allowed to identify a "QTL hotspot" region on chromosome 4 (Varshney et al., 2014). Furthermore, this genomic region was narrowed down to ~300 kb region containing 26 genes. Functional validation of these genes indicated differential expression of genes, viz., E3 ubiquitin ligase, serine/threonine protein kinases, and homocysteine S-methyltransferase, under water stress, confirming their significant role in conferring drought tolerance in chickpea. While identifying genomic regions conferring drought and heat tolerance in chickpea through genome wide-association mapping in a global set of 300 chickpea germplasm, allele mining was performed for a selected set of five drought stress–responsive candidate genes, namely, abscisic acid stress and ripening (ASR), CAP2 gene, ERECTA, sucrose synthase (SUSY), and sucrose phosphate synthase (SPS), in the given set of chickpea genotypes (Thudi et al., 2014). A total of 34 SNPs for ASR gene and 33 SNPs for ERECTA gene were identified. Roorkiwal et al. (2014) also found 79 SNPs and 41 indels in a set of ten drought stress–responsive candidate genes, including the five genes previously mentioned. Further haplotype analysis indicated alleles of the given set of genes were represented by more than two haplotype blocks, except alleles of CAP2 and SUSY gene (Roorkiwal et al., 2014). Similarly, employing

GWAS in association with the SNP derived from sequencing of 132 genotypes of chickpea indicated several significant associations of SNPs in auxin efflux carrier protein (PIN3), p-glycoprotein, and nodulin MtN21/EamA-like transporter related genes under drought stress in chickpea (Li et al., 2018). Further, Manchikatla et al. (2021) employed the MutMap technique to find candidate genes for early flowering and high seed weight, two essential features linked to drought resistance. Between 9.76 and 12.96 Mb, 102 potential SNPs were found on Ca6. The amplification of a total of 12 allele-specific primer pairs was confirmed in 8 different genotypes of chickpeas, although only Ca6HSDW10099486, Ca6HSDW9890335, Ca6HSDW9828083, and Ca6EF10509893 markers displayed allele-specific amplification. Further, these four allele-specific markers were validated in 82 chickpea germplasm with early flowering and high seed weight trait. Thus, these allele-specific markers could be used for identifying early and high seed weight chickpea lines associated with drought tolerance. Recently, Kumar et al. (2020) identified allelic variation in drought-responsive dehydrin (DHN) gene chickpea. Sequence alignment showed different number of SNPs in the given candidate gene among the chickpea genotypes used. It was noted that the indels were only present in the candidate gene from genotypes Pusa1103, Pusa362, ILC0 (Kumar et al., 2020). Similarly, considering ASR gene attributing to water stress tolerance, Sachdeva et al. (2020) found that ASR genes are conserved in chickpea, except a few variations. The role of this gene conferring drought tolerance in chickpea was also examined and validated at functional level in BGD72, ICCV10, and ICCV331 (drought-tolerant) and ICCV2, Pusa 362, and Pusa1003 (drought-sensitive). Thus, with DHN, ASR genes could be potentially used for designing superior chickpea genotypes with improved yield under drought stress. Recently, Barmukh et al. (2022) demonstrated that the allelic variation of CaTIFY4b gene harboring in the "QTL hotspot" region could greatly help in improving yield under water stress by improving pre-anthesis water use, transpiration efficiency, root architecture, and canopy development. This yield advantage in plants under drought stress due to the presence of CaTIFY4b-ICC 4958 allele has been established by comparing ICC 1882, ICC 4958, ICC 1882 ↔ Qh (possessing the ICC 1882 allele of the "QTL hotspot"), and ICC1882 + Qh (possessing the ICC 4958 allele of the "QTL hotspot") plants (Barmukh et al., 2022). The functional analysis of this gene indicated that CaTIFY4b may control plant organ size under drought stress by modulating the expression of "GRF- INTERACTING FACTOR1 (GIF1)" (Barmukh et al., 2022a). Further, to capture the various candidate genes underlying the "QTL hotspot" regions transferred in the various introgression lines through breeding process, Barmukh et al. performed the whole genome re-sequencing of the introgression lines (ILs) and parental chickpea lines (Barmukh et al., 2022b). Across ten genotypes, a total of 941,432 SNPs were detected, of which 923,401 were discovered to be bi-allelic. BGM 10216, DIBG 205, and BGM 10218 each had an introgressed region marked by SNPs with the donor genotype (ICC 4958) and shared five common SNPs in the coding or non-coding sections of four genes inside the region, according to analysis of the "QTL hotspot" region in five ILs (Barmukh et al., 2022b). Thus, the several alleles existing across the "QTL hotspot" regions have been introgressed in BGM 10216, BGM 10218, BG 4005, and BG 3097 lines, showing superior seed yield to their respective recurrent parents at multiple locations under drought stress environment. Furthermore, introgression of superior allele of CabHLH10harboring in the CaqDYI(P/H)1.1 QTL region into nearly isogenic lines developed from involving the following parents: ICC 4958, ICCV 93954, ICCV 10, and annigeri, ICC12968, ICC92311, and ICC 17160. These showed superior yield, improved root and shoot biomass, and high efficiency of photosynthesis gene(s) under drought stress (Thakro et al., 2023).

2.4.2 Salinity Stress

Salinity stress is becoming one of the major challenges for crop yield across the globe due to land degradation, excessive irrigation, and other related activities (Jha et al., 2019). Nearly 80 million hectares of the world's arable land are predicted to be under salinity stress (Flowers et al., 2010). Chickpea is sensitive to salinity stress (Samineni et al., 2011; Kumar et al., 2021). Salinity stress is the second most important abiotic stress causing serious yield limitation in chickpea, inflicting 8–10% yield loss in chickpea (Flowers et al., 2010). To understand the genetic basis of salinity stress tolerance, substantial work related to biparental-based QTL analysis has been done (Pushpavalli et al., 2015; Soren et al., 2020; Atieno et al., 2021). Previously, several QTLs contributing to yield and yield-related traits were elucidated under salinity stress on CaLG03 and CaLG06 (Soren et al., 2020), and six QTLs conferring resistance against salinity were identified on CaLG04, CaLG05, and CaLG06 (Atieno et al., 2021).

Flowering time might have a confounding effect on salinity tolerance when assessing tolerance to it (Atieno et al., 2021). Aiming at mapping QTL for salinity tolerance, Atieno et al. (2021) mapped flwqtl.1loci on CaLG05 and flwqtl.2 location on CaLG03 in the glasshouse and in Merredin2017 and Merredin2018 field experiments. The authors also found that locus flwqtl.1 corresponded to the CaELF3a gene reported by Ridge et al. (2017). Genesis836 and Rupali parents had contrasting alleles for this gene, with Rupali carrying the early flowering mutated form, caelf3a (Atieno et al., 2021). Thus, this gene could be used for marker-assisted selection for developing salinity-tolerant chickpea genotype.

Among the various genes, Trehalose-6-phosphate synthase (TPS) gene is an important gene contributing to salinity stress tolerance in chickpea. Allelic variation of TPS gene for salinity tolerance was noted in selected chickpea genotypes. Kumar et al. (2022) first reported allelic variation for TSP gene from chickpea genotypes. They also investigated that TPS gene sequences were conserved among the selected chickpea genotypes. Further, investigating the allelic variation of this gene in various chickpea genotypes indicated that Pusa 362, Pusa 1103, and IG5856 chickpea genotypes were promising genotypes for salinity tolerance (Kumar et al., 2022). Thus, these genotypes could be harnessed for developing salinity tolerance in chickpea genotypes.

2.4.3 Heat Tolerance

Given the global climate change, high-temperature stress is becoming an important abiotic stress causing serious yield loss in various crops, including chickpea. To develop heat-tolerant chickpea cultivar, substantial molecular work has been initiated. Based on biparental QTL mapping approach and genome-wide association mapping, several QTLs/genomic regions related to physiological traits, phenological traits, and yield contributing traits conferring heat tolerance were identified (Paul et al., 2018; Jha et al., 2021a, 2021b).

A comprehensive list of 71 QTLs (37 major QTLs and 40 minor QTLs) related to various phenological, morpho-physiological, and yield and yield-related traits was uncovered by subjecting phenotypic assessment of recombinant inbred line–based mapping population under normal and late sown conditions (Jha et al., 2021a). Additionally, a total of 32 candidate genes in the QTL regions encoding heat shock protein and heat shock transcription factors regulating pollen-specific genes were unveiled. In the future, these candidate genes could be validated in various natural chickpea population or chickpea genotypes for confirming their role in heat tolerance.

2.4.4 Cold Stress

Chickpea suffers from low-temperature stress during severe winter conditions (0–12°C). In north India, it is estimated to cause up to 15–20% of yield loss (Chaturvedi et al., 2009). Low-temperature stress, particularly below 15°C, during flowering leads to flower drop and pod abortion, causing yield penalty (Ali et al., 2005). So there is a need to identify genotypes which can set pods and show minimum yield reduction at severely low temperatures. Earlier works on screening methods for chilling temperature have been developed by ICRISAT (1989). Using a scale of 1 to 9, seedlings were evaluated for freezing (Singh et al., 1989). Toker (2005) discovered that a few wild species of *Cicer* were more resistant to freezing than well-known cultivars of the plant. According to Kanouni et al. (2015), seed yield and the number of secondary branches for freezing cold endurance were significantly correlated. Significant correlation was also found between 100-seed weight and cold tolerance rate. These agronomic characteristics are crucial for cold and freezing resistance. Photosystem is harmed by cold stress, and chloroplast stability is decreased at 4°C as well; triphenyl tetrazolium (TTC) content fell, and electrolyte leakage increased, ultimately leading to the relative root development rate stopping as well. Comparing cold-acclimated seedlings to non-acclimated ones, ABA content increased (Nayyar et al., 2005). There is no evidence for reported QTLs for cold tolerance in chickpea, and the chromosomal locations of the genes and QTLs are still unknown. Finding molecular markers linked to genes controlling various stress-related phenotypes may thereby improve the effectiveness of marker-assisted breeding for abiotic challenges.

2.5 CONCLUSION

The main objective of the chickpea breeding program is to develop cultivars that are superior to the existing one and have better tolerance to biotic and abiotic constraints. To achieve these breeding tasks, it is important to scan the available germplasm and wild relatives for more robust and responsive alleles coping up with these challenges. Abiotic stresses, like drought, salt, heat, and cold, cause major modulation in plant physio-biochemical and molecular responses. There is a need to explore the effect of combined stress and its responses in chickpea. The chickpea has been shown to be resistant to various biotic stresses, including AB and fusarium wilt, and breeding for this disease resistance is progressing; this could be possible due to the discovery of new resistance genes. However, further studies are the need of the hour in chickpea to better comprehend the relationships among economically important traits and yield and their effect on grain quality under stress conditions. Additionally, it has been possible to pinpoint QTLs for introgression into the background of elite cultivars for a number of traits that are resilient to biotic and abiotic stress. Combining molecular tools with traditional breeding techniques can hasten the introgression of genes or QTLs. On linkage maps, key QTLs that confer resistance to biotic or abiotic stresses have been associated with molecular markers. These molecular markers can be employed for effective pyramiding of the desired traits.

REFERENCES

Abbo S, Redden RJ, Yadav SS (2007) Utilization of wild relatives. In: Yadav SS, Redden RJ, Chen W, Sharma B (eds) *Chickpea Breeding and Management*. CAB International, Wallingford, pp. 338–354.

Ahmad F (1999) Random amplified polymorphic DNA (RAPD) analysis reveals genetic relationships among the annual *Cicer* species. *Theor Appl Genet* 98(3):657–663.

Ahmad S, Khan MA, Sahi ST, Ahmad R (2013) Evaluation of chickpea germplasm against *Ascochyta rabiei* (pass) lab. *J Anim Plant Sci* 23(2):440–443.

Ali M, Kumar S (2005) Chickpea (*Cicer arietinum*) research in India: Accomplishments and future strategies. *Indian J Agric Sci* 75(3):125–133.

Anbessa Y, Taran B, Warkentin TD, Tullu A, Vandenberg A (2009) Genetic analyses and conservation of QTL for ascochyta blight resistance in chickpea (*Cicer arietinum* L.). *Theor Appl Genet* 119(4):757–765.

Anuradha C, Gaur PM, Pande S, Gali KK, Ganesh M et al. (2011) Mapping QTL for resistance to botrytis grey mould in chickpea. *Euphytica* 182(1):1–9.

Asthana RD (1957) Some observations on the incidence of *Uromyces ciceris-arietini* (Grognon) Jakz. and Boyer on *Cicer arietinum* L. *Nagpur Agricultural College Magazine* 31:1–4.

Atieno J, Colmer TD, Taylor J, Li Y, Quealy J et al. (2021) Novel salinity tolerance loci in chickpea identified in glasshouse and field environments. *Front Plant Sci* 12:667910.

Badami PS, Mallikarjuna N, Moss JP (1997) Interspecific hybridization between *Cicer arietinum* and *C. pinnatifidum*. *Plant Breed* 116(4):393–395.

Barmukh R, Roorkiwal M, Dixit GP, Bajaj P, Kholova J et al. (2022a) Characterization of 'QTL-hotspot' introgression lines reveals physiological mechanisms and candidate genes associated with drought adaptation in chickpea. *J Exp Bot* 73(22):7255–7272.

Barmukh R, Roorkiwal M, Garg V, Khan AW, German L et al. (2022b) Genetic variation in CaTIFY4b contributes to drought adaptation in chickpea. *Plant Biotechnol J* 20(9):1701–1715. doi:10.1111/pbi.13840.

Bharadwaj C, Jorben J, Rao A, Roorkiwal M, Patil BS, Jayalakshmi, Ahammed SK, Saxena DR, Yasin M, Jahagirdar JE, Sontakke PL (2022) Development of high yielding Fusarium wilt resistant cultivar by pyramiding of "genes" through marker-assisted backcrossing in chickpea (Cicer arietinum L.). *Front Genet* 13:924287.

Bharadwaj C, Tripathi S, Soren KR, Thudi M, Singh RK et al. (2021) Introgression of "QTL-hotspot" region enhances drought tolerance and grain yield in three elite chickpea cultivars. *Plant Genome* 14(1):e20076.

Castillo P, Navas-Cortés JA, Landa BB, Jiménez-Díaz RM, Vovlas N (2008) Plant-parasitic nematodes attacking chickpea and their in-planta interactions with rhizobia and phytopathogenic fungi. *Plant Dis* 92(6):840–853.

Castro P, Pistón F, Madrid E, Millan T, Gil J, Rubio J (2010) Development of chickpea near-isogenic lines for Fusarium wilt. *Theor Appl Genet* 121:1519–1526.

Chaturvedi SK, Mishra DK, Vyas P, Mishra N (2009) Breeding for cold tolerance in chickpea. *Trends Biosc* 2(2):1–4.

Chen W, Sharma HC, Muehlbauer FJ (2011) *Compendium of Chickpea and Lentil Diseases and Pests*. American Phytopathol Society (APS) Press, MN, USA, 165 p.

Chongo G, Buchwaldt L, Gossen BD, Lafond GP, May WE et al. (2003) Foliar fungicides to manage ascochyta blight [*Ascochyta rabiei*] of chickpea in Canada. *Can J Plant Pathol* 25(2):135–142.

Cobos MJ, Fernández MJ, Rubio J, Kharrat M, Moreno MT et al. (2005) A linkage map of chickpea (*Cicer arietinum* L.) based on populations from Kabuli× Desi crosses: Location of genes for resistance to fusarium wilt race 0. *Theor Appl Genet* 110(7):1347–1353.

Davidson JA, Pande S, Bretag TW, Lindbeck KD, Krishna-Kishore G (2007) Biology and management of *Botrytis* spp. in legume crops. In: Elad Y, Williamson B, Tudzynski P, Delen N (eds) *Botrytis: Biology, Pathology and Control*. Springer, Dordrecht, pp. 295–318.

Dubey SC, Priyanka K, Singh V, Singh B (2011a) Race profiling and molecular diversity analysis of *Fusarium oxysporum* f. sp. *Cicer* is causing wilt in chickpea. *J Phytopathol* 160(10):576–587.

Dubey SC, Tripathi A, Bhavani R, Singh B (2011b) Evaluation of seed dressing and soil application formulations of Trichoderma species for integrated management of dry root rot of chickpea. *Biocontrol Sci Technol* 21(1):93–100.

Eker T, Erler F, Adak A, Imrek B, Guven H, Tosun HS, Sari D, Sari H, Upadhyaya HD, Toker C, Ikten C (2018) Screening of chickpea accessions for resistance against the pulse beetle, Callosobruchus chinensis L.(Coleoptera: Bruchidae). *J Stored Prod Res* 76:51–57.

FAOSTAT (2020) http://www.fao.org/faostat/en/#data/QC

Flowers TJ, Gaur PM, Gowda CL, Krishnamurthy L, Samineni S et al. (2010) Salt sensitivity in chickpea. *Plant Cell Environ* 33(4):490–509.

Garg T, Mallikarjuna BP, Thudi M, Samineni S, Singh S, Sandhu JS, Kaur L, Singh I, Sirari A, Basandrai AK, Basandrai D (2018) Identification of QTLs for resistance to Fusarium wilt and Ascochyta blight in a recombinant inbred population of chickpea (Cicer arietinum L.). *Euphytica* 214:1–1.

Gaur PM, Jukanti AK, Varshney RK (2012) Impact of genomic technologies on chickpea breeding strategies. *Agronomy* 2(3):199–221.

Ghosh R, Sharma M, Telangre R, Pande S (2013) Occurrence and distribution of chickpea diseases in central and southern parts of India. *Am J Plant Sci* 4(4):940–944.

Golla SK, Rajasekhar P, Akbar SM, Sharma HC (2018) Proteolytic activity in the midgut of *Helicoverpa armigera* (Noctuidae: Lepidoptera) larvae fed on wild relatives of chickpea, *Cicer arietinum*. *J Econ Entomol* 111(5):2409–2915.

Gowda SJ, Radhika P, Kadoo NY, Mhase LB, Gupta VS (2009) Molecular mapping of wilt resistance genes in chickpea. *Mol Breed* 24(2):177–183.

Greco N, Di Vito M, Saxena MC, Reddy MV (1988) Effect of *Heterodera ciceri* on yield of chickpea and lentil and development of this nematode on chickpea in Syria. *Nematologica* 34(1):98–114.

Grewal JS, Pal M, Rewal N (1992) Botrytis graymold of chickpea in India. In: Haware MP, Faris DG, Gowda CLL (eds) *Botrytis Graymold of Chickpea*. ICRISAT, Patancheru, AP, India, pp. 6–8.

Gunes A, Inal A, Adak MS, Bagci EG, Cicek N et al. (2008) Effect of drought stress implemented at pre-or post-anthesis stage on some physiological parameters as screening criteria in chickpea cultivars. *RussJ Plant Physiol* 55(1):59–67.

Haware MP (1998) Diseases of chickpea. In: Allen DJ, Lenne JM (eds) *The Pathology of Food and Pasture Legumes*. ICARDA, CAB International, Wallingford, UK, pp. 473–516.

Haware MP, McDonald D (1992) Integrated management of botrytis gray mold of chickpea. In: Haware MP, Faris DG, Gowda CLL (eds) *Botrytis Gray Mold of Chickpea. Summary Proceedings of the BARI/ICRISAT Working Group Meeting to Discuss Collaborative Research on Botrytis Gray Mold of Chickpea.* ICRISAT, Patancheru, India, pp. 3–6.

Hegde VS, Tripathi S, Bharadwaj C, Agrawal PK, Choudhary AK (2018) Genetics and genomics approaches to enhance adaptation and yield of chickpea (*Cicer arietinum* L.) in semi-arid environments. *SABRAO J Breed Genet* 50(2):217–241.

Horn NM, Makkouk KM, Kumari SG, Van den Heuvel JF, Reddy DV (1995) Survey of chickpea (*Cicer arientinum* L.) for chickpea stunt disease and associated viruses in Syria, Turkey and Lebanon. *Phytopathol Mediter* 1:192–198.

Horn NM, Reddy SV, Roberts IM, Reddy DV (1993) Chickpea chlorotic dwarf virus, a new leafhopper-transmitted Gemini virus of chickpea in India 1. *Ann Appl Biol* 122(3):467–479.

ICARDA (1998) *Germplasm Program: Legumes Annual Report for 1998.* ICARDA, Aleppo.

ICRISAT (1989) Chickpea kabuli variety ICCV6. In: ICRISAT (ed) *Plant Material Description ICRISAT No. 124.* ICARISAT, Hyderabad, India.

Jaganathan D, Thudi M, Kale S, Azam S, Roorkiwal M et al. (2015) Genotyping-by-sequencing based intra-specific genetic map refines a "QTL-hotspot" region for drought tolerance in chickpea. *Mol Genet Genom* 290(2):559–571.

Jendoubi W, Bouhadida M, Boukteb A, Béji M, Kharrat M (2017) Fusarium wilt affecting chickpea crop. *Agriculture* 7(3):23.

Jha UC, Bohra A, Jha R (2019) Omics' of salinity stress response in major grain legumes and breeding for improving crop yields in salt stressed environments. *Plant Cell Rep* 38:255–277.

Jha UC, Jha R, Thakro V, Kumar A, Gupta et al. (2021a) Discerning molecular diversity and association mapping for phenological, physiological and yield traits under high temperature stress in chickpea (*Cicer arietinum* L.). *J Genet* 100(1):1–5.

Jha UC, Palakurthi R, Nayyar H, Jha R, Valluri VK et al. (2021b) Major QTLs and potential candidate genes for heat stress tolerance identified in chickpea (*Cicer arietinum* L.). *Front Plant Sci* 12:1241.

Joshi S (1992) Botrytis gray mold of chickpea in Nepal. In: Haware MP, Faris DG, Gowda CLL (eds) *Botrytis Gray Mold of Chickpea.* ICRISAT, Patancheru, AP, India, pp. 12, 13.

Jukanti AK, Gaur PM, Gowda CL, Chibbar RN (2012) Nutritional quality and health benefits of chickpea (*Cicer arietinum* L.): A review. *Brit J Nutr* 108(S1):S11–S26.

Kaiser WJ, Hannan RM (1988) Seed transmission of *Ascochyta rabiei* in chickpea and its control by seed-treatment fungicides. *Seed Sci Technol* 16(3):625–637.

Kale SM, Jaganathan D, Ruperao P, Chen C, Punna R et al. (2015) Prioritization of candidate genes in "QTL-hotspot" region for drought tolerance in chickpea (*Cicer arietinum* L.). *Sci Rep* 5(1):1–4.

Kanouni H, Farayedi Y, Saeid A, Sabaghpour SH (2015) Stability analyses for seed yield of chickpea (*Cicer arietinum* L.) genotypes in the western cold zone of Iran. *J Agric Sci* 7(5):219.

Knights EJ, Southwell RJ, Schwinghamer MW, Harden S (2008) Resistance to Phytophthora medicaginis Hansen and Maxwell in wild *Cicer* species and its use in breeding root rot resistant chickpea (*Cicer arietinum* L.). *Aust J Agric Res* 59(4):383–387.

Kumar N, Bharadwaj C, Satyavathi CT, Pal M, Kumar et al. (2017) Morpho physiological characterization and grouping (SAHN) of chickpea genotypes for salinity tolerance. *Vegetos* 7:30:33.

Kumar N, Bharadwaj C, Soren KR, Meena MC, Priya PS et al. (2021a) Response of shoot ionic (Na^+/K^+) distribution on yield and agro-physiological traits in chickpea (*Cicer arietinum* L.) recombinant inbred lines population under salinity. *Indian J Genet* 81(4):590–593.

Kumar N, Soren KR, Bharadwaj C, PR SP, Shrivastava AK, Pal M et al. (2021b) Genome-wide transcriptome analysis and physiological variation modulates gene regulatory networks acclimating salinity tolerance in chickpea. *Environ Exp Bot* 187:104478.

Kumar T, Tiwari N, Bharadwaj C, Roorkiwal M, Reddy SP et al. (2022) A comprehensive analysis of Trehalose-6-phosphate synthase (TPS) gene for salinity tolerance in chickpea (*Cicer arietinum* L.). *Sci Rep* 12(1):1–2.

Kumar T, Tiwari N, Bharadwaj C, Sarker A, Pappula SP et al. (2020) Identification of allelic variation in drought responsive dehydrin gene based on sequence similarity in chickpea (*Cicer arietinum* l.). *Front Genet* 1579.

Kumari SG, Makkouk KM, Attar N (2006) An improved antiserum for sensitive serologic detection of Chickpea chlorotic dwarf virus. *J Phytopathol* 154(3):129–133.

Kushwah A, Bhatia D, Rani U, Yadav IS, Singh I, Bharadwaj C, Singh S (2021) Molecular mapping of quantitative trait loci for ascochyta blight and botrytis grey mould resistance in an inter-specific cross in chickpea (Cicer arietinum L.) using genotyping by sequencing. *Breed Sci* 71(2):229–239.

Li Y, Ruperao P, Batley J, Edwards D, Davidson J et al. (2017) Genome analysis identified novel candidate genes for ascochyta blight resistance in chickpea using whole genome re-sequencing data. *Front Plant Sci* 8:359.

Li Y, Ruperao P, Batley J, Edwards D, Khan T et al. (2018) Investigating drought tolerance in chickpea using genome-wide association mapping and genomic selection based on whole-genome resequencing data. *Front Plant Sci* 9:190.

Madrid E, Rubiales D, Moral A, Moreno MT, Millán T et al. (2008) Mechanism and molecular markers associated with rust resistance in a chickpea interspecific cross (*Cicer arietinum*× *Cicer reticulatum*). *Eur J Plant Pathol* 121(1):43–53.

Malhotra RS, Singh KB, Di Vito M, Greco N, Saxena MC (2002) Registration of ILC 10765 and ILC 10766 chickpea germplasm lines resistant to cyst nematode (Registrations of Germplasm). *Crop Sci* 42(5):1756–1657.

Manchikatla PK, Kalavikatte D, Mallikarjuna BP, Palakurthi R, Khan AW, Jha UC, Bajaj P, Singam P, Chitikineni A, Varshney RK, Thudi M (2021) MutMap approach enables rapid identification of candidate genes and development of markers associated with early flowering and enhanced seed size in chickpea (*Cicer arietinum* L.). *Front Plant Sci* 12:688694.

Mayer MS, Tullu A, Simon CJ, Kumar J, Kaiser WJ et al. (1997) Development of a DNA marker for Fusarium wilt resistance in chickpea. *Crop Sci* 37(5):1625–1629.

Merga B, Haji J (2019) Economic importance of chickpea: Production, value, and world trade. *Cogent Food Agric* 5(1):1615718.

Millan T, Clarke HJ, Siddique KH, Buhariwalla HK, Gaur PM, Kumar J, Gil J, Kahl G, Winter P (2006) Chickpea molecular breeding: New tools and concepts. *Euphytica* 147:81–103.

Ministry of Agriculture and Farmers Welfare, Government of India. Fourth Advance Estimates of Production of Food grains (2021–22) *Directorate of Economics and Statistics*. https://eands.dacnet.nic.in/Advance_Estimates.htm

Nayyar H, Bains T, Kumar S (2005) Low temperature induced floral abortion in chickpea: Relationship to abscisic acid and cryoprotectants in reproductive organs. *Environ Exp Bot* 53(1):39–47.

Nene YL, Reddy MV, Haware MP, Ghanekar AM, Amin KS et al. (2012) *Field Diagnosis of Chickpea Diseases and Their Control. Information Bulletin no. 28 (revised)*. ICRISAT, Patancheru, AP, India.

Nene YL, Sheila VK (1996) *A World List of Chickpea and Pigeonpea Pathogens*. ICRISAT, Patancheru, AP, India.

Nguyen TT, Taylor PW, Redden RJ, Ford R (2004) Genetic diversity estimates in *Cicer* using AFLP analysis. *Plant Breed* 123(2):173–179.

Pande SK, Siddique KH, Kishore GK, Bayaa B, Gaur PM et al. (2005) Ascochyta blight of chickpea (*Cicer arietinum* L.): A review of biology, pathogenicity, and disease management. *Aust J Agric Res* 56(4):317–332.

Patil BS, Basavarajappa MP, Chellapilla B, Jayant B (2016) Screening and Identification of resistance to rust in chickpea genotypes. In: *National Conference on Genetics and Cytogenetics*. UAS, Dharwad, India, pp. 116, 117.

Paul PJ, Samineni S, Thudi M, Sajja SB, Rathore A et al. (2018) Molecular mapping of QTLs for heat tolerance in chick-pea. *Int J Mol Sci* 19(8):pii:E2166.

Pushpavalli R, Krishnamurthy L, Thudi M, Gaur PM, Rao MV et al. (2015) Two key genomic regions harbour QTLs for salinity tolerance in ICCV 2× JG 11 derived chickpea (Cicer arietinum L.) recombinant inbred lines. *BMC Plant Biol* 15(1):1–5.

Rao GV, Rameshwar Rao V, Ghaffar MA (2013) Handbook on Chickpea and Pigeonpea Insect Pests Identification and Management. In: *Information Bulletin No. 57*. ICRISAT, Patancheru, Andhra Pradesh India, p. 96.

Rao VK, Krishnappa K (1995) Reaction of some chickpea cultivars to Meloidogyne incognita in the field. *Indian J Nematol* 25(2):217.

Ratnaparkhe MB, Santra DK, Tullu A, Muehlbauer FJ (1998) Inheritance of inter-simple-sequence-repeat polymorphisms and linkage with a fusarium wilt resistance gene in chickpea. *Theor Appl Genet* 96(3):348–353.

Redden RJ, Berger JD (2007) History and origin of chickpea. In: Yadav SS, Redden RJ, Chen W, Sharma B (eds) *Chickpea Breeding and Management*. CAB International, Wallingford, 1:1–3.

Ridge S, Deokar A, Lee R, Daba K, Macknight RC et al. (2017) The chickpea Early Flowering 1 (Efl1) locus is an ortholog of Arabidopsis ELF3. *Plant Physiol* 175(2):802–815.

Roorkiwal M, Bharadwaj C, Barmukh R, Dixit GP, Thudi M et al. (2020) Integrating genomics for chickpea improvement: Achievements and opportunities. *Theor Appl Genet* 133(5):1703–1720.

Roorkiwal M, Nayak SN, Thudi M, Upadhyaya HD, Brunel D et al. (2014) Allele diversity for abiotic stress responsive candidate genes in chickpea reference set using gene based SNP markers. *Front Plant Sci* 5:248.

Sabbavarapu MM, Sharma M, Chamarthi SK, Swapna N, Rathore A et al. (2013) Molecular mapping of QTLs for resistance to Fusarium wilt (race 1) and Ascochyta blight in chickpea (*Cicer arietinum* L.). *Euphytica* 193(1):121–133.

Sachdeva S, Bharadwaj C, Singh RK, Jain PK, Patil BS et al. (2020) Characterization of ASR gene and its role in drought tolerance in chickpea (*Cicer arietinum* L.). *PLoS One* 15(7):e0234550.

Samineni S, Siddique KH, Gaur PM, Colmer TD (2011) Salt sensitivity of the vegetative and reproductive stages in chickpea (*Cicer arietinum* L.): Podding is a particularly sensitive stage. *Environ Exp Bot* 71(2):260–268.

Santra DK, Tekeoglu M, Ratnaparkhe M, Kaiser WJ, Muehlbauer FJ (2000) Identification and mapping of QTLs conferring resistance to ascochyta blight in chickpea. *Crop Sci* 40(6):1606–1612.

Sharma KD, Muehlbauer FJ (2007) Fusarium wilt of chickpea: Physiological specialization, genetics of resistance and resistance gene tagging. *Euphytica* 157:1–4.

Sharma KK, Lavanya M, Anjaiah V (2006) Agrobacterium-mediated production of transgenic pigeonpea (*Cajanus cajan* L. Millsp.) expressing the synthetic Bt cry1Ab gene. *In Vitro Cell Dev Biol-Plant* 42(2):165–173.

Sharma M, Ghosh R, Krishnan RR, Nagamangala UN, Chamarthi SK (2012) Molecular and morphological diversity in *Rhizoctonia bataticola* isolates causing dry root rot of chickpea (*Cicer arietinum* L.) in India. *Afr J Biotechnol* 11(37):8949–8959.

Sharma M, Pande S (2013) Unravelling effects of temperature and soil moisture stress response on development of dry root rot [*Rhizoctonia bataticola* (Taub.)] Butler in Chickpea. *Am J Plant Sci* 4(3):584–589.

Shimray PW, Bharadwaj C, Patil BS, Sankar SM, Kumar N et al. (2022) Evaluation and identification of stable chickpea lines for yield-contributing traits from an association mapping panel. *Agronomy* 12(12):3115.

Sillero JC, Moreno-Alías I, Rubiales D (2012) Identification and characterization of resistance to rust (Uromyces ciceris-arietini (Grognot) Jacz. & Boyd) in a germplasm collection of *Cicer* spp. *Euphytica* 188:229–238.

Singh AK, Singh RV, Bharati RC, Chandra N, Kumar C et al. (2010) Introduction of wild and weedy relatives of crop plants in India. *EnvironEcol* 28(3A):1715–1721.

Singh G, Chen W, Rubiales D, Moore K, Sharma YR, Gan Y (2007) Diseases and their management. In: Yadav SS, Redden RJ, Chen W, Sharma B (eds) *Chickpea Breeding and Management*. CAB International, Wallingford, pp. 497–519.

Singh KB, Malhotra RS, Saxena MC (1989) Chickpea evaluation for cold tolerance under field conditions. *Crop Sci* 29(2):282–285.

Singh KB, Ocampo B (1997) Exploitation of wild Cicer species for yield improvement in chickpea. *Theor Appl Genet* 95(3):418–23.

Singh KB, Ocampo B, Robertson LD (1998) Diversity for abiotic and biotic stress resistance in the wild annual *Cicer* species. *Genet Resour Crop Evol* 45(1):9–17.

Singh M, Bisht IS, Dutta M, Kumar K, Basandrai AK et al. (2014) Characterization and evaluation of wild annual *Cicer* species for agro-morphological traits and major biotic stresses under northwestern Indian conditions. *Crop Sci* 54(1):229–239.

Solomon S, Qin D, Manning M, Averyt K, Marquis M et al. (2007) *Climate Change 2007-the Physical Science Basis: Working Group I Contribution to the Fourth Assessment Report of the IPCC*. Cambridge University Press.

Soren KR, Madugula P, Kumar N, Barmukh R, Sengar MS et al. (2020) Genetic dissection and identification of candidate genes for salinity tolerance using Axiom® Cicer SNP array in chickpea. *Int J Mol Sci* 21(14):5058.

Srivastava SK, Singh SN, Khare MN (1984) Assessment of yield losses in some promising gram cultivars due to fusarium wilt. *Indian J Plant Protec* 12:125–126.

Stephens A, Lombardi M, Cogan NO, Forster JW, Hobson K et al. (2014) Genetic marker discovery, intraspecific linkage map construction and quantitative trait locus analysis of Ascochyta blight resistance in chickpea (*Cicer arietinum* L.). *Mol Breed* 33(2):297–313.

Tapan K, Bharadwaj C, Rizvi AH, Ashutosh S, Shailesh T et al. (2015) Chickpea landraces: A valuable and divergent source for drought tolerance. *Int J Trop Agric* 33(2):633–638.

Taran B, Warkentin TD, Tullu A, Vandenberg A (2007) Genetic mapping of ascochyta blight resistance in chickpea (Cicer arietinum L.) using a simple sequence repeat linkage map. *Genome* 50(1):26–34.

Taran B, Warkentin TD, Vandenberg A, Holm FA (2010) Variation in chickpea germplasm for tolerance to imazethapyr and imazamox herbicides. *Can J Plant Sci* 90(1):139–142.

Taxak PC, Khanna SM, Bharadwaj C, Gaikwad K, Kaur S, Chopra M, Tandon G, Jaiswal S, Iquebal MA, Rai A, Kumar D (2017) Transcriptomic signature of Fusarium toxin in chickpea unveiling wilt pathogenicity pathways and marker discovery. *Physiol Mol Plant Pathol* 100:163–77.

Tekeoglu M, Tullu A, Kaiser WJ, Muehlbauer FJ (2000) Inheritance and linkage of two genes that confer resistance to Fusarium wilt in chickpea. *Crop Sci* 40(5):1247–1251.

Thakro V, Malik N, Basu U, Srivastava R, Narnoliya L et al. (2023) A superior gene allele involved in abscisic acid signaling enhances drought tolerance and yield in chickpea. *Plant Physiol* 191(3):1884–1912.

Thompson JP, Greco N, Eastwood R, Sharma SB, Scurrah M (2000) Integrated control of nematodes of cool season food legumes. In: Knight R (eds) *Linking Research and Marketing Opportunities for Pulses in the 21st Century*. Springer, Dordrecht, pp. 491–506.

Thompson JP, Reen RA, Clewett TG, Sheedy JG, Kelly AM et al. (2011) Hybridisation of Australian chickpea cultivars with wild *Cicer* spp. increases resistance to root-lesion nematodes (*Pratylenchus thornei* and *P. neglectus*). *Australa Plant Pathol* 40(6):601–611.

Thudi M, Upadhyaya HD, Rathore A, Gaur PM, Krishnamurthy L et al. (2014) Genetic dissection of drought and heat tolerance in chickpea through genome-wide and candidate gene-based association mapping approaches. *PLos One* 9(5):e96758.

Toker C (2005) Preliminary screening and selection for cold tolerance in annual wild *Cicer* species. *Genet Resour Crop Evol* 52(1):1–5.

Tullu A, Kaiser WJ, Kraft JM, Muehlbauer FJ (1999) A second gene for resistance to race 4 of Fusarium wilt in chickpea and linkage with a RAPD marker. *Euphytica* 109:43–50.

USDA (2021) https://fdc.nal.usda.gov/fdc-app.html#/food-details/174288/nutrients

Van der Maesen LJG, Maxted N, Javadi F, Coles S, Davies AMR (2007) Taxonomy of the genus *Cicer* revisited. In: Yadav SS, Redden RJ, Chen W, Sharma B (eds) *Chickpea Breeding and Management*. CAB International, Wallingford, pp. 14–46.

Varshney RK, Ribaut JM, Buckler ES, Tuberosa R, Rafalski JA et al. (2012) Can genomics boost productivity of orphan crops? *Nat Biotechnol* 30(12):1172–1176.

Varshney RK, Song C, Saxena RK, Azam S, Yu S et al. (2013) Draft genome sequence of chickpea (*Cicer arietinum*) provides a resource for trait improvement. *Nat Biotechnol* 31(3):240–246.

Varshney RK, Thudi M, Nayak SN, Gaur PM, Kashiwagi J et al. (2014) A genetic dissection of drought tolerance in chickpea (*Cicer arietinum* L.). *Theor Appl Genet* 127(2):445–462.

Westerlund FV, Campbell RN, Kimble KA (1974) Fungal root rots and wilt of chickpea in California. *Phytopathology* 64(4):432–436.

Winter P, Benko-Iseppon AM, Hüttel B, Ratnaparkhe MI, Tullu AB et al. (2000) A linkage map of the chickpea (*Cicer arietinum* L.) genome based on recombinant inbred lines from a *C. arietinum* × *C. reticulatum* cross: Localization of resistance genes for fusarium wilt races 4 and 5. *Theor Appl Genet* 101(7):1155–1163.

3 Prospects and Potentials of Allele Mining in Chickpea for Qualitative and Quantitative Traits

Sarita Pandey, Abhishek Anand, Abhishek Rathore, Satya Taddi, and Chittaranjan Kole

3.1 INTRODUCTION

The chickpea, a leading cool-season legume, holds immense significance for resource-limited farmers cultivating marginal soils in semi-arid regions around the globe. Originating in south-eastern Turkey, this crop was domesticated in the Middle East and gradually spread across the Mediterranean, India, and Ethiopia (Ladizinsky, 1975, van der Maesan, 1987). With a diploid karyotype, mostly self-pollinating tendencies, and occasional insect-mediated cross-pollination, chickpea encompasses 34 perennials and nine annual wild relatives. Characterized by a diploid state of $2n = 2x = 16$ chromosomes and a genome size of about 750 Mbp (Arumuganathan and Earle, 1991), chickpea not only contributes to soil fertility but also stands as a primary protein source for human populations. Historical and plant-based records unveil its consumption by early "hunter-gatherer societies" as a means of sustenance and communal aid, predating even 10,000 BC (Ladizinsky, 1975). Acknowledging the pivotal role of pulses in safeguarding food and nutritional security, the Food and Agriculture Organization (FAO) designated the year 2016 as the International Year of Pulses. Up until 2005, chickpea languished as an orphan crop, hindered by a lack of genomic resources that impeded the advancement of genomics-driven breeding when contrasted with conventional methodologies (Varshney et al., 2005). However, through concerted national and international efforts, chickpea has now emerged as a genomics-rich resource (Thudi et al., 2014a; Varshney, 2016). In the realm of cool-season legumes, chickpea holds the distinction of being the third most crucial grain legume, trailing only behind dry beans and peas. Among them, dry beans reign supreme, cultivated on a vast 34.5 million hectares with a production of 30.4 million tons, followed by chickpeas at 17.8 million hectares and cowpeas (dry) at 12.5 million hectares (FAO.org). In 2021, the total harvested area under chickpea was 15,004,885 ha, with production of 15,871,845.65 t (FAOstat.org, 2023) (Figure 3.1). The major chickpea-producing countries are India, Australia, and Myanmar (Figure 3.2) (DES, 2023, MOAF&W, GoI). The categorization of chickpea seeds based on their color further divides them into desi type (characterized by dark-hued seeds) or kabuli type (characterized by beige-colored seeds). Distinguishing features set "desi" chickpeas apart from their distinct traits: seeds characterized by angular contours, a deep-hued seed coat, flowers tinted pink, anthocyanin pigmentation present on stems, a textured seed surface, and a growth habit that is semi-erect or semi-spreading. Conversely, "kabuli" variants showcase owl-like seed shapes, seeds with a beige hue, flowers of a white shade, a smooth seed surface, the absence of anthocyanin pigmentation, and a growth habit that is semi-spreading. Pundir et al. (1985) and Singh

DOI: 10.1201/9781003385059-3

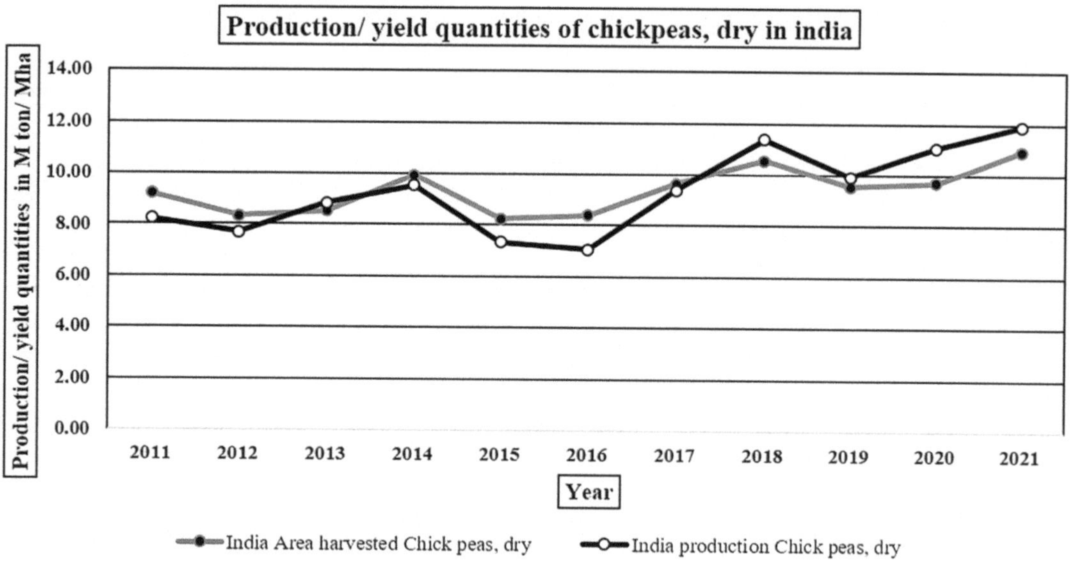

FIGURE 3.1 Production of chickpea from 2011 to 2021.

Source: FAOstat.org, curated on July 12, 2023.

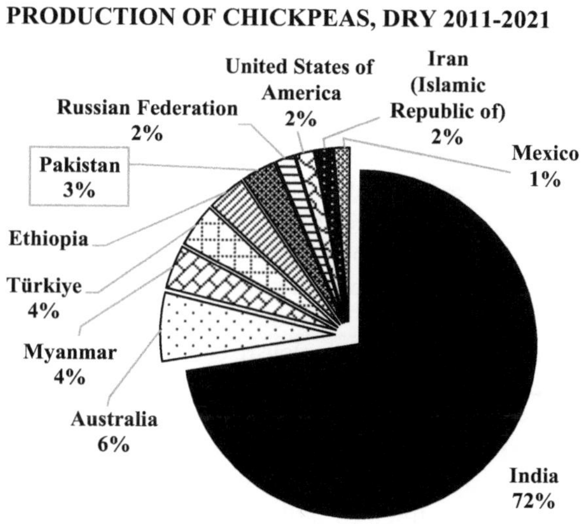

FIGURE 3.2 Share of chickpea production by different countries.

Source: Updated data from FAO, curated on July 12, 2023.

et al. (2013) provide further insight into these distinctions. Chickpea stands out with its exceptional nutritional profile among grain legumes, boasting around 23% protein content while avoiding the anti-nutritional or toxic factors often encountered in other legumes. Remarkably, up to 80% of its nitrogen requirements are met through symbiotic nitrogen fixation.

3.1.1 Origin and Distribution

Nikolai Vavilov, the Russian botanist and geneticist, delineated four primary centers of diversity for chickpea – the Mediterranean, Central Asia, the Near East, and India. Additionally, a secondary center is recognized in Ethiopia. *Cicer* species are categorized into three gene pools. The primary genetic reservoir (GP1) comprises the cultivated species *C. arietinum*, in tandem with two wild annual counterparts, namely, *C. echinospermum* and *C. reticulatum* (the proposed wild progenitor). Additionally, certain researchers extend the association to encompass the perennial *C. anatolicum* within this core genetic pool. The subsequent genetic grouping (GP2), designated as the secondary tier, encompasses a trio of annual species: *C. echinispermum*, *C. judaicum*, and *C. pinnatifidum*. Lastly, the tertiary genetic repository (GP3), representing the most distantly related segment, encompasses three annual species – *C. yamashitae*, *C. chorassanicum*, and *C. cuneatum* – alongside a collection of 34 perennial species. Desi chickpea types, predominantly found in India and Ethiopia, exhibit small angular colored seeds with a rough coat. They are believed to have closer genetic ties to the probable wild ancestor *C. reticulatum*. These varieties exhibit a bushy growth habit and boast blue-violet flowers. In contrast, kabuli types, which are primarily cultivated in the Mediterranean region, feature large beige-colored seeds shaped like owl heads, sporting a smoother coat. The plants of kabuli types tend to have an upright growth habit and white flowers (Gaur et al., 2019).

3.1.2 Morphological Characteristics and Concept of Core Collection

Chickpea taxonomically classified within the monogeneric tribe *Cicereae* of the Fabaceae family comprises nine annual species and 34 perennial species within the genus *Cicer.* The cultivated chickpea is an herb with diverse growth habits (varied from prostrate to erect). The presence of glandular and non-glandular hairs on the plant's surface results in the release of a unique acidic secretion, serving as a natural defense mechanism against piercing and sucking pests. The stem of the chickpea plant features primary, secondary, and tertiary branches, all bearing leaves and flowers. The compound leaf, commonly observed, possesses five to seven pairs of leaflets. The typical papilionaceous flower structure includes a prominent standard petal, two wing petals, and two keel petals. Diadelphous stamens and a stigma with one to four ovules constitute the flower, where anthers open a day before blooming, promotes to self-pollination. Around four weeks post-pollination, the pod matures, usually containing one to three seeds. Importantly, chickpea seeds lack dormancy. Efforts to utilize chickpea resources have resulted in the development of core collections within the legume field. Significantly, Hannan et al. (1994) meticulously curated a comprehensive core selection comprising 505 chickpea accessions, distilled from a broader collection of 7,613 specimens under the stewardship of the USDA's Western Regional Plant Introduction Station (WRPIS). Furthermore, ICRISAT forged an essential subset encompassing 1,956 accessions, with a concentrated emphasis on geographical provenance and morphological attributes (Upadhyay and Ortiz, 2001). Subsequent work by Upadhyay et al. (2007) identified 28 early-maturing chickpea lines with broad geographical distribution. A mini-core set of 211 accessions was created for improved trait evaluation (Upadhyaya and Ortiz, 2001), leading to the identification of promising traits like water use efficiency and stress tolerance (Upadhyaya et al., 2010) (Table 3.1). In 2012, the complete chickpea assemblage of the Indian National Gene Bank, totaling 14,651 accessions, underwent comprehensive characterization, revealing valuable agronomic traits (Archak et al., 2016), including unique morph types. To enhance the utility of valuable germplasm, unique traits are registered with the Germplasm Registration and Information System (GRIS). These traits aid in the development of superior plant types with high yield, such as the determinate phenotype seen in BGD 9971. Noteworthy traits, like multiple flowers per node, were identified for improving chickpea plant type and yield. Furthermore, the curation of chickpea germplasm is managed by ICARDA (International Center for Agricultural Research in the Dry Areas) and ICRISAT (International Crops Research Institute for the Semi-Arid Tropics) centers

TABLE 3.1
Details about Core Collections in Chickpea

Crop	Core/Mini-Core Collection Size	Base Accessions	Number of Studied Morphological Traits	References
Chickpea	1,956	16,991	13	Upadhyaya et al. (2001)
	211	1,956	22	Upadhyaya and Ortiz (2001)
	1,103	14,651	20	Archak et al. (2016)

(Plekhanova et al., 2017). Despite these efforts, the current tally of unique accessions stemming from both the primary and secondary gene reservoirs remains limited, as highlighted by Berger and Turner (2003). However, through recent global partnerships, the repositories of wild relatives have been broadened, as indicated by von Wettberg et al. (2018).

3.1.3 Biotic and Abiotic Stresses

In chickpea cultivation, prominent diseases include fusarium wilt, collar rot, Ascochyta blight, dry root rot, and Botrytis gray mold, while the primary pest is the pod borer. While chickpea germplasm screening has revealed numerous sources resistant to fusarium wilt, strong resistant donors for other diseases (dry root rot, collar rot, Botrytis gray mold, and pod borer) are scarce. The National Bureau of Plant Genetic Resources assessed over 2,500 accessions for resistance to certain diseases, yielding only a few moderately resistant candidates due to the narrow genetic diversity within cultivated germplasm. Notably, resistance sources have been identified for Ascochyta blight disease (Gayacharan et al., 2020). Additional valuable sources were recognized at the ICARDA and the ICRISAT (Singh, 1997) (Table 3.1). Chickpea's potential yield is hindered by both biotic and abiotic stresses, which have been exacerbated by recent climate changes, resulting in increased droughts, pest infestations (pod borer), and disease occurrences (fusarium wilt, Ascochyta blight, dry root rot, etc.). Despite past breeding efforts, the complex nature of these stresses and a limited understanding of the genetic traits involved have hampered successful results. Chickpea faces with notable abiotic pressures, such as concluding drought, ultimate heat stress, and cold spells during the advanced vegetative phases. The Northern Plains of India, once a fertile domain for chickpea yield, have experienced a downturn owing to escalating minimum nocturnal temperatures. The hurdle of terminal heat stress obstructs the extension of chickpea cultivation onto rice-fallow terrain. As a solution to these concerns, scientists are actively exploring novel avenues of tolerance to foster the development of short-duration and heat-resistant chickpea varieties. Several highly heat-tolerant chickpea lines have been identified, and a reference collection for heat stress tolerance has been established. Screening for drought stress has led to the discovery of five tolerant accessions, and a germplasm line (ICC4958) has been used for breeding drought-resistant varieties. For salinity stress, highly tolerant accessions were identified, and promising genotypes for cold tolerance were recognized. Chickpea's potential for winter cultivation is constrained by freezing temperatures during flowering. Genes for cold tolerance are sought to improve yields during this season. Cold tolerance is controlled by several genes, with both additive and non-additive effects influencing inheritance. A field screening method and a scoring scale have been developed to assess cold tolerance, aiding breeding efforts. Some wild annual *Cicer* species, such as *C. bijugum* and *C. reticulatum*, have exhibited significant cold tolerance. Additionally, chickpea's potential for winter cultivation is hampered by susceptibility to Ascochyta blight. To improve productivity, the combination of cold tolerance and resistance to Ascochyta blight is crucial. Notably, *C. bijugum* and

C. reticulatum have emerged as valuable sources for enhancing cold tolerance and combating this disease, aiding in the development of robust chickpea genotypes for winter seasons.

3.1.4 Nutritional Quality

Chickpeas are recognized for their nutritionally dense grains, serving as another source of supplementary nutrients. The seeds comprise 63% overall carbohydrates, 21% protein, and a fat content ranging from 2.70% to 6.48%. It is worth highlighting their abundance in vital minerals like potassium, phosphorus, magnesium, and calcium. Moreover, they contain vitamins C, B2, B3, B5, γ–tocopherol, E (α–tocopherol), and folic acid, while comprehensive nutritional profiling for chickpea has been recently done to identify nutrient-specific donors with high mineral content, like Zn, Ca, and P, and the identified lines could be used for enhancing the nutritional content of modern chickpea varieties (Roorkiwal et al., 2022). The vitamins, minerals, and fibers found in chickpea grains contribute to their potential health benefits.

3.1.5 Germplasm and Genetic Diversity

The study by Upadhyaya et al. (2011) explored the genetic variation, population composition, and interconnections among 48 chickpea genotypes, comprising 19 indigenous Iranian landraces and 29 globally sourced lines and cultivars. Thirty-eight simple sequence repeat (SSR) markers were used for seven distinctive morphological traits. The morphological traits exhibited high diversity and coefficients of variation (Upadhyaya et al., 2011). Few publications have extensively reviewed the germplasm conserved in the ICRISAT repository (Upadhyaya et al., 2011). For the enhancement of germplasm management and preservation, the proposition of core and mini-core collections has emerged (Upadhyaya and Ortiz, 2001), alongside the identification of germplasm tailored to specific traits for breeding and genomics-driven selection (Upadhyaya et al., 2012). Molecular-level characterization of chickpea germplasm has been conducted in multiple investigations (Iruela et al., 2002; Croser et al., 2003; Nguyen et al., 2004; Rao et al., 2007; Upadhyaya et al., 2008; Sefera et al., 2011; Choudhary et al., 2012) complementing traditional phenotypic assessments (Upadhyaya et al., 2012). However, trait mapping remains essential for understanding complex traits like drought tolerance. Mini-core collection evaluation helped identify the germplasm lines with prolific root systems for terminal drought tolerance. Recombinant inbred line (RIL) mapping populations were developed to study various biotic and abiotic stresses (Gaur et al., 2008, 2013). For the functional verification of drought-responsive genes, a "target-induced local lesions in genome" (TILLING) population was formulated, comprising 3,072 M2 chickpea lines (ICRISAT, data not published). The landscape of chickpea has experienced a profound shift, evolving from an "orphan legume" to a "genomic resource-rich" crop, courtesy of recent strides in genomic tools, such as genome sequencing and allele mining (Varshney et al., 2012; Jain et al., 2013). Farhani et al. (2019) conducted a study where they utilized Diversity Arrays Technology (DArT) seq-based single-nucleotide polymorphism (SNP) markers to genotype 186 chickpea genotypes, aiming to investigate the genetic diversity within the population.

3.1.6 Phenotyping and Phenomics

Evaluating the diversity present within germplasm holds paramount significance for plant breeders endeavoring to enhance crops and gene bank curators aiming to proficiently manage their collections. A comprehensive characterization of an expansive chickpea germplasm collection encompassing an array of morph physiological and reproductive attributes has been conducted at ICRISAT, situated in Patancheru, India. Through the scrutiny of 16,820 accessions and 13 distinct traits, the diversity assessment yielded an intriguing revelation: notable variations in means and heterogeneous variances for agronomic traits across different regions (Upadhyaya, 2003). Akin diversity assessments within pigeon peas (involving 11,402 accessions and 26 traits) also unveiled noteworthy disparities in means and heterogeneous variances among varying regions (Xu and Crouch, 2008; Passioura, 2010). The landscape of plant phenotyping has undergone a transformative shift, transitioning from invasive or damaging methods to the realm of high-throughput, precise, and non-destructive imaging techniques. A diverse range of phenomics platforms is now globally available, providing comprehensive solutions for phenotyping requirements. These encompass a variety of tools, including infrared cameras for monitoring temperature and transpiration patterns, fluorescent microscopy and spectroscopy for evaluating photosynthesis and photosynthetic rates, three-dimensional cameras for finely capturing growth responses, lidars (light detection and ranging) for measuring growth rates, and magnetic resonance imaging for exploring root and leaf physiology (Finkel, 2009, Gupta et al., 2012). The advent of digital imaging has granted the capacity to observe, quantify, and trace various aspects of plant development, functionality, and health that were previously unimaginable using conventional measurement techniques. A suite of software programs has emerged to facilitate data extraction from digital images encompassing roots, shoots, leaves, seeds, and grains (Sozzani and Benfey, 2011, Cobb et al., 2013). These burgeoning phenomics tools, in conjunction with anticipated future advancements, are poised to facilitate the rapid scanning of thousands of plants within a single operational day, mirroring the high-throughput nature of DNA sequencing in the realm of genomics. The precision and accuracy of the data generated through these advanced facilities hold profound import and utility, contributing to the meaningful dissection of genetics and the application of genotype–environment interactions (GE) for crop enhancement, encompassing the realms of chickpea amelioration.

3.1.7 Biochemical Diversity

Isozymes have emerged as a valuable tool for delving into the genetic diversity present within crop populations and unraveling the intricate relationships that interweave different species. An in-depth exploration led by Kazan and Muehlbauer (1991) orchestrated an encompassing analysis, scrutinizing nine annual chickpea species alongside one perennial variant. Their focus was directed at 19 enzymes that serve as proxies for 30 potential gene loci. The results of this comprehensive inquiry yielded the stratification of these species into four distinct clusters, delineated by their unique isozyme profiles. The initial linkage map of chickpea was created using isozyme markers (Gaur and Slinkard, 1990), while the interconnections between annual *Cicer* species were elucidated through phylogenetic analyses (Kazan and Muehlbauer, 1991). To be precise, Group I encompassed *C. arietinum*, *C. reticulatum*, *C. echinospermum*, and *C. anatolicum*, while Group II enfolded *C. judaicum*, *C. bijugum*, and *C. pinnatifidum*. Moving onward, Group III comprised *C. chorassanicum* and *C. yamashitae*, while Group IV embraced *C. cuneatum*. Subsequent investigations involving annual chickpea species continued to underscore this analogous pattern (Labdi et al., 1996, Tayyar and Waines, 1996) (Table 3.2). Nevertheless, a measure of inconsistency emerged regarding the positioning of *C. chorassanicum* relative to *C. yamashitae* within the isozyme-based classification. This dissonance was further magnified by studies involving interspecific hybridization (Kazan and Muehlbauer, 1991) and scrutiny of seed storage protein profiles (Ahmad et al., 1992), both of which assigned *C. chorassanicum* and *C. yamashitae* to separate groupings. Intriguingly, the evaluation of species relationships through the lens of random amplified polymorphic DNA (RAPD) markers yielded outcomes that converged with the previously established isozyme-based cluster arrangement. Of notable interest is the revelation that, despite the reproductive isolation characterizing *C. chorassanicum* and *C. yamashitae*, the evolution of reproductive barriers does not invariably mirror the divergence seen within isozyme loci (Ahmad et al., 1992). Among the perennial species, a grouping emerged that united *C. anatolicum* and *C. soongaricum*, sharing a cluster alongside *C. chorassanicum* and *C. yamashitae*, as ascertained by Tayyar and Waines (1996).

TABLE 3.2
Information on the Status of Collections of Chickpea and Their Wild Relatives Available in the Indian National Gene Bank, Along with Production and Yield Status of Chickpea Cultivated Worldwide

Botanical Name		Chromosome Number and Ploidy Level	Production: Top 3 Major Producing Countries (Million Tons)			Top 3 Countries with Highest Yield (kg/ha)			Total World Production (MT)
			1st	2nd	3rd	1st	2nd	3rd	
Cicer arietinum	2,961 (exotic) 11,452 (indigenous) 14,413 (total)	2n = 2x = 16	India (7.06)	Australia (1.32)	Myanmar (0.56)	China (5,177)	Israel (4,148)	Sudan (4,048)	14.6
Cicer wild species: *C. bijugum* (31), *C. chorassanicum* (2), *C. cuneatum* (6), *C. echinospermum* (18), *C. judaicum* (54), *C. microphyllum* (35), *C. pinnatifidum* (27), *C. reticulatum* (18), *C. yamashitae* (4), unknown species (9)	148 (exotic) 56 (indigenous) 204 (total)	–	–	–	–	–	–	–	–

3.1.8 Molecular Diversity

While conventional breeding techniques have successfully enhanced yield and abiotic stress tolerance in chickpea, they are limited in their applicability to highly heritable and visually discernible traits (Maphosa et al., 2020). The adaptive traits crucial for abiotic stress tolerance and yield, however, are often influenced by multiple genes, possess low heritability, and are profoundly impacted by environmental factors (Dormatey et al., 2020). Given the challenges posed by climate change and its extreme conditions, conventional breeding alone may not ensure the necessary grain yield levels for global food security, particularly in chickpea (Maphosa et al., 2020). To address this, modern omics-based breeding methods offer a valuable complement by enhancing selection efficiency, reducing breeding duration, and pinpointing specific genes governing desired traits in legumes (Jacob et al., 2016). An integrated approach that combines conventional and omics-based techniques holds the key to developing resilient, high-yielding chickpea varieties capable of withstanding climate change impacts. The various omics disciplines, encompassing genomics, transcriptomic, proteomics, and metabolomics, have been harnessed to scrutinize plant responses spanning diverse molecular strata when confronted with adverse environmental circumstances. Throughout the preceding decade, substantial advancements have been achieved in augmenting the genomic reservoirs of chickpea, facilitating the deconstruction of the genetic foundations underpinning pivotal traits associated with tolerance and yield performance (Jacob et al., 2016, Gaur et al., 2019). Genomics encompasses the formulation of molecular markers that facilitate the assessment of genetic diversity, construction of genetic maps, and pinpointing of genome segments, recognized as quantitative trait loci (QTLs), which correlate with desirable agricultural attributes (Reddy et al., 2012). The chickpea domain has observed the inception of a multitude of molecular markers, including RAPD (Tullu et al., 1998), SSR (Sethy et al., 2006), DArT (Thudi et al., 2011), and SNP (Hiremath et al., 2012) (Table 3.3). The emergence of next-generation sequencing (NGS) and high-throughput genotyping technologies has paved the way for the development of the high-density "Axiom®CicerSNP Array," boasting 50,590 unique SNPs (Roorkiwal et al., 2018a). These markers have substantially

TABLE 3.3
List of Markers Along with Traits Used in Chickpea

Molecular Marker	Trait	No. of Makers	Reference
ASAP	Wilt resistance	1	Jukte et al. (2017)
ISSR	Agronomic traits	23	Yadav et al. (2022)
	Agro-morphological traits	45	Janghel et al. (2021)
CDDP	Morphological traits	11	Hajibarat et al. (2016)
SCoT	Morphological traits	9	Hajibarat et al. (2016)
STMS	Agronomic traits	35	Bharadwaj et al. (2010)
	Fusarium wilt resistance	5	Rajput et al. (2023)
	Wilt resistance	3	Jukte et al. (2017)
InDels	Biotic and abiotic stress	8,307	Varshney et al. (2019)
RAPD	Seed trait	9	Begum et al. (2019)
	Morphological traits	10	Varshney et al. (2019)
	Genotype	42	Farahani et al. (2016)
SCAR	Genotype	42	Farahani et al. (2016)
SNPs	Agronomic traits	71	Nayak et al. (2010)
	Drought tolerance	828	Jaganathan et al. (2015)

(Continued)

TABLE 3.3 *(Continued)*
List of Markers Along with Traits Used in Chickpea

Molecular Marker	Trait	No. of Makers	Reference
SSRs	Agronomic traits	147	Nayak et al. (2010)
	Tolerant/resistant to different abiotic and biotic stresses	33	Ningwal et al. (2023)
	Drought tolerance	279	Jaganathan et al. (2015)
	Drought tolerance	15	Yadav et al. (2019)
	Seed trait	10	Begum et al. (2019)
	Agro-morphological trait	31	Singh et al. (2021)
	Morphological trait	10	Khelurkar et al. (2022)
EST-SSR	Drought tolerance	14	Jaganathan et al. (2015)
GMM	Drought tolerance	4	Jaganathan et al. (2015)
DArT	Drought tolerance	21	Jaganathan et al. (2015)
	Agronomic traits	15,360	Sharma et al. (2013)
STMS	Agronomic traits	35	Bharadwaj et al. (2010)
	Fusarium wilt resistance	5	Rajput et al. (2023)

Note: *ASAP*, allele-specific associated primers; *ISSR*, inter-simple sequence repeat; *CDDP*, conserved DNA-derived polymorphism; *SCoT*, start codon targeted; *STMS*, sequence-tagged microsatellite sites; *InDels*, insertions, or deletions; *RAPD*, random amplified polymorphic DNA; *SCAR*, sequence-characterized amplified region; *SNPs*, single-nucleotide polymorphisms; *SSRs*; *EST-SSR*, expressed sequence tag–simple sequence repeat; *GMM*, genic molecular markers; *DArT*, Diversity Arrays Technology; *STMS*, sequence-tagged microsatellite sites.

eased the generation of diverse genetic maps, encompassing those with high-density profiles (Kujur et al., 2015, Gaur et al., 2015, and Barmukh et al., 2021). Furthermore, the accessibility of draft genome sequences (Gupta et al., 2017; Varshney et al., 2021) extensive re-sequencing endeavors (Thudi et al., 2016) and an exhaustive map documenting genetic variation within 3,171 cultivated and 195 wild chickpea accessions (Varshney et al., 2021) has reinforced the chickpea genomic landscape. The integration and effective utilization of these genomic assets harbor the potential to amplify precision and efficacy within chickpea breeding programs. Significantly, various genomic regions or QTLs linked to abiotic stress tolerance (notably drought, extreme temperatures, and salinity) and yield-associated traits have been pinpointed through QTL and association mapping within the chickpea realm (Arriagada et al., 2022). The accompanying table presents a condensed overview of the markers deployed in chickpea research.

3.1.9 Association Mapping and Marker–Trait Association

In a comprehensive study, Thudi et al. (2014a) extensively characterized the chickpea reference set utilizing 16,046 markers, which comprised 35 SSRs, 15,360 DArT features, and 651 SNPs. The analysis of SSR markers unveiled 917 alleles exhibiting polymorphism information content (PIC) values spanning from 0.48 to 0.96. Notably, Roorkiwal et al. (2014) amplified ten drought-responsive candidate genes within the reference set/mini-core collection. Subsequent population structure analysis employed 85 evenly distributed DArT loci across the chickpea genome on the reference set (Thudi et al., 2014b). Diapari et al. (2014) identified two specific SNPs, on chromosome number 1 and on chromosome number 4, associated with iron and zinc concentrations augmenting the previously identified SNPs for these traits. Exploring drought-related characteristics in a sparse SSR marker-based biparental mapping population (ILC588 × ILC 3279) led to the discovery of 15 and 93 QTLs (Rehman et al., 2011, Hamwieh et al., 2013). Notably, a significant "QTL hotspot" on chromosome 4 was pinpointed through conventional QTL analysis utilizing both SSR and GBS markers, subsequently narrowed down to an approximate ~300 kb region encompassing 26 genes (Varshney et al., 2014b; Jaganathan et al., 2015; Kale et al., 2015). Employing the GWAS approach, Thudi et al. (2014a) revealed over 200 markers associated with drought-related traits using SSR, DArT, and SNP markers. In response to the challenge of marker-assisted selection for complex traits like drought tolerance, genomic selection (GS) emerged as an alternative method (Collins et al., 2008). In a recent GS study, Roorkiwal et al. (2016) reported diminished prediction accuracies for rainfed yield compared to irrigated conditions in chickpea.

3.2 GENOMICS-ASSISTED BREEDING

Genomic-assisted breeding (GAB) integrates genomic tools to enhance selection precision and efficiency in breeding. Key strategies within GAB include genomics, proteomics, and transcriptomics, which aid in identifying closely linked molecular markers associated with economically important traits. These markers facilitate phenotype prediction from genotypic data. High-throughput genotyping made feasible by next-generation sequencing (NGS) enables the development of extensive genome-wide markers. For the simultaneous integration of multiple QTLs into a specific genetic background, marker-assisted backcrossing (MABC) is employed, particularly for significant improvements in traits governed by major genes/QTLs. However, due to the polygenic nature of many economically important traits, marker-assisted recurrent selection (MARS) is favored for enhancing polygenic characteristics. The genome-wide selection (GWS) or GS approach has emerged as a potent method for selecting favorable progenies resulting from advantageous crosses (Jannink et al., 2010). Another technique, the advanced backcross QTL (AB-QTL) approach, has been harnessed to simultaneously identify and transfer beneficial alleles from wild species or relatives into elite lines, thereby enhancing productivity and stress resistance. This technique has proven effective for introgressing desirable traits, including disease resistance, from *C. reticulatum* into chickpea (Singh et al., 2005). Enhanced prediction accuracies can be achieved by adopting a GS approach that incorporates G × E effects and considers multiple variables simultaneously within chickpea breeding programs (Roorkiwal et al., 2018b). The integration of GS techniques into pre-breeding initiatives holds significant promise, aiding in the efficient screening of accessions for subsequent introgression (Crossa et al., 2017). Varshney et al. (2018) have outlined a preliminary framework for sequence-based breeding utilizing the GS approach. The availability of the chickpea genome draft has paved the way for the application of genomics-assisted breeding methods to enhance drought tolerance. Strategies such as MABC, marker-assisted recurrent selection, and GS have been implemented (Thudi et al., 2014b). MABC has facilitated the successful incorporation of QTL hotspots into modern elite cultivars, such as JG11, Chefe, and KAK2, accomplished through multiple selection rounds employing SSR and AFLP markers (Varshney et al., 2013). MARS has been employed to accumulate favorable alleles, elevating drought tolerance by augmenting the frequency of desirable alleles (Varshney et al., 2012). ICRISAT utilized MARS to enhance drought tolerance in elite "desi" chickpea genotypes (Thudi et al., 2014a). Trait mapping and germplasm characterization efforts have also played a role in enhancing drought tolerance. Researchers have evaluated trait-specific germplasm collections and employed genomics-assisted selection techniques (Upadhyaya et al., 2012). Validation of drought-responsive genes was achieved through TILLING utilizing next-generation sequencing (Thudi et al., 2014a). Leveraging advanced techniques such as the multiparent advanced generation intercross (MAGIC) population in next-generation multiple mapping has proven beneficial for unraveling intricate traits and identifying novel genes (Glaszmann et al., 2010). Despite major abiotic and biotic stresses such as drought and fusarium wilt (FW), progress has been made in developing chickpea varieties with enhanced drought tolerance and FW resistance through approaches like MABC (Thudi et al., 2014a). Notably, select varieties have been released and are contributing to improved chickpea production (Thudi et al., 2014b).

3.3 FUNCTIONAL GENOMICS

The rapid advancement of genome science has expanded its scope beyond model organisms, facilitating in-depth investigations of traits of interest across diverse living organisms. Next-generation re-sequencing techniques are rapidly advancing the sequencing of crop plant genomes, producing high-density SNP genotyping that facilitates the comprehension of genetic diversity and genes associated with traits. By merging phenotyping and genomic databases, the investigation into intricate genetic traits has been empowered, culminating in the discovery of novel genes and QTLs, consequently amplifying genetic advancements for traits characterized by reduced heritability (Bernardo, 2004). Genomic tools have enabled prediction of gene function, location, and interactions, yet precise validation through phenotyping using next-generation phenotyping techniques is crucial. Accurate phenotyping data is essential for identifying relevant QTLs or chromosomal regions and establishing phenotype–genotype associations. The combination of high-resolution genotype and phenotype data provides an avenue to dissect complex quantitative traits, particularly valuable for crop genomics and practical breeding. To overcome challenges posed by low heritability and Type II errors, large-scale phenotyping across various environments enhances QTL detection probability and accuracy. GWS leverages molecular profiling and precise phenotyping, eliminating the need for explicit QTL identification and leading to improved breeding outcomes (Heffner et al., 2009 and Jansen et al., 1995). The utilization of high-throughput phenomics platforms has played a pivotal role in uncovering genes and QTLs linked to a wide array of traits in crops, such as rice, wheat, barley, and mustard (Topp et al., 2013). These methodologies are applied to survey mutation populations with sequenced genomes, facilitating the identification of mutants displaying modified root system architecture (Shorinola et al., 2018). The convergence of genomics and phenomics is catalyzing a transformation in crop breeding, with compelling evidence showcasing that the integration of genomics and marker-assisted selection has yielded significant genetic advancements compared to conventional phenotypic selection methods (Battenfield et al., 2016). Even within less extensively studied crops like food legumes, the integration of molecular markers and phenotyping techniques has been employed to introduce disease resistance genes into chickpea (Varshney et al., 2014a; Pratap et al., 2017). The streamlined and cost-effective acquisition of genomic data has catalyzed the development of GS strategies, especially for complex traits, offering promising avenues for enhancing genetic improvement. Precise phenotyping in training populations allows the dissection of complex traits into contributions from genotype, phenotype, and environment, enabling breeders to adopt GS strategies for effective trait improvement (Cabrera-Bosquet et al., 2012; Heslot et al., 2012). This integrated approach holds the promise of revolutionizing the way new crop varieties are developed and bred, with the synergy between genomics and phenomics poised to unlock novel advancements in crop improvement (Araus et al., 2014)

3.3.1 Transcriptomics

Extensive exploration of the chickpea transcriptome has yielded valuable insights into a myriad of biological processes. Employing a diverse range of techniques, including EST sequencing, SAGE, and notably, NGS, researchers have meticulously dissected the transcriptomes of various tissues, such as root, shoot, flower, seed, and nodules. These investigations have been driven by the quest to unravel tissue-specific developmental processes and functions. In a prior study focused on root transcriptomics, an EST library was meticulously constructed through the application of subtractive suppressive hybridization (SSH) between two closely related chickpea varieties. This endeavor resulted in the identification of over 2,800 ESTs, which were subsequently harnessed to create the "Chickpea Root Expressed Sequence Tag Database" (Jayshree et al., 2005). Building upon this foundation, several comprehensive transcriptome analyses, underpinned by NGS, have been conducted within the chickpea realm (Garg et al., 2011b; hiremath et al., 2011). Pioneering the NGS realm, a seminal study utilized Illumina sequencing to scrutinize the transcriptome of the ICC 4958 chickpea genotype, spanning root and shoot tissues. The resultant de novo assembly bore fruit in the form of 53,409 transcripts. Further enriching the chickpea transcriptome compendium, Garg et al. (2011a) conducted a Roche/454-based analysis encompassing diverse tissue types, including "shoot," "root," "mature leaf," "flower bud," and "young pod" from the ICC 4958 chickpea genotype. This holistic dataset coalesced into the creation of the "Chickpea Transcriptome Database" (CTDB), a repository endowed with a user-friendly interface to facilitate access to chickpea transcriptome data (Garg et al., 2011b). In a broader taxonomic context, the transcriptome of the wild progenitor of chickpea, the *Cicer reticulatum* PI489777 variety, underwent scrutiny under the meticulous analysis of Jhanwar et al. (2012). Furthermore, transcriptomes drawn from kabuli, desi, and wild chickpeas were juxtaposed, yielding a refined iteration of the Chickpea Transcriptome Database (V2.0) (Verma et al., 2015). Delving deeper into developmental dynamics, the intricacies of flower development were deciphered through an Illumina sequencing–based transcriptome analysis. This endeavor unearthed 1,118 genes marked by differential regulation during distinct flower bud and flower developmental stages (Singh et al., 2013). The dynamic landscape of chickpea seed development was similarly explored using NGS-based transcriptomics, shedding light on distinct developmental stages (Pradhan et al., 2014). The realm of chickpea embryo development also witnessed scrutiny, with the sequencing of ESTs culminating in the identification of 1,480 expressed unigenes pivotal in embryo development (Gupta et al., 2015). Venturing beneath the soil, a DeepSuperSAGE-based investigation unearthed 71 genes characterized by differential expression within root nodules (Afonso-Grunz F et al., 2014). Environmental challenges such as drought and salinity stress were met head-on through transcriptomic scrutiny, resulting in the generation of a vast repertoire of ESTs. Among these, a subset of 105 ESTs exhibited differential expression in response to the stresses (Deokar et al., 2011). A comparative analysis between drought-tolerant (ICC 4958) and drought-resistant (ICC 1882) chickpea varieties further underscored the genetic underpinnings of drought responsiveness, unearthing 5,494 drought-responsive ESTs (Deokar et al., 2011). A microarray-based expedition into the transcriptome of chickpea roots and leaves, subjected to drought stress, identified 4,815 genes exhibiting differential expression (Garg et al., 2015). Shifting focus to salt tolerance, a combination of SuperSAGE and 454 sequencing unveiled the transcriptional dynamics of the salt-tolerant variety INRAT-93, highlighting 363 upregulated and 106 downregulated genes within root and nodule tissues (Garg et al., 2015). The comprehensive landscape of stress responses in chickpea was encapsulated in a study by Garg et al. (2011a) which delved into the transcriptomic shifts occurring within root and shoot tissues under the duress of desiccation, salinity, and cold stress (Garg et al., 2015). Building on this, an in-depth transcriptome analysis of distinct chickpea varieties (drought-tolerant, drought-sensitive, salinity-tolerant, and salinity-sensitive) revealed a plethora of transcripts (18,462), each emblematic of the complex interplay between genotypic variations and stress conditions (Garg et al., 2016). To dissect the intricacies of cold stress response, AFLP-based transcript profiling (cDNA-AFLP) elucidated the dynamics of 102 transcript-derived fragments (TDFs) within cold-tolerant chickpea plants, reflective of their adaptive strategies (Dinari et al., 2013). The transcriptomic retort of chickpea to *A. rabiei*, a notorious pathogen, was unveiled through the sequencing of an EST library derived from ICC 3996, a blight-resistant chickpea variety, upon infection with *A. rabiei* (Coram et al., 2005). Concomitantly, a wealth of genomic markers was identified, comprising 303 polymorphic SSRs, 14,462 SNPs, and 1,864 insertions/deletions (InDels). Notably, many of these markers were intricately linked to defense-related genes, underscoring the interplay between genetics and defense mechanisms. The convergence of biotic and abiotic stress responses came into focus through a meticulous microarray analysis of chickpea ICC 3996, unraveling the shared genetic repertoire underlying responses to drought, cold, high salinity, and biotic stress *A. rabiei* infection (Mantri et al., 2010). Broadening the scope, the small RNA landscape of chickpea was probed through the sequencing of small RNA libraries derived from normal tissues and those subjected to diverse stress conditions (Jain et al., 2014Kohli et al., 2014 and Srivastava et al., 2015). The genetic landscape was further enriched by mining lncRNA from RNA-seq data spanning 11 distinct chickpea tissues (Khemka et al., 2016).

Marker development emerged as an integral facet of transcriptome analysis, evident in the generation of 106 EST-based markers sourced from a root tissue EST library (Buhariwalla et al., 2005). The pursuit of marker development extended to EST-SSR markers, exemplified by the utilization of 2,131 ESTs to create 246 EST-SSR markers (Choudhary et al., 2012). Beyond SSRs, a diverse array of markers emerged, encompassing ESTPs, PIPs, and EST-SNPs. For instance, Choudhary et al. (2012) harnessed transcriptome data to yield 125 EST-SSRs, 109 ESTPs, 102 SNPs, and 151 ITPs. Gupta et al. (2015) made their mark through the development of 367 novel EST-derived functional markers, comprising 187 EST-SSRs, 130 potential intron polymorphisms (PIPs), and 50 expressed sequence tag polymorphisms (ESTPs) (Gupta et al., 2015). In a bid to fortify genetic tools, 71 gene-based SNP markers were fashioned, drawing upon candidate chickpea transcripts (Nayak et al., 2010). Notable contributions to the genetic atlas of chickpea also include the sequencing of the ICCV2 transcriptome, yielding the identification of 5,409 SSRs (Nayak et al., 2010). Similarly, a transcriptome-based approach led to the identification of 51,632 genic SNPs (Srivastva et al., 2016). A monumental endeavor involving RNA sequencing of the wild perennial *Cicer microphyllum* yielded the identification of 11,621 differentially expressed genes in the context of root vs. shoot tissues.

3.3.2 Metabolomics

While transcriptomic and proteomic investigations represent crucial strides in unraveling the intricate biological mechanisms associated with abiotic stress tolerance, they offer only a partial glimpse into the comprehensive cellular response exhibited by plants under such stressors. This limitation stems from the fact that many underlying biological processes are ultimately orchestrated by cellular metabolites (Janská et al., 2010). In this context, metabolomics emerges as a burgeoning field within the realm of "omics" research, centering on the high-throughput characterization of small-molecule metabolites within biological matrices (Steinfath et al., 2010). Functioning as a molecular bridge between genotype and phenotype, metabolomics assumes a pivotal role in elucidating the intricate metabolic changes that underlie plant development and responses to applied stressors. These metabolic insights provide a more accurate reflection of biological end points compared to transcriptomic or proteomic analyses (Steinfath et al., 2010). Chickpea's metabolomic landscape under diverse abiotic stress conditions has been a relatively underexplored territory. For instance, Nisa et al. (2020) conducted an analysis on desi and kabuli chickpea genotypes grown under rainfed conditions, revealing notable differences in metabolite profiles. Osmoprotectants like oxalic acid, threonic acid, inositol, maltose, and L-proline exhibited significant variations between the evaluated genotypes, with higher levels observed in the desi genotype under rainfed conditions. This underscores the involvement of inositol phosphate metabolism as a defense mechanism against water scarcity during chickpea production (Schauer et al., 2006). Similarly, Khan et al. reported distinct changes in metabolite levels between drought-sensitive and drought-tolerant (Návarová et al., 2012) chickpea cultivars under drought stress. These findings were echoed in a subsequent study by Khan et al. wherein drought-tolerant genotypes demonstrated heightened accumulation of five metabolites (L-proline, L-arginine, L-histidine, L-isoleucine, and tryptophan) in response to drought stress. Salinity stress, as investigated by Dias et al., induced a reduction in arabinose concentration but a substantial increase in tricarboxylic acid (TCA) cycle metabolites – such as isocitrate, cis-aconitate, citrate, fumarate, and malate – in the salt-sensitive Rupali genotype. Incorporating metabolomics into the broader omics framework, as outlined in preceding sections, allows for the identification of QTLs, genes, proteins, and metabolites that collectively contribute to the expression of specific phenotypes. This integrative approach furnishes a holistic comprehension of the intricate molecular networks governing plant responses to abiotic stresses, ultimately influencing chickpea yields. A GC-MS-based untargeted metabolomics approach in chickpea revealed significant differences in several metabolites, as reported by Nisa et al. (2020).

3.3.3 Proteomics

The chloroplast, a distinctive energy organelle exclusive to plant cells, stands as a dynamic entity that orchestrates a diverse array of metabolic pathways. It serves as the primary site for energy conversion, influencing the entire ecological hierarchy. The exponential growth in sequence data and advancements in mass spectrometry techniques have ushered in fresh avenues for the opportune exploration of the global proteome housed within this organelle. In this investigation, our focus was directed towards creating a comprehensive inventory of chloroplast proteins in a grain legume, specifically the chickpea. Our efforts culminated in the establishment of a definitive reference map for the proteome. To ensure precise attribution of the identified proteins, we meticulously monitored the purity of the chloroplast-enriched fraction. This was accomplished through rigorous assessment using a variety of chemical and immunological indices, in addition to pigment and enzyme analyses. The outcome of our proteome analysis unveiled the presence of 2,451 proteins, encompassing 27 isoforms, which encompass both predicted and novel constituents of the chloroplast. Stringent sequence analysis was employed to validate the identity of these proteins. Our comprehensive prediction of protein localization, based on sequence data, indicated that over 50% of the proteins were confidently associated with the chloroplast, as indicated by multiple distinct algorithms. Delving into the chromosomal distribution of these identified proteins across both the nuclear and chloroplast genomes revealed the presence of 55 chloroplast-encoded genes. A thorough comparative analysis between our dataset and the existing non-redundant pool of identified chloroplast proteins from diverse species not only uncovered novel candidates but also revealed significant overlaps. In a parallel context, Li et al. (2014) conducted a comprehensive review of plant phenotyping techniques, discussing their applicability, advantages, and limitations. When addressing drought tolerance in chickpea, a combination of phenotyping, genetic dissection, and molecular breeding is employed (Upadhyaya et al., 2012).

3.3.4 Prospects of Allele Mining in Chickpea

Allele mining presents an effective tactic for unearthing a trove of latent, yet substantial, natural and functionally diverse allelic variations harbored within the realm of wild and cultivated genetic resources. This strategic approach propels the augmentation of crop genetics, thus elevating global agriculture's productivity and sustainability (Figure 3.3). Thoroughly understanding the richness of natural allelic diversity and unraveling the complexities of genetic structure and linkage disequilibrium (LD) patterns within wild germplasm accessions emerge as a crucial endeavor. To accomplish this, extensive genotyping utilizing informative microsatellite and SNP markers becomes essential, thereby laying the foundation for advancing chickpea genetics.

In a research endeavor undertaken by Kujur et al. (2015), a notable tally of 44,844 high-quality SNPs was unveiled by sequencing a collection of 93 distinct cultivated chickpea accessions. This significant discovery was made possible by employing a dual approach encompassing both reference genome–guided and de novo–based genotyping-by-sequencing (GBS) methodologies, which were systematically mapped onto the eight chromosomes of desi and kabuli chickpea variants. These groundbreaking findings echoed the substantial efforts previously dedicated to molecular mapping of plant height-related QTLs, both within and between different genetic linkage maps of chickpea (Thudi et al., 2014a; Roorkiwal et al., 2018b; Barmukh et al., 2022). This concerted progress paves the way for efficient genetic applications. To further in this line, the focus has shifted toward the identification of genes hosting high-resolution QTLs on a genome-wide scale. This ambitious goal is pursued by constructing SNP marker-based ultra-high-resolution genetic linkage maps through GBS assays. Subsequent validation via trait association analysis and differential expression profiling promises to be a promising avenue for driving chickpea genetic improvement (Figure 3.3). The interest toward mutagenesis as a potent method for deciphering gene function and elevating

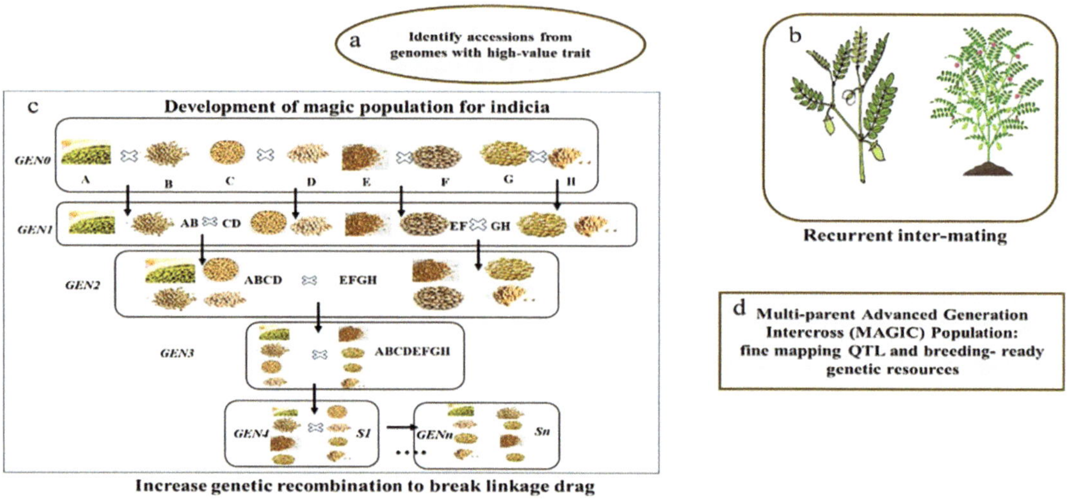

FIGURE 3.3 Schematic representation for mating design to convert genetic diversity to genetic resources. The design involves four phases: (1) identifying new accessions for desired traits validated by phenotyping; (2) creation of a MAGIC population; (3) breaking unfavorable linkages; and (4) producing superior resources.

traits has increased. TILLING, a reverse genetics technique, serves as an efficient early screening platform for pinpointing point mutations in genes of significance (Kurowska et al., 2011; Hunter et al., 2014). It is noteworthy that chemical mutagens have gained prominence for their ability to generate a diverse spectrum of mutations, aligning with the method's resolution for single nucleotide and small InDel variations. The strategic utilization of low-cost TILLING entailed genotypic screening within the same M2 generation. A curated selection of genes important for tackling major chickpea diseases like Ascochyta blight and fusarium wilt, alongside those associated with early flowering pathways, was subjected to this scrutiny. Leveraging TILLING-by-sequencing (TBS), a potent technique to unearth novel alleles/mutations within candidate genes, a population referred to as "CaTILL" encompassing 3,164 M3 lines was developed. A chickpea genotype, ICC 4958, was treated with 0.2% ethyl methane sulfonate (EMS) to create this invaluable resource. With a focus on 23 key candidate genes involved in various stress responses, flowering, symbiosis, nodulation, and plant development, the TBS approach uncovered a total of 91 novel alleles/mutations within these 23 genes.

Efficiently uncovering novel gene-based allelic variants within the natural chickpea population is pivotal for association mapping, aimed at identifying molecular markers governing crucial agronomic traits. This research employs an innovative and resource-efficient EcoTILLING strategy, coupled with agarose gel detection, for the identification of 1,133 new SNP allelic variants across a wide range of coding and regulatory sequences of 1,133 transcription factor (TF) genes within a diverse collection of 192 desi and kabuli chickpea accessions, forming a panel associated with seed weight. To unlock the full genetic potential of chickpea, it becomes imperative to comprehensively characterize these foundational germplasm resources at both genotypic and phenotypic levels, with a specific focus on pivotal stress tolerance and yield-related traits (Upadhyaya et al., 2008, 2011; Varshney et al., 2013; Saxena et al., 2014). Recognizing the significance of allele mining for crop improvement, EcoTILLING can effectively identify novel allelic variants in important candidate and known genes associated with diverse agricultural traits in chickpea. While the agarose gel-based EcoTILLING method usually pools DNA from two diverse accessions, it can be expanded to multiple accessions in chickpea, known for its limited genetic diversity, facilitating robust allele mining (Raghavan et al., 2007). In the face of challenges, this methodology proves to be a valuable tool for revealing natural allelic variations driven by pressures of domestication, especially among diverse desi, kabuli, and wild chickpea accessions thriving in various agro-climatic conditions. In a study conducted by Bajaj et al. (2016), researchers adopted an innovative and efficient pool-based EcoTILLING approach, combined with agarose gel detection, which not only saved time but also reduced labor requirements and costs. Through this approach, they successfully identified 1,133 novel SNP allelic variants located in a diverse range of coding and regulatory sequences of 1,133 TF genes. This extensive genotyping was performed across 192 diverse desi and kabuli chickpea accessions, forming a seed weight association panel.

3.4 CONCLUSION

In the future, the prospects of allele mining in chickpea present a dynamic and promising trajectory. As genetic technologies continue to advance, the refinement of allele mining strategies is anticipated. The integration of high-throughput sequencing, bioinformatics, and phenotypic data will enhance the precision and efficiency of allele discovery. Furthermore, the synthesis of allele mining with other cutting-edge techniques like genome editing holds potential for targeted manipulation of specific alleles to achieve desired agronomic traits. The chapter underscores the importance of collaborative efforts among international research institutes to synergistically unravel chickpea's genetic diversity and exploit its potential for crop improvement. As allele mining becomes increasingly sophisticated, its applications are expected to transcend traditional breeding approaches, facilitating the development of chickpea cultivars with enhanced nutritional content, disease resistance, and yield. The future landscape of chickpea agriculture is poised to benefit significantly from the insights and innovations stemming from allele mining, ushering in a new era of precision breeding and sustainable crop production.

REFERENCES

Afonso-Grunz F, Molina C, Hoffmeier K, Rycak L, Kudapa H et al. (2014) Genome-based analysis of the transcriptome from mature chickpea root nodules. *Front Plant Sci* 5:325.

Ahmad F, Gaur PM, Slinkard AE (1992) Isozyme polymorphism and phylogenetic interpretations in the genus *Cicer* L. *Theor Appl Genet* 83:620–627.

Araus JL, Cairns JE (2014) Field high-throughput phenotyping: The new crop breeding frontier. *Trends Plant Sci* 19:52–61.

Archak S, Tyagi RK, Harer PN, Mahase LB, Singh N et al. (2016) Characterization of chickpea germplasm conserved in the Indian National Gene bank and development of a core set using qualitative and quantitative trait data. *Crop J* 4(5):417–424.

Arumuganathan K, Earle ED (1991) Nuclear DNA content of some important plant species. *Plant Mol Biol Rep* 9:208–218.

Arriagada O, Cacciuttolo F, Cabeza RA, Carrasco B, Schwember AR (2022) A Comprehensive review on chickpea (*Cicer arietinum* L.) breeding for abiotic stress tolerance and climate change resilience. *Int. J. Mol. Sci.* 23(12):6794.

Bajaj D, Srivastava R, Nath M, Tripathi S, Bharadwaj C et al. (2016) Eco TILLING-based association mapping efficiently delineates functionally relevant natural allelic variants of candidate genes governing agronomic traits in chickpea. *Front Plant Sci* 7:450.

Barmukh R, Roorkiwal M, Dixit GP, Bajaj P, Kholova J et al. (2022) Characterization of 'QTL-hotspot' introgression lines reveals physiological mechanisms and candidate genes associated with drought adaptation in chickpea. *J Exp Bot* 73(22):7255–7272.

Barmukh R, Soren KR, Madugula P, Gangwar P, Shanmugavadivel PS et al. (2021) Construction of a high-density genetic map and QTL analysis for yield, yield components and agronomic traits in chickpea (*Cicer arietinum* L.). *PLoS One* 16:0251669.

Battenfield SD, Guzmán C, Gaynor RC, Singh RP, Peña RJ et al. (2016) Genomic selection for processing and end-use quality traits in the CIMMYT spring bread wheat breeding program. *The Plant Genome* 9(2):01.

Begum KN, Alam SS (2019) Genetic diversity in nine chickpea (*Cicer arietinum* L.) varieties based on different molecular markers. *Bang J Bot* 48(1):195–203.

Berger J, Abbo S, Turner NC (2003) Eco geography of annual wild Cicer species: The poor state of the world collection. *Crop Sci* 43(3):1076–1090.

Bernardo R (2004) what proportion of declared QTL in plants are false? *Theor Appl Genet* 109:419–424.

Bharadwaj C, Chauhan SK, Rajguru G, Srivastava R, Tara Satyavathi C et al. (2010) Diversity analysis of chickpea (*Cicer arietinum* L.) cultivars using MS markers. *Indian J Agric Sci* 80(11):947.

Buhariwalla HK, Jayashree B, Eshwar K, Crouch JH (2005) Development of ESTs from chickpea roots and their use in diversity analysis of the Cicer genus. *BMC Plant Biol* 5(1):1.

Cabrera-Bosquet L, Crossa J, von Zitzewitz J, Serret MD, Luis Araus J (2012) High-throughput phenotyping and genomic selection: The frontiers of crop breeding converge. *J Integr Plant Biol* 54:312–320.

Choudhary S, Gaur R, Gupta S (2012) EST-derived genic molecular markers: Development and utilization for generating an advanced transcript map of chickpea. *Theor Appl Genet* 124(8):1449–1462.

Cobb JN, DeClerck G, Greenberg A, Clark R, McCouch S (2013) Next-generation phenotyping: Requirements and strategies for enhancing our understanding of genotype–phenotype relationships and its relevance to crop improvement. *Theor Appl Genet* 126:867–887.

Collins NC, Tardieu F, Tuberosa R (2008) Quantitative trait loci and crop performance under abiotic stress: Where do we stand? *Plant Physiol* 147(2):469–486.

Coram TE, Pang ECK (2005) Isolation and analysis of candidate ascochyta blight defence genes in chickpea. Part I. Generation and analysis of an expressed sequence tag (EST) library. *Physiol Mol Plant Pathol* 66(5):192–200.

Croser JS, Ahmad F, Clarke HJ, Siddique KHM (2003) Utilisation of wild Cicer in chickpea improvement progress, constraints, and prospects. *Aust J Agric Res* 54(5):429–44.

Crossa J, Pérez-Rodríguez P, Cuevas J, Montesinos-López O, Jarquín D et al. (2017) Genomic selection in plant breeding: Methods, models, and perspectives. *Trends Plant Sci* 22:961–975.

Deokar AA, Kondawar V, Jain PK, Karuppayil SM, Raju NL et al. (2011 Comparative analysis of expressed sequence tags (ESTS) between drought-tolerant and susceptible genotypes of chickpea under terminal drought stress. *BMC Plant Biol* 11(1):70 1621.

DES (2023) Directorate of economics and statistics. Crop production statistics information system. Ministry of Agriculture & Farmers Welfare (MOAF&W), GoI, India.

Diapari M, Sindhu A, Bett K, Deokar A, Warkentin TD et al. (2014) Genetic diversity and association mapping of iron and zinc concentrations in chickpea (*Cicer arietinum* L.). *Genome* 57(8):459–468.

Dinari A, Niazi A, Afsharifar AR, Ramezani A (2013) Identification of upregulated genes under cold stress in cold-tolerant chickpea using the cDNA-AFLP approach. *PLoS One* 8(1):52757.

Dormatey R, Sun C, Ali K, Coulter JA, Bi Z, Bai J (2020) Gene pyramiding for sustainable crop improvement against biotic and abiotic stresses. *Agronomy* 10:1255.

FAOstat.org (2023) *FAOSTAT Statistical Database*. Food and Agriculture Organization of the United Nations (FAO), Rome.

Farahani S, Morid B, Maleki M (2016) Identification of resistant chickpea genotypes to races 1, 2, 3, 4 and 5 of *Fusarium oxysporum* f. sp *ciceri*, the cause of fusarium wilt using molecular markers. *Seed Plant Improv J* 32(1)

Farahani S, Talebi R, Maleki M, Mehrabi R, Kanouni H (2019) Pathogenic diversity of Ascochyta rabiei isolates and identification of resistance sources in core collection of chickpea germplasm. *Plant Pathol J* 35(4):321.

Finkel E (2009) Scientists seek easier access to seed banks. *Science* 324:1376.

Garg R, Bhattacharjee A, Jain M (2015) Genome-scale transcriptomic insights into molecular aspects of abiotic stress responses in chickpea. *Plant Mol Biol Rep* 33(3):388–400.

Garg R, Patel RK, Jhanwar S, Priya P, Bhattacharjee A et al. (2011a) Gene discovery and tissue-specific transcriptome analysis in chickpea with massively parallel prose-quenching and web resource development. *Plant Physiol* 156(4):1661–1678.

Garg R, Patel RK, Tyagi AK, Jain M (2011b) De novo assembly of chickpea transcriptome using short reads for gene discovery and marker identification. *DNA Res* 18(1):53–63.

Garg R, Shankar R, Thakkar B, Kudapa H, Krishnamurthy L et al. (2016) Transcriptome analyses reveal genotype- and developmental stage-specific molecular responses to drought and salinity stresses in chickpea. *Sci Rep* 6:19228.

Gaur PM, Chamarthi SK, Krishnamurthy L, Tripathi, S, Kashiwagi J et al. (2013) Fast-track introgression of "QTL-hotspot" for root traits and other drought tolerance traits in JG 11, an elite and leading variety of chickpea. *The Plant Genome* 6(3):07.

Gaur PM, Krishnamurthy L, Kashiwagi J (2008) Improving drought-avoidance root traits in chickpea (*Cicer arietinum* L.) current status of research at ICRISAT. *Plant Prod Sci* 11(1):3–11.

Gaur PM, Samineni S, Thudi M, Tripathi S, Sajja SB et al. (2019) Integrated breeding approaches for improving drought and heat adaptation in chickpea (*Cicer arietinum* L.). *Plant Breed* 138(4):389–400.

Gaur PM and Slinkard AE (1990) Inheritance and linkage of isozyme coding genes in chickpea. *J Hered* 81:455–461.

Gaur R, Jeena G, Shah N, Gupta S, Pradhan S et al. (2015) High density linkage mapping of genomic and transcriptomic SNPs for synteny analysis and anchoring the genome sequence of chickpea. *Sci Rep* 5:13387.

Gayacharan Rani U, Singh S, Basandrai AK, Rathee VK (2020) Identification of novel resistant sources for ascochyta blight (*Ascochyta rabiei*) in chickpea. *PLoS One* 15(10):0240589.

Glaszmann JC, Kilian B, Upadhyaya HD, Varshney RK (2010) Accessing genetic diversity for crop improvement. *Curr Opin Plant Biol* 13(2):167–173.

Gupta S, Garg V, Bhatia S (2015) A new set of ESTs from chickpea (*Cicer arietinum* L.) embryo reveals two novel F-box genes, CarF-box_PP2 and CarF-box_LysM, with potential roles in seed development. *PLoS One* 10(3):0121100.

Gupta S, Nawaz K, Parween S, Roy R, Sahu K et al. (2017) Draft genome sequence of (*Cicer reticulatum* L.) the wild progenitor of chickpea provides a resource for agronomic trait improvement. *DNA Res* 24:1–10.

Gupta SC, Sahu S (2012) Response of chickpea to micronutrients and biofertilizers in vertisol. *Legume Research-An Int J* 35(3):248–251.

Hajibarat Z, Saidi, A, Talebi R (2016) Evaluation of genetic diversity of chickpea (*Cicer arietinum* L.) genotypes based on morphological traits and CDDP and SCoT markers. *Seed Plant Improv J* 32(2)

Hamwieh A, Imtiaz M, Malhotra RS (2013) Multi-environment QTL analyses for drought-related traits in a recombinant inbred population of chickpea (*Cicer arientinum* L.). *Theor Appl Genet* 126:1025–1038.

Hannan RM, Kaiser WJ, Muehlbauer FJ et al. (1994) Experience with Ascochyta blight of chickpea in the United States. In: Muehlbauer FJ and Kaiser WJ (eds) *Expanding the Production and Use of Cool Season Food Legumes: A Global Perspective of Persistent Constraints and of Opportunities and Strategies for Further Increasing the Productivity and Use of Pea, Lentil, Faba Bean, Chickpea and Grasspea in Different Farming Systems*. Kluwer Academic Publishers, Dordrecht, pp. 849–858.

Heffne EL, Sorrells ME, Jannink JL (2009) Genomic selection for crop improvement. *Crop Sci* 49(1):1–12.

Heslot N, Yang HP, Sorrells ME, Jannink JL (2012) Genomic selection in plant breeding: A comparison of models. *Crop Sci* 52:146–160.

Hiremath PJ, Farmer A, Cannon SB, Woodward J, Kudapa H et al. (2011) Large-scale transcriptome analysis in chickpea (*Cicer arietinum* L.) an orphan legume crop of the semi-arid tropics of Asia and Africa. *Plant Biotechnol* 9(8):922–931.

Hiremath PJ, Kumar A, Penmetsa RV, Farmer A, Schlueter JA et al. (2012) Large-scale development of cost-effective SNP marker assays for diversity assessment and genetic mapping in chickpea and comparative mapping in legumes. *Plant Biotechnol J* 10:716–732.

Hunter DJ, Schofield D, Callander E (2014) The individual and socioeconomic impact of osteoarthritis. *Nat Rev Rheumatol* 10(7):437–441.

Iruela M, Rubio J, Cubero JI, Gil J, Millan T (2002) Phylogenetic analysis in the genus Cicer and cultivated chickpea using RAPD and ISSR markers. *Theor Appl Genet* 104(4):643–651.

Jacob C, Carrasco B, Schwember AR (2016) Advances in breeding and biotechnology of legume crops. *Plant Cell Tiss Org Cult* 127:561–584.

Jaganathan D, Thudi M, Kale S, Azam S, Roorkiwal M et al. (2015) Genotyping-by-sequencing based intra-specific genetic map refines a "QTL-hotspot" region for drought tolerance in chickpea. *Mol Genet Genom* 290:559–571.

Jain M, Chevala VV, Garg R (2014) Genome-wide discovery and differential regulation of conserved and novel microRNAs in chickpea via deep sequencing. *J Exp Bot* 65(20):5945–5958.

Jain M, Misra G, Patel RK, Priya P, Jhanwar et al. (2013) A draft genome sequence of the pulse crop chickpea (*Cicer arietinum* L.). *Plant J* 74(5):715–729.

Janghel DK, Kumar K, Kumar M, Chhabra AK (2021) Genetic diversity assessment in chickpea (*Cicer arietinum* L.) through agro-morphological and ISSR molecular markers. *Legume Res* 44(7):751–758.

Jannink JL, Lorenz AJ, Iwata H (2010) Genomic selection in plant breeding: From theory to practice. *Brief Funct Genom* 9:166–177.

Jansen RC, Van Ooijen JW, Stam P, Lister C, Dean C (1995) Genotype-by-environment interaction in genetic mapping of multiple quantitative trait loci. *Theor Appl Genet* 91:33–37.

Janská A, Marsík P, Zelenková S, Ovesná J (2010) Cold stress and acclimation – What is important for metabolic adjustment? *Plant Biol* 12:395–405.

Jayshree B, Buhariwalla HK, Shinde S, Crouch JH (2005) A legume genomics resources. The chickpea root expressed sequence tag database resources. *Elect J Biotechnol* 8(2):128–133.

Jhanwar S, Priya P, Garg R, Parida SK, Tyagi AK (2012) Transcriptome sequencing of wild chickpea as a rich resource for marker development. *Plant Biotechnol J* 10(6):690–702.

Jutke SR, Mane SS, Tatte R, Surnar S (2017) Marker assisted selection for chickpea wilt resistance by using ASAP and MS markers. *Trends Biosci* 10(31):6458–6461.

Kale SM, Jaganathan D, Ruperao P, Chen C, Punna R et al. (2015) Prioritization of candidate genes in "QTL-hotspot" region for drought tolerance in chickpea (*Cicer arietinum* L.). *Sci Rep* 5(1):1–14.

Kazan K, Muehlbauer FJ (1991) Allozyme variation and phylogeny in annual species of *Cicer* (*Leguminosae*). *Plant Syst Evol* 175:11–21.

Khelurkar VC, Moharil MP, Sakhare SB, Jadhav PV, Potdukhe NR et al. (2022) Utilization of SSR markers for hybridity assessment in F1's chickpea plants. *Pharm Innov J* 11(3):625–628.

Khemka N, Singh VK, Garg R, Jain M (2016) Genome-wide analysis of long intergenic non-coding RNAs in chickpea and their potential role in flower development. *Sci Rep* 6:33297.

Kohli D, Joshi G, Deokar AA, Bhardwaj AR, Agarwal M, Katiyar-Agarwal S et al. (2014) Identification and characterization of Wilt and salt stress-responsive microRNAs in chickpea through high-throughput sequencing. *PLoS One* 9(10):108851.

Kujur A, Upadhyaya HD, Shree T, Bajaj D, Das S, Saxena MS, Badoni S et al. (2015) Ultra-high density intra-specific genetic linkage maps accelerate identification of functionally relevant molecular tags governing important agronomic traits in chickpea. *Sci Rep* 5:9468.

Kurowska-Stolarska M, Alivernini S, Ballantine LE, Asquith DL, Millar NL et al. (2011) MicroRNA-155 as a proint flammatory regulator in clinical and experimental arthritis. *Proc Natl Acad Sci USA* 108(27):11193–11198.

Labdi M, Robertson LD, Singh KB, Charrier A (1996) Genetic diversity and phylogenetic relationships among the annual Cicer species as revealed by isozyme polymorphism. *Euphytica* 88:181–188.

Ladizinsky G (1975) A new *Cicer* from Turkey. *Notes Roy Bot Gard Edinb* 34:201–202.

Li L, Zhang Q, Huang D (2014) A review of imaging techniques for plant phenotyping. *Sensors* 14(11) 20078–20111.

Mantri NL, Ford R, Coram TE, Pang ECK (2010) Evidence of unique and shared responses to major biotic and abiotic stresses in chickpea. *Environ Exp Bot* 69(3):286–292.

Maphosa L, Richards MF, Norton SL, Nguyen GN (2020) Breeding for abiotic stress adaptation in chickpea (*Cicer arietinum* L.): A comprehensive review. *Crop Breed Genet Genom* 4(3).

Návarová H, Bernsdorff F, Döring AC, Zeier J (2012) Pipecolic acid, an endogenous mediator of defense amplification and priming, is a critical regulator of inducible plant immunity. *Plant Cell* 24(12):5123–5141.

Nayak SN, Zhu H, Varghese N, Datta S, Choi HK, Horres R et al. (2010) Integration of novel SSR and gene-based SNP marker loci in the chickpea genetic map and establishment of new anchor points with *Medicago truncatula* genome. *Theor Appl Genet* 120:1415–1441.

Nguyen NT, Ho SS, Low CLN (2004) A polymeric micro-gripper with integrated thermal actuators. *J Micromechan Microeng* 14(7):969.

Nisa ZU, Arif A, Waheed MQ, Shah TM, Iqbal A et al. (2020) comparative metabolomic study on desi and kabuli chickpea (*Cicer arietinum* L.) genotypes under rainfed and irrigated field conditions. *Sci Rep* 10(1):13919.

Ningwal R, Tripathi MK, Tiwari S, Asati R, Yadav RK, Tripathi N (2023) Identification of polymorphic SSR marker and diversity analysis in a set of desi chickpea genotypes. *BFAIJ* 15(3):45–51.

Passioura JB, Angus JF (2010) Improving productivity of crops in water-limited environments. *Adv Agron* 106:37–75.

Plekhanova E, Vishnyakova MA, Bulyntsev S, Chang PL, Carrasquilla-Garcia N (2017) Genomic and phenotypic analysis of Vavilov's historic landraces reveals the impact of environment and genomic islands of agronomic traits. *Sci Rep* 7(1):4816.

Pradhan S, Bandhiwal N, Shah N, Kant C, Gaur R et al. (2014) Global transcriptome analysis of developing chickpea (*Cicer arietinum* L.) seeds. *Front Plant Sci* 5:698.

Pratap A, Chaturvedi SK, Tomar R, Rajan N, Malviya N et al. (2017) Marker-assisted introgression of resistance to fusarium wilt race 2 in Pusa 256, an elite cultivar of desi chickpea. *Mol Genet Genom* 292:1237–1245.

Pundir RPS, Singh RB (1985) Biosystematic relationships among *Cajanus, Atylosia,* and *Rhynchosia* species and evolution of pigeon pea (*Cajanus cajan* (L.) Millsp.). *Theor Appl Genet* 69:531–534.

Raghavan C, Naredo MEB, Wang HH, Atienza G, Liu B et al. (2007) Rapid method for detecting SNPs on agarose gels and its application in candidate gene mapping. *Mol Breed* 19:87–101.

Rehman AU, Malhotra RS, Bett K, Tar'An B, Bueckert R, Warkentin TD (2011) Mapping QTL associated with traits affecting grain yield in chickpea (Cicer arietinum L.) under terminal drought stress. *Crop Science*, 51(2): 450–463.

Rajput S, Jain S, Tiwari S, Sikarwar RS, Tripathi N et al. (2023) Evaluation of gene-based markers against Fusarium wilt disease in chickpea (*Cicer arietinum* L.). *Curr J Appl Sci Technol* 42(6):26–32.

Rao LS, Usha Rani P, Deshmukh PS, Kumar PA, Panguluri SK (2007) RAPD and ISSR fingerprinting in cultivated chickpea (*Cicer arietinum* L.) and its wild progenitor Cicer reticulatum Ladizinsky. *Genet Resour Crop Evol* 54:1235–1244.

Reddy DS, Bhatnagar-Mathur P, Vadez V, Sharma KK (2012) Grain legumes (soybean, chickpea, and peanut): Omics approaches to enhance abiotic stress tolerance. In: Tuteja N, Gill SS, Tiburcio AF, Tuteja R (eds) *Improving Crop Resistance to Abiotic Stress*. Wiley, Weinheim, Germany, pp. 995–1032.

Roorkiwal M, Jain A, Kale SM, Doddamani D, Chitikineni A (2018a) Development and evaluation of high-density Axiom®*CicerSNP* Array for high-resolution genetic mapping and breeding applications in chickpea. *Plant Biotechnol J* 16:890–901.

Roorkiwal M, Jarquin D, Singh MK, Gaur PM, Bharadwaj C et al. (2018b) Genomic-enabled prediction models using multi-environment trials to estimate the effect of genotype x environment interaction on prediction accuracy in chickpea. *Sci Rep* 8:11701.

Roorkiwal M, Rathore A, Das RR, Singh MK, Jain A et al. (2016) Genome-enabled prediction models for yield related traits in chickpea. *Front Plant Sci* 7:1666.

Roorkiwal M, Sachdeva S, Bharadwaj C, Patil BS, Pal M et al. (2022) Agronomic performance of chickpea affected by drought stress at different growth stages. *Agronomy* 12(5):995.

Roorkiwal M, Von Wettberg EJ, Upadhyaya HD, Warschefsky E, Rathore A, Varshney RK (2014) Exploring germplasm diversity to understand the domestication process in Cicer spp. using SNP and DArT markers. *PLoS One* 9(7):e102016.

Saxena MS, Bajaj D, Kujur A, Das S, Badoni S, Kumar V (2014) Natural allelic diversity, genetic structure and linkage disequilibrium pattern in wild chickpea. *PLoS One* 9(9):e107484.

Schauer N, Semel Y, Roessner U, Gur A, Balbo I et al. (2006) Comprehensive metabolic profiling and phenotyping of interspecific introgression lines for tomato improvement. *Nat Biotechnol* 24:447–454.

Sefera T, Abebie B, Gaur PM, Assefa K, Varshney RK (2011) Characterizations and genetic diversity analysis of selected chickpea cultivars of nine countries using simple sequence repeat (SSR) markers. *Crop Pasture Sci* 62(2):177–187.

Sethy NK, Shokeen B, Edwards KJ, Bhatia S (2006) Development of microsatellite markers and analysis of intraspecific genetic variability in chickpea (*Cicer arietinum* L.). *Theor Appl Genet* 112:1416–1428.

Sharma S, Upadhyaya HD, Gowda CLL, Kumar S, Singh S (2013) Genetic analysis for seed size in three crosses of chickpea (Cicer arietinum L.). *Can J Plant Sci* 93(3):387–395.

Shorinola O, Kaye R, Golan G, Peleg Z, Kepinski S et al. (2018) Isolation and characterisation of mutants with altered seminal root numbers in hexaploid wheat. *BioRxiv* 364018.

Singh A (1997) financial liberalization, stock markets and economic development. *Econ J* 107442:771–782.

Singh D, Pushpavalli SNCVL, Vanisri S, Kumar GA (2021) Assessment of genetic diversity in chickpea genotypes (Cicer arietinum L.) using agro-morphological and SSR markers. *Pharm Innov J* 10(8):784–791.

Singh S, Gumber RK, Joshi N, Singh K (2005) Introgression from wild *Cicer reticulatum* to cultivated chickpea for productivity and disease resistance. *Plant Breed* 124:477–480.

Singh VK, Garg R, Jain M (2013) A global view of transcriptome dynamics during flower development in chickpea by deep sequencing. *Plant Biotechnol J* 11(6):691–701.

Sozzani R and Benfey PN (2011) High-throughput phenotyping of multicellular organisms: Finding the link between genotype and phenotype. *Genome Biol* 12:1–7.

Srivastava S, Zheng Y, Kudapa, H, Jagadeeswaran G, Hivrale V et al. (2015) High throughput sequencing of small RNA component of leaves and inflorescence revealed conserved and novel miRNAs as well as phasiRNA loci in chickpea. *Plant Sci* 235:46–57.

Srivastva R, Bajaj D, Malik A, Singh M (2016) Transcriptome landscape perennial wild Cicer microphyllum uncovers functionally relevant molecular tags regulating agronomic traits in chickpea. *Sci Rep* 6:33616.

Steinfath M, Strehmel, Peters R, Schauer N, Groth D et al. (2010) Discovering plant metabolic biomarkers for phenotype prediction using an untargeted approach. *Plant Biotechnol J* 8:900–911.

Tayyar R, Federici CV, Waines GJ (1996) Natural outcrossing in chickpea (*Cicer arietinum* L.). *Crop Sci* 36(1):203–205.

Thudi M, Bohra A, Nayak SN, Varghese N, Shah T et al. (2011) Novel SSR markers from BAC-end sequences, DArT arrays and a comprehensive genetic map with 1,291 marker loci for chickpea (*Cicer arietinum* L.). *PLoS One* 6:27275.

Thudi M, Gaur PM, Krishnamurthy L, Mir RR, Kudapa H et al. (2014a) Genomics-assisted breeding for drought tolerance: A dream comes true in chickpea! *Funct Plant Biol* 41:1178–1190.

Thudi M, Khan AW, Kumar V, Gaur PM, Katta K et al. (2016) Whole genome re-sequencing reveals genome-wide variations among parental lines of 16 mapping populations in chickpea (*Cicer arietinum* L.). *BMC Plant Biol* 16 (Suppl 1):53–64.

Thudi M, Upadhyaya HD, Rathore A, Gaur PM, Krishnamurthy L et al. (2014b) Genetic dissection of drought and heat tolerance in chickpea through genome-wide and candidate gene-based association mapping approaches. *PLoS One* 9(5):96758.

Topp CN, Iyer-Pascuzzi AS, Anderson JT, Lee CR, Zurek PR et al. (2013) 3D phenotyping and quantitative trait locus mapping identify core regions of the rice genome controlling root architecture. *Proc Natl Acad Sci USA* 110: E1695–E1704.

Tullu A, Muehlbauer FJ, Simon CJ, Mayer MS, Kumar J et al. (1998) Inheritance and linkage of a gene for resistance to race 4 of fusarium wilt and RAPD markers in chickpea. *Euphytica* 102:227–232.

Upadhyaya HD (2003) Geographical patterns of variation for morphological and agronomic characteristics in the chickpea germplasm collection. *Euphytica* 132:343–352.

Upadhyaya HD, Dwivedi SL, Baum M, Varshney RK, Udupa SM et al. (2008) Genetic structure, diversity, and allelic richness in composite collection and reference set in chickpea (*Cicer arietinum* L.). *BMC Plant Biol* 8(1):1–12.

Upadhyaya HD, Kashiwagi J, Varshney RK, Gaur PM, Saxena KB et al. (2012) Phenotyping chickpeas and pigeonpeas for adaptation to drought. *Front Physiol* 3:179.

Upadhyaya HD and Ortiz R (2001) A mini core subset for capturing diversity and promoting utilization of chickpea genetic resources. *Theor Appl Genet* 102:1292–1298.

Upadhyaya HD, Salimath PM, Gowda CLL, Singh S (2007) New early-maturing germplasm lines for utilization in chickpea improvement. *Euphytica* 157:195–208.

Upadhyaya HD, Thudi M, Dronavalli N, Gujaria N, Singh S et al. (2011) Genomic tools and germplasm diversity for chickpea improvement. *Plant Genet Resour* 9(1):45–58.

Upadhyaya HD, Yadav D, Dronavalli N, Gowda CLL, Singh S (2010) Mini core germplasm collections for infusing genetic diversity in plant breeding programs. *Elect J Plant Breed* 1(4):1294–1309.

Van der Maesen LJG (1987) *Cicer* L. Origin, history and taxonomy of chickpea. In: Saxena MC, Singh KB (eds) *The Chickpea*. CAB International Cambrian News Ltd, Aberystwyth, UK, pp. 11–34.

Varshney RK (2016) Exciting journey of 10 years from genomes to fields and markets: Some success stories of genomics-assisted breeding in chickpea, pigeon pea and groundnut. *Plant Sci* 242:98–107.

Varshney RK, Graner A, Sorrells ME (2005) Genomics-assisted breeding for crop improvement. *Trends Plant Sci* 10:621–630.

Varshney RK, Mir RR, Bhatia, S, Thudi M, Hu Y et al. (2014a) Integrated physical, genetic and genome map of chickpea (*Cicer arietinum* L.). *Funct Integr Genom* 14:59–73.

Varshney RK, Mohan SM, Gaur PM, Chamarthi SK, Singh VK et al. (2014b) Marker-assisted backcrossing to introgress resistance to Fusarium wilt race 1 and Ascochyta blight in C 214, an elite cultivar of chickpea. *Plant Genome* 7(1):2013–2010.

Varshney RK, Ribaut JM, Buckler ES, Tuberosa R, Rafalski JA, Langridge P (2012) Can genomics boost productivity of orphan crops? *Nat Biotechnol* 30(12):1172–1176.

Varshney RK, Roorkiwal M, Sun S, Bajaj P, Chitikineni A et al. (2021) A chickpea genetic variation map based on the sequencing 3366 genomes. *Nature* 599:622–627.

Varshney RK, Song C, Saxena of RK, Azam S, Yu S et al. (2013) Draft genome sequence of chickpea (*Cicer arietinum*) provides a resource for trait improvement. *Nat Biotechnol* 31:240–246.

Varshney RK, Thudi M, Pandey MK, Tardieu F, Ojiewo C et al. (2018) Accelerating genetic gains in legumes for the development of prosperous smallholder agriculture: Integrating genomics, phenotyping, systems modelling and agronomy. *J Exp Bot* 69:3293–3312.

Varshney RK, Thudi M, Roorkiwal M, He W, Upadhyaya HD (2019) Resequencing of 429 chickpea accessions from 45 countries provides insights into genome diversity, domestication and agronomic traits. *Nat Genet* 51(5):857–864.

Verma M, Kumar V, Patel RK, Garg R, Jain M (2015) CTDB: An integrated chickpea transcrip- tome database for functional and applied genomics. *PLoS One* 10(8):0136880.

von Wettberg EJ, Chang PL, Başdemir F, Carrasquila-Garcia N, Korbu LB et al. (2018) Ecology and genomics of an important crop wild relative as a prelude to agricultural innovation. *Nat Commun* 9(1):649.

Xu Y, Crouch JH (2008) Marker-assisted selection in plant breeding: From publications to practice. *Crop Sci* 48(2):391–407.

Yadav G, Jayaswal D, Jayaswall K, Bhandawat A, Singh A et al. (2022) Identification and characterization of chickpea genotypes for early flowering and higher seed germination through molecular markers. *Mol Biol Rep* 49(7):6181–6188.

Yadav S, Shah V, Mod B (2019) Genetic diversity analysis between different varieties of chickpea (Cicer arietinum L.) using SSR markers. *Int J Appl Sci Biotechnol* 7(2):236–242.

4 Allele Mining for Crop Genome Designing in Cowpea

H. A. Bhargavi, B. S. Harshitha, Sunaina Yadav, Karthik Kumar M., Brijesh K. Mehta, Padmavati G. Gore, Kuldeep Tripathi, Ravi Prakash Saini, P. S. Basavaraj, and M. C. Keerthi

4.1 INTRODUCTION

The primary issue concerning agricultural modernization is genetic uniformity, which leads to narrowing a genetic base with reduced genetic diversity; crop improvement programs will be challenging. The accumulation of advantageous alleles from the extensive plant genetic resources throughout the world has advanced plant breeding in terms of the generation of superior and high-yielding varieties of crops. However, a sizable part of these superior alleles was never used because they were lost during evolution and domestication. The unexploited genetic variation existing in wild relatives and landraces of crop plants could be exploited fruitfully for the development of agronomically superior cultivars. To further improve the genetic potential of crop varieties for different agronomic traits, it is necessary to search the huge germplasm resources again for novel alleles (Sammour R H, 2011).

Cowpea is one of the most important grain legumes grown in dry tropical areas of America, South Asia, and Africa. It is mainly used as dry grain and fodder since it has excellent nutritional and nutraceutical properties (da Silva et al., 2018). Through its contributions to nutritional security, income production, and soil fertility enhancement, it strongly supports the livelihood of small-scale farmers. The low productivity of cowpea is mainly due to numerous abiotic and biotic stresses. The major abiotic stress factor comprises drought, heat, salinity, and low soil fertility, while biotic stresses include rust, anthracnose, powdery mildew, bacterial blight, cowpea mosaic virus (CMV), aphid, pod bugs, and leaf minor (Sahay et al., 2022; Mahesha et al., 2022). All these constraints lead to limited access to quality seeds of improved varieties by farmers for planting. In the cowpea hybridization programs worldwide, thousands of breeding lines and hundreds of elite cultivars are developed yearly. The accumulation of advantageous alleles from the vast plant genetic resources that exist throughout the world primarily made it feasible to form these breeding lines.

Allele mining is a promising method for identifying allelic variation present in germplasm; hence, genetic base can be broadened. The discovery of single-nucleotide polymorphisms (SNPs) or insertions/deletions (InDels) at coding regions or exons of the gene was the only focus of the early investigations of allele mining since it was anticipated that these changes would influence the structure and function of the encoded protein. However, according to recent studies, nucleotide alterations in non-coding regions (5′ UTR), such as the promoter, introns, and 3′ UTR, also significantly affect the synthesis and accumulation of transcripts, which, in turn, affects how traits are expressed (Kumar et al., 2010). The generation of allele-specific molecular markers, the facilitation of the introduction of novel alleles by marker-assisted selection (MAS), and molecular discrimination across related species are further potential applications of allele mining.

DOI: 10.1201/9781003385059-4

Compared to many other crops, cowpea genomic resources have just recently been established. Accelerated identification and annotation of new genes can be anticipated with the rapid accumulation of sequence data and expression data in multiple genomic databases, allowing for the creation of allele-specific markers. Based on gene and genome sequences, polymerase chain reaction (PCR)–based strategies are devised to amplify beneficial alleles of genes from a wide range of genotypes and recognize nucleotide variance in the alleles (Kumar et al., 2010). This method would help us access the various novel alleles associated with biotic and abiotic tolerance, nutritional quality, and enhanced yield in cowpea. However, genome-wide association studies (GWAS) and deep re-sequencing analyses are promising approaches to detect nucleotide variations in all genes controlling for various traits.

4.2 COWPEA GENETIC RESOURCES

Plant genetic resources for food and agriculture (PGRFA) are defined as "any genetic material of plant origin of actual or potential value for food and agriculture" (FAO, 2009). Genetic diversity is the basis for crop breeding programs and is an insurance against unforeseen threats to agriculture, like biotic and abiotic stress and climate changes (Gepts, 2006). Cowpea genetic resources are being conserved in the form of landraces (or primitive cultivars): dynamic populations of a cultivated plant that have historical origin, have distinct identity, and lack formal crop improvement, as well as often being genetically diverse, locally adapted, and associated with traditional farming systems (Camacho et al., 2005). *Old cultivars*, sometimes known as *obsolete cultivars*, refer to cultivated varieties which have fallen into disuse. Farmers variety, registered under the PPV&FR Act, have been traditionally cultivated and evolved by farmers in their fields or those that are wild relatives or landraces of a variety about which farmers possess common knowledge. Modern cultivated varieties (modern cultivars) are agronomic varieties in current use and newly developed varieties. Special stocks include advanced breeding lines (i.e., pre-released varieties developed by plant breeders), mapping populations. Crop wild relatives (CWR) are wild plant species that are genetically related to cultivated crops. CWR are the wild ancestors of the domesticated plant and other, more distantly related species. Plant genetic resource conservation aims to ensure that the maximum possible allelic genetic diversity, and therefore potentially valuable traits for breeding of a crop, is maintained and is available and accessible for utilization.

Genetic resources of cowpea are being conserved under long-term storage at –20°C as base collection and in medium-term storage at 5°C for the active collection, from which seeds will be distributed. The largest collection of cowpea genetic resources (15,004 accessions) is held by the Genetic Resources Center (GRC) of the International Institute of Tropical Agriculture (IITA), Ibadan, Nigeria. IITA conserved germplasm belonging to 89 countries, and 7,912 accessions belonged to West Africa, the center of diversity for the crop (Mahalakshmi et al., 2007). Other important collections are conserved in the USA with the US Department of Agriculture (USDA), Griffin (8,244 accessions), and with the University of California–Riverside (UCR) (5,600 accessions) (Table 4.1).

4.2.1 Cowpea Core Collection

A large number of germplasm collection accessions could hinder their full exploitation, evaluation, and utilization in crop breeding programs. Evaluation of such large collections in detail would be expensive and time-consuming. This task could be more straightforwardly fulfilled by developing subsets of the whole collection, called *active working collections* by Harlan (1972) and *core collections* by Frankel and Brown (1984). Brown (1989) suggested that the core collection should include a maximum of the genetic variation of the whole collection with minimum repetitiveness, ideally conserving at least 70% of the alleles in the entire collection. IITA has developed a "core collection" of 2,062 accessions of cowpea using geographical, agronomic, and botanical descriptors (Mahalakshmi et al., 2007). Core collection of 2,062 accessions of cowpea was constituted of a

TABLE 4.1
Cowpea Genetic Resources Conserved Ex Situ at National, Regional, and International Holders

Country	Institute/Gene Bank	No. of Accessions
Nigeria	International Institute of Tropical Agriculture (IITA)*	15,004
USDA-ARS	Plant Genetic Resources Conservation Unit, USDA-ARS**	8,244
UC–Riverside	University of California, Riverside	5,600
India	National Gene Bank of India, ICAR-National Bureau of Plant Genetic Resources***	4,009
Russia	N.I. Vavilov All-Russian Institute of Plant Industry of RAAS	1,945
South Korea	National Agrobiodiversity Center	910
South Africa	Agricultural Research Council (ARC)-grain	886
Kenya	National Genebank of Kenya	875
Spain	Instituto Nacional de Investigation y Technologia Agraria Alimentaria	466
Tanzania	National Plant Genetic Resources Center of Tanzania	386
Nigeria	National Center for Genetic Resources and Biotechnology	384
Belgium	Belgium (National Botanical Garden),	331
AVRDC-Taiwan	Taiwan (AVRDC, the World Vegetable Center),	322
Zambia (SADC)	Regional Plant Genetic Resources Centre	305
Germany	IPK Leibniz Institute of Plant Genetics and Crop Plant Research	291
Angola (SADC)	Regional Plant Genetic Resources Centre	172
Benin	Institut National des Recherches Agricoles du Benin	155
Cote d'Ivoire	Centre National de Recherche Agronomiques	126
Togo	Institut Togolais de Recherche Agronomique	100
Namibia (SADC)	Regional Plant Genetic Resources Centre	57
Swaziland (SADC)	Regional Plant Genetic Resources Centre	45

Source: https://www.croptrust.org/.
* IITA: (http://singer.cgiar.org/index.jsp).
** USDA-ARS: https://www.genesys-pgr.org/wiews/USA016.
***NBPGR: http://www.nbpgr.ernet.in/Research_Projects/Base_Collection_in_NGB.aspx.

TABLE 4.2
Composition of Global Cowpea Core Developed by IITA, Nigeria

Sr. No.	Genetic Resources	Total Accessions (Number)	IITA Core Set (Number)
1	Landraces	10,227	1,701
2	Advance lines	1,422	225
3	Unknown biological status	838	130
4	Wild and weedy	64	06
Total			**2,062**

total of 1,701 accessions of landraces, 225 accessions of advanced cultivars and breeding or research lines, 130 accessions that either lack information on origin or biological status, and six accessions of wild and weedy forms from the available 64 wild and weed accessions were added. A core collection of cowpea provides an opportunity for the exploitation of the cowpea germplasm for crop improvement (Table 4.2).

4.2.2 Registration of Trait-Specific Germplasm of Cowpea

Recognizing the importance of the issue, a mechanism for "registration of plant germplasm" was instituted in 1996 at the ICAR-National Bureau of Plant Genetic Resources (NBPGR), New Delhi, ICAR. This mechanism is envisaged to serve as a recognized tool for the registration of PGRFA at the national level. This would also facilitate access to the developed or identified potentially valuable germplasm for utilization in crop improvement programs. Registered germplasm of cowpea is conserved at the National Gene Bank for long-term storage, accompanied by passport and gene bank–related information (Kak and Gupta, 2021). Till now, seven accessions of cowpea have been registered for different traits (Table 4.3).

4.3 APPROACHES OF ALLELE MINING

Genetic variations are the stepping stone toward crop improvement. Wild races, traditional cultivars, and landraces are a rich source of valuable genes that can be used to develop high-yielding, nutritionally sound, and stress-tolerant varieties. Allele mining is a tool for finding newer alleles of a known gene or locus governing an important trait (agronomical, yield, and stress-related) in the concerned crop species and has a potential in future crop improvement programs. Identification of sequence polymorphisms for a specific gene in naturally occurring populations can be accomplished using two main methods. EcoTILLING (ecotype target-induced local lesion in the genome) and sequencing-based allele mining. These two approaches are briefly described in this section, and steps involved in each approach are illustrated in Figure 4.1.

TABLE 4.3
Cowpea Genetic Stocks Registered in India

SN	National Identity	Donor Identity	Year	Developing Institute	Novel Unique Features
1.	IC299972	CAZC-B	2004	ICAR–Central Arid Zone Research Institute (CAZRI), Jodhpur, Rajasthan	Black seed coat color
2.	IC202803	KM-5501	2008	ICAR–National Bureau of Plant Genetic Resources (NBPGR), Pusa Campus, New Delhi	Bold seed
3.	IC519745	KDRS-205	2008	ICAR–National Bureau of Plant Genetic Resources (NBPGR) Regional Station, Rajendranagar, Hyderabad, Andhra Pradesh	Resistance to black-eyed cowpea mosaic virus
4.	EC724523	EC724523	2019	ICAR–National Bureau of Plant Genetic Resources, Pusa Campus, New Delhi	Resistant to root-knot nematode (*Meloidogyne incognita*)
5.	EC723686	EC723686	2019	ICAR–National Bureau of Plant Genetic Resources, Pusa Campus, New Delhi	Resistant to root-knot nematode (*Meloidogyne incognita*)
6.	EC725122	EC725122	2019	ICAR–National Bureau of Plant Genetic Resources, Pusa Campus, New Delhi	Resistance to root-knot nematode (*Meloidogyne incognita*)
7.	EC762384	EC762384	2021	ICAR–National Bureau of Plant Genetic Resources, Pusa Campus, New Delhi	Dense pubescence

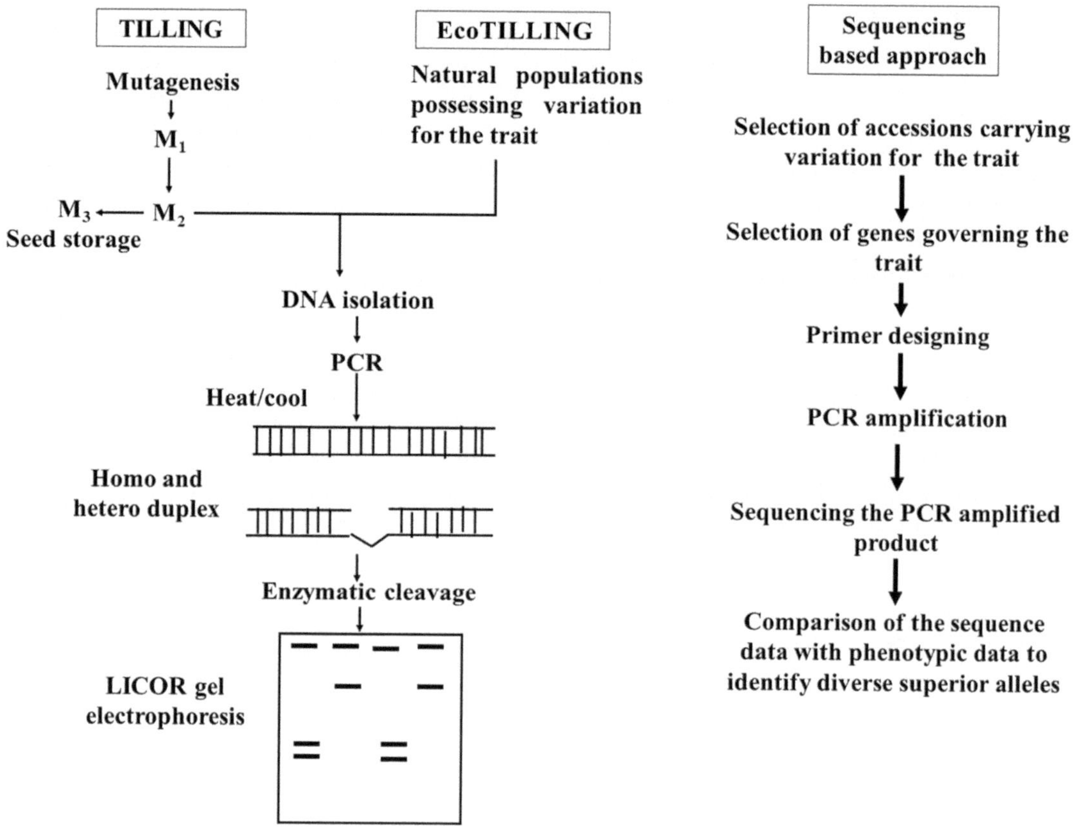

FIGURE 4.1 Approaches of allele mining.

4.3.1 TILLING Approach

It is a reverse genetic method in which heteroduplex analysis is used to pinpoint the polymorphisms brought on by induced mutations in a target gene. It offers a way to assess the degree of variation in mutations that have been artificially produced. EcoTILLING is similar to targeted induced local lesions in genomes, but naturally occurring polymorphism is detected as opposed to induced mutations. EcoTILLING is advantageous for those crop species wherever it is difficult to carry out chemical mutagenesis (Comai et al., 2004). EcoTILLING helps in the detection of point mutations or InDels in the gene(s) of interest in a high-throughput manner through heteroduplex formation and evaluation by using mismatch specific endonuclease enzyme (Till et al., 2003). Due to its precision in detecting the site of mutation in a specific gene or genomic region and without the involvement of genetic engineering, EcoTILLING approach has taken an upper hand in comparison to other tools of functional genomics like RNAi-based gene silencing and T-DNA knockouts (Sabetta et al., 2011). EcoTILLING-based approaches have been used for identifying novel alleles in several crop species (Comai et al., 2004; Mejlhede et al., 2006; Ibiza et al., 2010; Negrao et al., 2011; Yu et al., 2012; Frerichmann et al., 2013; Zeng et al., 2016; Raja et al., 2017). TILLING is an ideal approach to identify novel alleles of a gene since it helps overcome the problem of screening a large number of germplasm. The diverse cowpea germplasm is available from collections in the International Institute of Tropical Agriculture (IITA), University of California, Riverside, CA (USA); in the USDA repository in Griffin, GA (USA); and in India's National Bureau of Plant Genetic Resources (NBPGR) in New Delhi (Muñoz-Amatriaín et al., 2017). These collections contain variability related to pests, pathogens, nutritional attributes, seed traits, plant architecture, and adaptation to harsh environments. So these germplasm accessions can be assessed for molecular polymorphism through TILLING and EcoTILLING for detection of SNPs in germplasm.

4.3.2 Sequencing-Based Allele Mining

Sequencing-based allele mining involves PCR-based amplification of alleles of a gene in a diverse set of genotypes, followed by DNA sequencing and identification of nucleotide variation in the alleles (Kumar et al., 2020). This novel genomic tool can identify multiple alleles of several genes in a diverse panel. Unlike EcoTILLING, sequencing-based allele mining is a cheap technique, does not involve tedious steps, and also does not require more sophisticated laboratory and instruments like Li-Corgenotyper. The haplotype structure and diversity analysis of individuals using this technique can help in establishing genetic association among plants (Min et al., 2021). This allows us to detect and locate the point mutations (SNPs) and InDels in the individuals to construct haplotypes (Chhabra et al., 2022). In addition, sequencing-based allele mining in identified accessions and their association with phenotypic variations provide a tremendous impetus to precision breeding programs in cowpea. The available germplasm repositories (core and mini-core) and reference SNP maps of cowpea will serve as valuable resources for mining allelic diversity for beneficial traits and further associational analysis, mapping, and cloning of novel genes (Ketema et al., 2020).

Drought-tolerant genotypes and associated SNPs are available in cowpea. The homozygous SNPs and InDels identified in *DHN* gene in chickpea genotypes suggest that the sequencing-based allele mining can help discover allelic variations in drought-tolerant cowpea genotypes (Kumar et al., 2020). Bajaj et al. (2015), using integrated reference genome- and *de novo*–based genotyping-by-sequencing (GBS) assay, identified functional allelic variants for domestication pattern and stress-responsive genes in wild chickpea accessions. This approach implicates that the allelic variants (SNPs) for these traits can also be identified in wild cowpea accessions and can be transferred into cultivated gene pools for genetic improvement. The multi-parent advanced generation intercross (MAGIC) lines developed in cowpea possessed tremendous genetic variation that can be used for detecting allelic variations in genes for drought, salt, and other beneficial key traits (Huynh et al., 2018; Ravelombola et al., 2021, 2022). Ogawa et al. (2018) used haplotype-based allele mining in the Japan-MAGIC rice population to detect the allelic variations for glutinous endosperm and culm length. They proposed that the same techniques can be used for allele mining new traits in rice and other crops. Over the last two decades, biparental mapping and GWAS have identified a wealth of quantitative trait loci (QTLs) for several traits in cowpea. The next step is to identify the candidate genes underlying the key QTLs of interest through the integration of genetic mapping, genomics, and precise phenotyping strategies. Once the candidate genes are identified, it can empower targeted modification using gene editing tools. However, identifying candidate genes and narrowing down the number of potential candidate genes to causative genes are the future challenges in cowpea that need to be unveiled. In this context, next-generation sequencing (NGS) data provides a valuable source to identify haplotypes, alleles, and causal variants.

The rapid and substantial cost reduction of NGS platforms has recently increased the feasibility of sequencing-based allele mining. Massive parallel sequencing has made it possible to sequence thousands of genes together, or even entire genome or transcriptome more efficiently and cost-effectively (Varshney et al., 2009; Kaboldt et al., 2013). With the advances in sequencing technologies, it is possible to find newer variations of candidate genes and candidate regions in large gene bank collections. We can even capture genetic diversity within and between germplasm pools, map genes, and QTLs and determine their functional roles. Therefore, these technologies are gaining importance in mining useful alleles in the available germplasm pools. Various large-scale NGS platforms include 454/Roche, Illumina/Solexa systems, Applied Biosystems/SOLiD, Heliscope (single-molecule sequencer), ion TORRENT, etc. These sequencing platforms vary from each other in their sequencing technique, read length, and error rates. Therefore, the right choice of sequencing platform depends upon the number of factors, like number of candidate genes, number of samples, availability of reference genome, etc. Cowpea is a highly self-pollinating crop with very limited chances of outcrossing. Most of the developed cultivars share common parents and therefore have a narrow genetic base. Due to poor genetic base, developed varieties are vulnerable against various kinds of biotic and abiotic stresses (Boukar et al., 2016). Therefore, the search for newer and more beneficial alleles of desirable genes in wild relatives and landraces of cowpea by applying the tool of allele mining is very much required.

4.4 GENOMIC RESOURCES AND BIOINFORMATICS TOOLS REQUIRED FOR ALLELE MINING

Cowpeas are a protein-rich legume that can be used as both grain and fodder. It is not only a main food but also a nutritional security for sub-Saharan Africa (SSA). It not only provides food but also provides strong support to small marginal farmers in many parts of the world, particularly African countries. Globally, more than 6.5 million metric tons of cowpea are produced annually from 14.5 million hectares. However, the productivity of the cowpea is low due to constant exposure to one or more biotic and abiotic stress factors. Besides this, there is a lack of quality seeds for farmers (Boukar et al., 2016). In the past, efforts at the national and international levels made some progress in cowpea improvement. More recently, sophisticated molecular breeding techniques could also help improve cowpea productivity. Recent advances in cutting-edge technologies, like NGS, GWAS, genomic selection (GS), and genome editing (GE), changed the face of pulse improvement to achieve food and nutritional security (Moose and Mumm, 2008). With the rapid advances in NGS, a lot of sequence information is made available in the public databases. It is vital to utilize this genomic information to identify and isolate the novel alleles, genes, and haplotypes of agronomically important traits from crop gene pools to deploy for the development of improved cultivars suitably. Allele mining is a promising approach to analyze naturally occurring allelic variation at candidate genes affecting important agronomic trait. Allele mining assists in the tracking of allele evolution, the discovery of novel haplotypes, and the creation of allele-specific markers for use in marker-assisted selection (Kumar et al., 2010). Several bioinformatic tools and genomic resources have been developed in many legume crops, including cowpea; they are described in this section. Chen et al. (2007) developed Cowpea Gene Space/Genomics Knowledge Base (CGKB) under the Cowpea Genome Initiative project, which is based on 298,848 cowpea gene space sequences (GSS) isolated by methylation filtering of genomic DNA. There are three knowledge bases under CGKB; they are GSS annotation and comparative genomics knowledgebase, GSS enzyme and metabolic pathway knowledge base, and GSS simple sequence repeats (SSRs) knowledge base for molecular marker discovery. The bioinformatic tools developed are summarized in Table 4.4.

4.5 MOLECULAR MAPPING

4.5.1 Biparental QTL Mapping

Quantitative trait locus (QTL) mapping aims to reveal the genetic architecture of quantitative traits and to develop molecular markers for genomics-assisted breeding (Bernardo, 2008). It relies on statistical linkage analysis between molecular markers and traits of interest in biparental mapping populations, such as F2, F2:3, recombinant inbred lines (RILs), near-isogenic lines (NILs) backcross (BC), and doubled haploids (DH) (Chhabra et al., 2019). The steps involved in QTL mapping

TABLE 4.4
Bioinformatics Tools and Databases Developed

Tools/Software	Description	Utility	Reference
HarvEST:Cowpea	Consists of a consensus genetic map derived from 5 mapping population	EST database with gene function analysis and primer design	Muchero et al. (2009a): http://harvest.ucr.edu/
Cowpea Gene Space/Genomics Knowledge Base (CGKB)	Genetic markers, gene space, metabolic pathways, mitochondrial, and chloroplast sequences	A tool that assists in gene discovery, enzyme and metabolic pathways of cowpea	Chen et al. (2007): http://cowpeagenomics.med.virginia.edu/ CGKB/
Physical Map	Contains information about nucleotide bases	Allele mining, gene discovery, comparative genomics, etc.	http://phymap.ucdavis.edu/cowpea/
SNPselector, KBioConverter, and Backcross Selector	Management of genotyping data	Genomic-assisted breeding of cowpea	http://breedit.org
CowpeaPan	CowpeaPan genome resource	Genome analysis, comparative genomics, marker development, GWAS, GS, etc.	https://phytozome-next.jgi.doe.gov/cowpeapan
CMap	Web-based tool for visualizing genetic and comparative maps	Comparative genomic study	https://cmap.legumeinfo.org/
HaploSNPer	A web-based tool to identify SNPs and allele	SNP mining, haplotype reconstruction	http://www.bioinformatics.nl/tools/haplosnper/
QualitySNP		SNP mining	Tang et al. (2006)
SNPServer	A web-based tool for SNP detection	Real-time discovery of SNPs associated with specific sequence	Savage et al. (2005)
novoSNP	A software package for allele mining in faster way	SNP and InDels mining	Weckx et al. (2005)
Phytozome	Web-accessible information resource providing genome sequences and annotations of various plant species	Comparative genomics studies	http://www.phytozome.net/Phytozome_info.php
Kevin's GATTACA World			http://kevin-gattaca.blogspot.com
Next-Gen Sequencing	A working guide to the rapidly developing world of next-generation DNA sequencing, with an emphasis on bioinformatics	Base calling, alignment to reference genome, de novo assembly, SNP discovery, etc.	http://nextgenseq.blogspot.com/
Lex Nederbragt	Blog having many software for bioinformatic works		http://lexnederbragt.com/blog
PlantGDB	Provides sequence data for >70,000 plant species, custom EST assemblies (PUT) for over 150 species, web tools and plant genome browsers, as well as an outreach portal for plant genomics	Discovery, annotation, and expression analysis of arginine/serine (SR) proteins	www.plantgdb.org/MtGDB/
Legume Base	Database of model legume species		legumebase.nbrp.j
Legume information system	Enables access to resources and tools, such as browsers, InterMine instances, organized primarily by genus	Integrates genetic and genomic information across many legume species, with tools for research	https://www.legumeinfo.org/

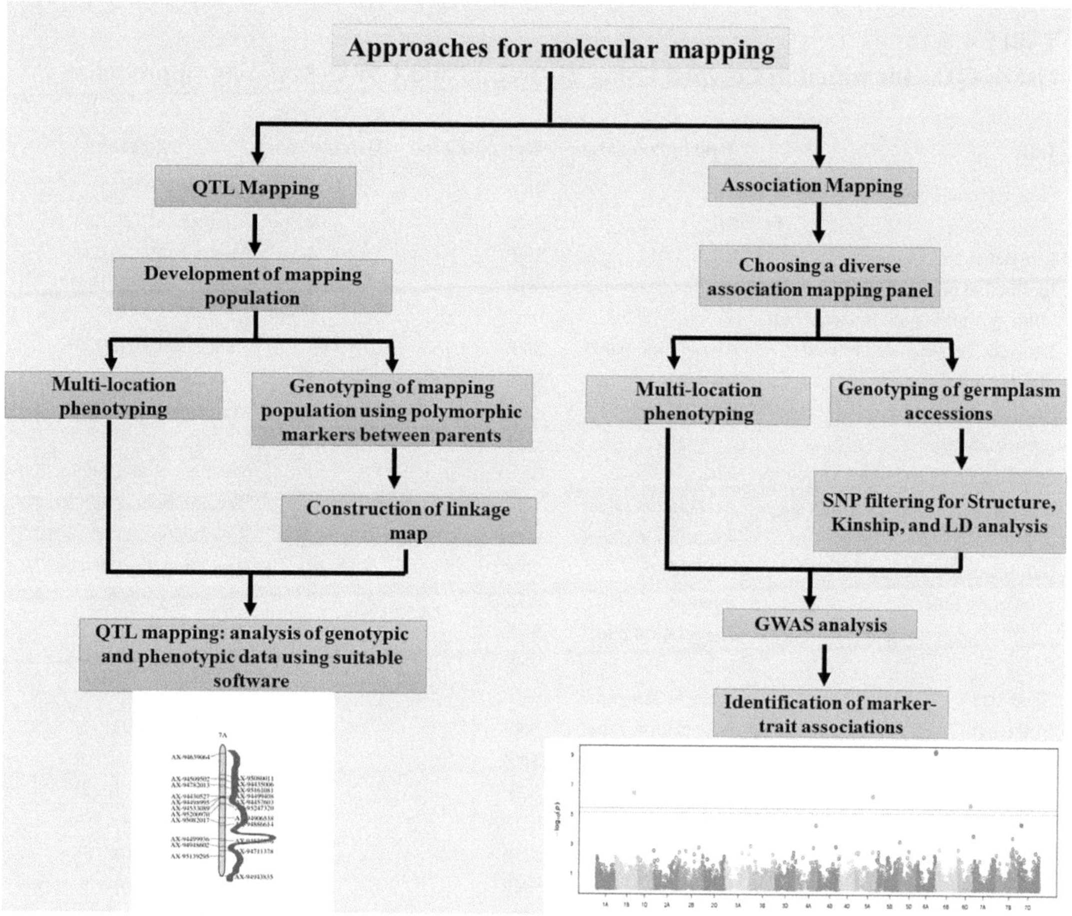

FIGURE 4.2 Flowchart of QTL mapping and association mapping in plants.

were briefly outlined in Figure 4.2. The genetic and genomic resources of cowpea are enriching every year due to continuous evaluation of germplasms and use of new high-throughput genotyping platforms. Several genetic linkage maps have been developed in cowpea for QTL discovery, diversity analysis, association mapping, map-based cloning, and marker-assisted breeding (Lucas et al., 2011). The main prerequisite for marker-assisted breeding of target trait is the availability of tightly liked molecular markers (QTLs). In this direction, several QTLs for desired traits have been identified in cowpea (Table 4.5). Flowering time in cowpea is an adaptive trait, which requires a particular range of temperature and light. Hitherto few QTLs have been identified for flowering time in cowpea. Ubi et al. (2000) discovered two major QTLs for days to flowering using interspecific RILs. Andargie et al. (2013) reported five QTLs for flower opening and three QTLs for days to flowering. Angira et al. (2020), using SNPs, identified one major QTL for days to flowering and one major QTL for plant height. QTL for domestication traits, perenniality, and floral scent have also been revealed in cowpea (Lo et al., 2018; Lo et al., 2020). These QTLs are valuable to efficiently exploit wild species in improving cowpea's adaptability and stress resilience.

Various abiotic and biotic stresses also hamper the production and productivity of cowpea. The impact of abiotic factors can be minimized by different management techniques, however not always successful, as the abiotic stresses cannot be predicted accurately. Chemicals can reduce the yield losses due to diseases and insect-pests, but it increases the cost and is not economically

TABLE 4.5
List of QTLs Identified in Cowpea Using Biparental and GWAS Mapping Approaches

Trait	Type of Population	Type of Marker	No. of QTLs Discovered	Reference
Heat tolerance	RIL	SNP	5	Lucas et al. (2013)
	RIL	SNP	9	Angira et al. (2022a)
Iron deficiency chlorosis	RIL	SNP	8	Angira et al. (2022b)
Ineffective nodulation and nitrogen utilization–related traits	F2	SNP	2	Ohlson et al. (2018a)
Drought tolerance and related traits	Association panel	SNP	65	Nkhoma et al. (2020)
	MAGIC	SNP	5–35	Ravelombola et al. (2021)
Seedling drought-induced senescence	RIL	AFLP	10	Muchero et al. (2009a, 2009b)
Seedling drought-induced maturity	RIL	AFLP	2	Muchero et al. (2009a, 2009b)
Salt tolerance at germination stage	Association panel	SNP	3	Ravelombola et al. (2018)
Salt tolerance at seedling stage	Association panel	SNP	7	Ravelombola et al. (2018)
Salt tolerance–related traits	MAGIC	SNP	1–23	Ravelombola et al. (2022)
Flowering time	Interspecific RIL	RAPD	2	Ubi et al. (2000)
	Association panel	SNP	7	Paudel et al. (2021)
	RIL	SNP	1	Angira et al. (2020)
Seed size	Association panel	SNP	17	Lo et al. (2019)
Pod length	Association panel	SNP	72	Xu et al. (2017)
Seed coat color	Association panel	SNP	1	Herniter et al. (2018)
Carotenoids content	Association panel	SNP	7	Sodedji et al. (2022)
Domestication traits	RIL	SNP	16	Lo et al. (2018)
Perenniality	RIL	SNP	1	Lo et al. (2020)
Floral scent	RIL	SNP	2	Lo et al. (2020)
Root architecture	Association panel	SNP	11	Burridge et al. (2017)
Bacterial blight resistance	Association panel	SNP	4	Shi et al. (2016)
	RIL	SNP	3	Agbicodo et al. (2010)
Golden mosaic virus	F2	AFLP	3	Rodrigues et al. (2012)
Striga resistance	F2	SCAR	2	Ouedraogo et al. (2012)
Macrophomina resistance	RIL	SNP	9	Muchero et al. (2011)
Foliar thrips	RIL	AFLP	3	Muchero et al. (2010)
Root-knot nematode	RIL and F2:3	SNP	1	Huynh et al. (2016)
	RIL	SNP	1	Santos et al. (2018)
Fusarium wilt resistance	RILs	SNP	3	Pottorff et al. (2014)
Brown blotch resistance	F2	AS-PCR	4	Ohlson et al. (2018b)
Flower bud thrips resistance	F2	SNP	3	Sobda et al. (2017)
Aphid resistance	RIL	SNP	2	Huynh et al. (2015)
	Association panel	SNP	5	Ongom et al. (2022)
Bruchid resistance	Association panel	SNP	14	Kpoviessi et al. (2022)

feasible for farmers. The climate stress-tolerant varieties are the feasible and sustainable method. The donor sources for tolerance/resistance to these traits have been identified in cowpea but, being quantitative traits, are difficult to introgress in cultivated varieties. In this direction, various QTLs have been identified, which can be effectively utilized using marker-assisted breeding. Lucas et al. (2013) and Angira et al. (2022a) identified five and nine SNPs, respectively, for heat tolerance using RIL populations. Ohlson et al. (2018a) discovered two SNPs for nitrogen utilization–related traits, and Angira et al. (2022b) identified eight SNPs for iron deficiency chlorosis in cowpea. High-density SNP linkage map has not been developed for drought and salt tolerance using biparental mapping, which is the major constraint for breeding drought-tolerant cowpea. However, few researchers (Muchero et al., 2009a, 2009b) developed AFLP linkage map and identified QTLs for drought tolerance. Further, association mapping has been used for identifying SNPs for drought and salt tolerance in cowpea (Ravelombola et al., 2018; Nkhoma et al., 2020; Ravelombola et al., 2021; Ravelombola et al., 2022). These abiotic stresses–tolerant QTLs can be utilized for development of climate-resilient cowpea varieties for environmental security. Likewise, QTLs have also been identified in cowpea for disease resistance, such as bacterial leaf blight, golden mosaic virus, *Striga*, *Macrophomina*, root-knot nematode, fusarium wilt, and brown blotch, and for insect-pest resistance, viz., foliar thrips, flower bud thrips, aphids, and bruchid (Table 4.5). These QTLs can be deployed for developing cowpea varieties resistant to major diseases and insect-pests.

4.5.2 Genome-Wide Association Studies (GWAS)

Biparental mapping has been used for QTL discovery of many useful traits in cowpea. However, the discovery of QTLs for complex traits in biparental mapping is limited by low-resolution power and use of specific mapping populations. GWAS provides fine resolution; therefore, it is considered to be a robust and powerful tool for dissection of complex traits (Cortes et al., 2021). It uses the diverse germplasm sets that have undergone many rounds of historical recombination, while biparental mapping accounts only those recombinations that have been generated during the development of mapping populations (Figure 4.2) (Yu and Buckler, 2006). GWAS depends on the linkage disequilibrium between marker and QTL; therefore, marker density should be high to narrowing down the location of causal variant (Flint-Garcia et al., 2003). In cowpea, core set and mini-core sets have been developed, and dense SNP maps are available to enrich genetic and genomic resources (Fatokun et al., 2018). With cowpea being a self-pollinated crop, application of GWAS in cowpea has an additional benefit of employing "genotyping once and phenotyping many times" strategy, as the genotypic data generated for an association panel can be reused to map many other phenotypic variations (Kumar et al., 2015). Besides QTL discovery, GWAS is also useful for identifying allelic variants in candidate genes of complex quantitative traits, including plant architecture, stress, and quality (Sahay et al., 2022). Recent advances in phenotyping, genotyping, and bioinformatics tools have made it possible to conduct GWAS for QTL discovery and marker development in many legume crops. GWAS has also been applied in natural collections of cowpea for root architecture, agronomic traits, yield-related traits, flowering time, nutritional quality, and resistance to various biotic and abiotic stresses (Table 4.5).

The special type of mapping populations, such as nested association mapping (NAM) and MAGIC, can also be used in GWAS. NAM offers high-resolution power while minimizing the marker density by accounting both historical and recent recombination events (McMullen et al., 2009). MAGIC involves multiple rounds of recombination between eight diverse parental lines, thereby increasing the frequency of rare alleles in the RILs (Huang et al., 2015). The first cowpea MAGIC population was developed by Huynh et al. (2018) by crossing eight diverse parental lines carrying many key traits relevant to cowpea. This population is constituted as an important genetic resource for high-resolution genetic mapping. Ravelombola et al. (2021) conducted GWAS in cowpea MAGIC population (305 RILs) and discovered 15 SNPs for drought-induced change in plant growth habit, and 5-35 SNPs for four drought tolerance indices. Ravelombola et al. (2022) identified 1-23 SNPs for nine salt tolerance–related traits using 234 MAGIC lines of cowpea. The aforementioned markers can be applied as selection tool in marker-assisted backcrossing (MABC), marker-assisted recurrent selection (MARS), and genomic selection for developing improved varieties of cowpea (Chhabra et al., 2020; Mehta et al., 2020).

4.6 APPLICATIONS OF ALLELE MINING IN COWPEA

"Allele mining" analyzes naturally occurring allelic variations or appropriate alleles of a candidate gene affecting important agronomic properties that may be improved in crops. It aids in the reconstruction of allele evolution, the discovery of novel haplotypes, and the creation of molecular markers distinct to each allele for MAS.

4.6.1 Allelic Variation Characterization in Gene Banks

Measures of allelic diversity, or the number of various allelic types segregating in the population, are also frequently used, especially in conservation genetics studies. For instance, it is understood that the proportion of alleles in a population that segregates provides fundamental details about previous variations in population size (Nei et al., 1975). To determine the genetic links between the accessions of the collection and to evaluate their genetic worthiness in terms of its potential to enhance a target trait, it is crucial to characterize the genetic diversity or allelic/genic diversity among them. Numerous international crop research organizations that preserve crop germplasm collections have started investigations to define the allelic diversity of crop plants after realizing the potential of allele mining in plant genetic resources (PGR) management applications. International institutes working on cowpea, such as the International Centre for Tropical Agriculture (CIAT), studied and characterized genetic diversity in various germplasm using phenotypic traits and molecular markers (Nkhoma et al., 2020). However, still there is scope that allele mining can be one of the best strategies to identify superior alleles and to broaden genetic diversity studies in cowpea.

4.6.2 Haplotype Identification

The *haplotype* is a set of a nucleotide sequence or a set of alleles that tends to be inherited by the next generation without showing recombination and thus forms a set of DNA variants. Haplotype-based allele mining approaches are used to detect allelic variation in the genes controlling target trait that can be used in crop breeding programs. Studies on association mapping would be based on knowledge of the most frequent haplotype modifications and their frequency in the populations. The great potential of haplotype-based allele mining for genetic dissection of important traits in cowpea still needs to be explored.

4.6.3 Allele-Specific Marker Development for Marker-Assisted Selection

In order to precisely introduce the detected "superior and/or unique" alleles to the appropriate genetic background, allele-specific marker assays will need to be developed. This will be made possible by the identification of sequence variation. The occurrence of sequence variation at important genes has been shown in various case studies reported in recent years, and some research have shown the value of these variations by creating allele-specific molecular markers for MAS. The study in cowpea highlights the significance of molecular marker deployment and corroborates the efforts toward optimization of the breeding program for increased genetic gain (Ongom et al., 2021). Allele mining might be a great choice in identifying allelic variation for most of the needed traits for crop improvement programs, and thus, identified allelic variations can be used in the development of allele-specific markers for marker-assisted selection.

4.6.4 Allelic Synteny and Evolution Study

Synteny, which occurs from descent from a common ancestor, is the retained order of alleles of genes on chromosomes of related species. It is possible to evaluate syntenic links among the identified loci/genes across the species/genera using the sequencing data gathered through allele mining investigations. Understanding about evolutionary relationships help in genetic diversity studies. Synteny-based functional genomics studies have been done in cowpea based on expressed sequence tags (ESTs)–derived SNPs, and results support evolutionary closeness between soyabean and cowpea (Muchero et al., 2009a). Thus, the study of allelic synteny and evolutionary relationships can be done through allele mining for better understanding about genomic framework of cowpea.

4.7 CONCLUSION

Cowpea is one of the most important grain legumes which support the livelihood of resource-poor farmers worldwide. Furthermore, it is also an important nutrient-rich fodder for livestock. The limited productivity of cowpea is due to a number of abiotic and biotic factors. The low availability of genomic resources for cowpea has contributed to the slow development of higher-yielding varieties adapted to various adverse environments. Thus, it is imperative to use the allele mining tool to quickly advance crop improvement by introducing alleles imparting resistance to biotic and abiotic stress and high yield potential from wild relatives and landraces of cowpea to elite cultivars. The availability of NGS tools is anticipated to profoundly alter the management of cowpea genetic resource collections and the services offered by gene banks to the user community in addition to revolutionizing genomics research.

REFERENCES

Agbicodo EM, Fatokun CA, Bandyopadhyay R (2010) Identification of markers associated with bacterial blight resistance loci in cowpea [*Vigna unguiculata* (L.) Walp.]. *Euphytica* 175:215–226.

Andargie M, Pasquet RS, Muluvi GM, Timko MP (2013) Quantitative trait loci analysis of flowering time related traits identified in recombinant inbred lines of cowpea (*Vigna unguiculata*). *Genome* 56(5):289–294.

Angira B, Zhang Y, Scheuring CF, Zhang Y, Masor L et al. (2020) Quantitative trait loci influencing days to flowering and plant height in cowpea, *Vigna unguiculata* (L.) Walp. *Mol Gen Genom* 295:1187–1195.

Angira B, Zhang Y, Scheuring CF, Zhang Y, Masor L et al. (2022a) Construction of a single nucleotide polymorphism linkage map and identification of quantitative trait loci controlling heat tolerance in cowpea, *Vigna unguiculata* (L.) Walp. *Mol Genet Genom* 297(6):1481–1493.

Angira B, Zhang Y, Zhang Y, Scheuring CF, Masor L et al. (2022b) Genetic dissection of iron deficiency chlorosis by QTL analysis in cowpea. *Euphytica* 218:38.

Bajaj D, Das S, Badoni S, Kumar V, Singh M et al. (2015) Genome-wide high-throughput SNP discovery and genotyping for understanding natural (functional) allelic diversity and domestication patterns in wild chickpea. *Sci Rep* 5:12468.

Bernardo R (2008) Molecular markers and selection for complex traits in plants: Learning from the last 20 years. *Crop Sci* 48(5):1649–1664.

Boukar O, Fatokun CA, Huynh B-L, Roberts PA, Close TJ (2016) Genomic tools in cowpea breeding programs: Status and perspectives. *Front Plant Sci* 7:757.

Brown AHD (1989) A case for core collections. In: Brown AHD, Frankel OH, Marshall DR, Williams JT (eds) *The Use of Plant Genetic Resources*. Cambridge University Press, Cambridge, pp. 136–156.

Burridge JD, Schneider HM, Huynh BL (2017) Genome-wide association mapping and agronomic impact of cowpea root architecture. *Theor Appl Genet* 130:419–431.

Camacho Villa T, Maxted N, Scholten M, Ford-Lloyd B (2005) Defning and identifying crop landraces. *Plant Genet Resour* 3:373–384.

Chen X, Laudeman TW, Rushton PJ, Spraggins TA, Timko MP (2007) CGKB: An annotation knowledge base for cowpea (*Vigna unguiculata* L.) methylation filtered genomic gene space sequences. *BMC Bioinform* 8:129.

Chhabra R, Hossain F, Muthusamy V, Baveja A, Mehta BK et al. (2019) Development and validation of breeder-friendly functional markers of *sugary1* gene encoding starch-debranching enzyme affecting kernel sweetness in maize (*Zea mays*). *Crop Pasture Sci* 70(10):868–875.

Chhabra R, Hossain F, Muthusamy V, Baveja A, Mehta BK et al. (2020) Development and validation of gene-based markers for *shrunken2*-reference allele and their utilization in marker-assisted sweet corn (*Zea mays Sachharata*) breeding programme. *Plant Breed* 139(6):1135–1144.

Chhabra R, Muthusamy V, Baveja A, Katral A, Mehta B et al. (2022) Allelic variation in *shrunken2* gene affecting kernel sweetness in exotic-and indigenous-maize inbreds. *PLoS One* 17:9.

Comai L, Young K, Till BJ, Reynolds SH, Greene EA et al. (2004) Efficient discovery of DNA polymorphisms in natural populations by Ecotilling. *Plant J* 37:778–786.

Cortes TL, Zhang Z, Yu J (2021) Status and prospects of genome-wide association studies in plants. *Plant Genome* 14:1.

da Silva AC, da Costa Santos D, Junior DLT, da Silva PB, dos Santos et al. (2018) Cowpea: A strategic legume species for food security and health. In: *Legume Seed Nutraceutical Research*. IntechOpen. http://dx.doi.org/10.5772/intechopen.79006.

Fatokun C, Girma G, Abberton M, Gedil M, Unachukwu N et al. (2018) Genetic diversity and population structure of a mini-core subset from the world cowpea (*Vigna unguiculata* (L.) Walp.) germplasm collection. *Sci Rep* 8(1):1–10.

Flint-Garcia SA, Thornsberry JM, Buckler IV ES (2003) Structure of linkage disequilibrium in plants. *Annu Rev Plant Biol* 54:357–374.

Food and Agriculture Organization of the United Nations (2009) *International Treaty on Plant Genetic Resources for Food and Agriculture*. Food and Agriculture Organization of the United Nations, Rome.

Frankel OH, Brown AHD (1984) Plant genetic resources today: A critical appraisal. In: Holden JHW, Williams JT (eds) *Crop Genetic Resources: Conservation and Evaluation*. George Allen & Unwin, London, pp. 249–257.

Frerichmann SL, Kirchhoff M, Müller AE, Scheidig AJ, Jung C et al. (2013) Eco TILLING in Beta vulgaris reveals polymorphisms in the FLC-like gene BvFL1 that are associated with annuality and winter hardiness. *BMC Plant Biol* 13(1):1–14.

Gepts P (2006) Plant genetic resources conservation and utilization: The accomplishments and future of a societal insurance policy. *Crop Sci* 46:2278–2292.

Harlan JR (1972) Genetic resources in sorghum. In: Rao NGP, House LR (eds) *Sorghum in the Seventies*. Oxford and IBH, New Delhi, pp. 1–13.

Herniter IA, Muñoz-Amatriaín M, Lo S, Guo YN, Close TJ (2018) Identification of candidate genes controlling black seed coat and pod tip color in cowpea (*Vigna unguiculata* [L.] Walp). *Genes Genomes Genet* 8:3347–3355.

Huang BE, Verbyla KL, Verbyla AP, Raghavan C, Singh VK et al. (2015) MAGIC populations in crops: Current status and future prospects. *Theor Appl Genet* 128(6):999–1017.

Huynh BL, Ehlers JD, Huang BE, Muñoz-Amatriaín M, Lonardi S et al. (2018) A multi-parent advanced generation inter-cross MAGIC population for genetic analysis and improvement of cowpea (*Vigna unguiculata* L. Walp.). *Plant J* 93:1129–1142.

Huynh BL, Ehlers JD, Ndeve A (2015) Genetic mapping and legume synteny of aphid resistance in African cowpea (*Vigna unguiculata* L. Walp) grown in California. *Mol Breed* 35:36.

Huynh BL, Matthews WC, Ehlers JD (2016) A major QTL corresponding to the *Rk* locus for resistance to root-knot nematodes in cowpea (*Vigna unguiculata* L. Walp.). *Theor Appl Genet* 129:87–95.

Ibiza VP, Cañizares J, Nuez F (2010) Eco TILLING in capsicum species: Searching for new virus resistances. *BMC Genom* 11:631.

Kak A, Gupta V (2021) Registration of trait specific germplasm. In: Gupta V, Gore Padmavati G, Malav Pavan K, Hooda KS, Singh K (eds) *Training Manual on "Virtual Training Programme on Plant Genetic Resources Management and Utilization"*. ICAR-National Bureau of Plant Genetic Resources, New Delhi, India, pp. 1–372.

Ketema S, Tesfaye B, Keneni G, AmsaluFenta B, Assefa E et al. (2020) DArTSeq SNP-based markers revealed high genetic diversity and structured population in Ethiopian cowpea [*Vigna unguiculata* (L.) Walp] germplasms. *PLoS One* 15:10.

Koboldt DC, Steinberg KM, Larson DE, Wilson RK, Mardis ER (2013) The next-generation sequencing revolution and its impact on genomics. *Cell* 155(1):27–38.

Kpoviessi AD, Agbahoungba S, Agoyi EE, Badji A, Assogbadjo AE et al. (2022) Application of multi-locus GWAS for the detection of bruchid resistance loci in cowpea (*Vigna unguiculata*). *Plant Breed* 141:439–450.

Kumar GR, Sakthivel K, Sundaram RM, Neeraja CN, Balachandran SM et al. (2010) Allele mining in crops: Prospects and potentials. *Biotechnol Adv* 28(4):451–461.

Kumar T, Tiwari N, Bharadwaj C, Sarker A, Pappula SPR et al. (2020) Identification of allelic variation in drought responsive dehydrin gene based on sequence similarity in chickpea (*Cicer arietinum* L.). *Front Genet* 1579.

Kumar V, Singh A, Mithra SA, Krishnamurthy SL, Parida SK et al. (2015) Genome-wide association mapping of salinity tolerance in rice (*Oryza sativa*). *DNA Res* 22(2):133–145.

Lo S, Fatokun C, Boukar O, Gepts P, Close TJ et al. (2020) Identification of QTL for perenniality and floral scent in cowpea (*Vigna unguiculata* [L.] Walp.). *PLoS One* 15(4).

Lo S, Muñoz-Amatriaín M, Boukar O, Herniter I, Cisse N et al. (2018) Identification of QTL controlling domestication-related traits in cowpea (*Vigna unguiculata* L. Walp). *Sci Rep* 8:6261.

Lo S, Muñoz-Amatriaín M, Hokin SA, Cisse N, Roberts PA et al. (2019) A genome-wide association and meta-analysis reveal regions associated with seed size in cowpea [*Vigna unguiculata* (L.) Walp]. *Theor Appl Genet* 132(11):3079–3087.

Lucas MR, Diop NN, Wanamaker S, Ehlers JD, Roberts PA et al. (2011) Cowpea–soybean synteny clarified through an improved genetic map. *Plant Genome* 4(3):218–225.

Lucas MR, Ehlers JD, Huynh BL (2013) Markers for breeding heat-tolerant cowpea. *Mol Breed* 31:529–536.

Mahalakshmi V, Ng Q, Lawson M, Ortiz R (2007) Cowpea [*Vigna unguiculata* (L.) Walp.] core collection defined by geographical, agronomical and botanical descriptors. *Plant Genet Resour* 5(3):113–119.

Mahesha HS, Keerthi MC, Shivakumar KV, Bhargavi HA, Saini RP et al. (2022) Development of biotic stress resistant cowpea. In: Kole C (ed) Genomic *Designing for Biotic Stress Resistant Pulse Crops*. Springer, Cham, pp. 213–251.

McMullen MD, Kresovich S, Villeda HS, Bradbury P, Li H et al. (2009) Genetic properties of the maize nested association mapping population. *Science* 325(5941):737–740.

Mehta BK, Muthusamy V, Zunjare RU, Baveja A, Chauhan HS et al. (2020) Biofortification of sweet corn hybrids for provitamin-A, lysine and tryptophan using molecular breeding. *J Cereal Sci* 96:103093.

Mejlhede N, Kyjovska Z, Backes G, Burhenne K, Rasmussen SK et al. (2006) Eco TILLING for the identification of allelic variation in the powdery mildew resistance genes mlo and Mla of barley. *Plant Breed* 125:461–467.

Min MH, Maung TZ, Cao Y, Phitaktansakul R, Lee GS et al. (2021) Haplotype analysis of BADH1 by next-generation sequencing reveals association with salt tolerance in rice during domestication. *Int J Mol Sci* 22(14):7578.

Moose SP, Mumm RH (2008) Molecular plant breeding as the foundation for 21st century crop improvement. *Plant Physiol* 147(3):969–977.

Muchero W, Diop NN, Bhat PR, Fenton RD, Wanamaker S et al. (2009a) A consensus genetic map of cowpea [*Vigna unguiculata* (L) Walp.] and synteny based on EST-derived SNPs. *Proc Natl Acad Sci* 106(43):18159–18164.

Muchero W, Ehlers JD, Close TJ (2011) Genic SNP markers and legume synteny reveal candidate genes underlying QTL for *Macrophomina phaseolina* resistance and maturity in cowpea [*Vigna unguiculata* (L) Walp.]. *BMC Genom* 12:8.

Muchero W, Ehlers JD, Close TJ, Roberts PA (2009b) Mapping QTL for drought stress-induced premature senescence and maturity in cowpea [*Vigna unguiculata*(L.) Walp.]. *Theor Appl Genet* 118(5):849–863.

Muchero W, Ehlers JD, Roberts PA (2010) QTL analysis for resistance to foliar damage caused by *Thrips tabaci* and *Frankliniella schultzei* (Thysanoptera: Thripidae) feeding in cowpea [*Vigna unguiculata* (L.) Walp.]. *Mol Breed* 25(1):47–56.

Muñoz-Amatriaín M, Mirebrahim H, Xu P, Wanamaker SI, Luo M et al. (2017) Genome resources for climate-resilient cowpea, an essential crop for food security. *Plant J* 89(5):1042–1054.

Negrao S, Courtois B, Ahmadi N, Abreu I, Saibo N et al. (2011) Recent updates on salinity stress in rice: From physiological to molecular responses. *Crit Rev Plant Sci* 30:329–377.

Nei M, Maruyama T, Chakraborty R (1975) The bottleneck effect and genetic variability in populations. *Evolution* 1–10.

Nkhoma N, Shimelis H, Laing MD, Shayanowako A, Mathew I (2020) Assessing the genetic diversity of cowpea [Vigna unguiculata (L.) Walp.] germplasm collections using phenotypic traits and SNP markers. *BMC Genet* 21(1):1–16.

Ogawa D, Yamamoto E, Ohtani T, Kanno N, Tsunematsu H et al. (2018) Haplotype-based allele mining in the Japan-MAGIC rice population. *Sci Rep* 8:4379.

Ohlson EW, Seido SL, Mohammed S, Santos CA, Timko MP (2018a) QTL mapping of ineffective nodulation and nitrogen utilization-related traits in the IC-1 mutant of cowpea. *Crop Sci* 58(1):264–272.

Ohlson EW, Thio GI, Sawadogo M (2018b) Quantitative trait loci analysis of brown blotch resistance in cowpea variety KN1. *Mol Breed* 38:110.

Ongom PO, Fatokun C, Togola A, Salvo S, Oyebode OG et al. (2021) Molecular fingerprinting and hybridity authentication in cowpea using single nucleotide polymorphism based Kompetitive allele-specific PCR assay. *Front Plant Sci* 2205.

Ongom PO, Togola A, Fatokun C, Boukar O (2022) A genome-wide scan divulges key loci involved in resistance to aphids (*Aphis craccivora*) in cowpea (*Vigna unguiculata*). *Genes* 13(11):2002.

Ouedraogo JT, Ouedraogo M, Gowda BS, Timko MP (2012) Development of sequence characterized amplified region (SCAR) markers linked to race-specific resistance to *Strigagesnerioides* in cowpea (*Vigna unguiculata* L.). *Afr J Biotechnol* 11:12555–12562.

Paudel D, Dareus R, Rosenwald J, Muñoz-Amatriaín M, Rios EF (2021) Genome-wide association study reveals candidate genes for flowering time in cowpea (*Vigna unguiculata* [L.] Walp.). *Front Genet* 1021.

Pottorff MO, Li G, Ehlers JD (2014) Genetic mapping, synteny, and physical location of two loci for *Fusarium oxysporum* f. sp. *tracheiphilum* race 4 resistance in cowpea [*Vigna unguiculata* (L.) Walp]. *Mol Breed* 33:779–791.

Raja RB, Agasimani S, Jaiswal S, Thiruvengadam V, Sabariappan R et al. (2017) EcoTILLING by sequencing reveals polymorphisms in genes encoding starch synthases that are associated with low glycemic response in rice. *BMC Plant Biol* 17(1):1–13.

Ravelombola W, Shi A, Huynh BL (2021) Loci discovery, network-guided approach, and genomic prediction for drought tolerance index in a multi-parent advanced generation intercross (MAGIC) cowpea population. *Hortic Res* 8:24.

Ravelombola W, Shi A, Huynh BL, Qin J, Xiong H et al. (2022) Genetic architecture of salt tolerance in a multiparent advanced generation inter-cross (MAGIC) cowpea population. *BMC Genom* 23:100.

Ravelombola W, Shi A, Weng Y (2018) Association analysis of salt tolerance in cowpea (*Vigna unguiculata* (L.) Walp) at germination and seedling stages. *Theor Appl Genet* 131:79–91.

Rodrigues MA, Santos CAF, Santana JRF (2012) Mapping of AFLP loci linked to tolerance to cowpea golden mosaic virus. *Genet Mol Res* 11(4):gmr1819.

Sabetta W, Alba V, Blanco A, Montemurro C (2011) sunTILL: A TILLING resource for gene function analysis in sunflower. *Plant Methods* 7(1):1–13.

Sahay G, Mehta BK, Dikshit N, Shashikumara P, Bhargavi HA et al. (2022) Developing climate-resilient cowpea (*Vigna unguiculata* [L.]Walp.) through genomics-assisted breeding approaches. In: Jha UC, Nayyar H, Agrawal SK, Siddique KHM (eds) *Developing Climate Resilient Grain and Forage Legumes*. Springer, Singapore, pp. 77–103.

Sammour RH (2011) Genetic diversity and allele mining in soybean germplasm. In: Krezhova D (ed) *Soybean-Genetics and Novel Techniques for Yield Enhancement*. IntechOpen, pp. 3–22.

Santos JR, Ndeve AD, Huynh BL, Matthews WC, Roberts PA (2018) QTL mapping and transcriptome analysis of cowpea reveals candidate genes for root-knot nematode resistance. *PLoS One* 13:1.

Savage D, Batley J, Erwin T, Logan E, Love CG et al. (2005) SNPServer: A real-time SNP discovery tool. *Nucleic Acids Res* 33:493–495.

Shi A, Buckley B, Mou B (2016) Association analysis of cowpea bacterial blight resistance in USDA cowpea germplasm. *Euphytica* 208:143–155.

Sobda G, Boukar O, Tongoona PB, Ayertey J, Offei KS (2017) Quantitative trait loci (QTL) for cowpea resistance to flower bud thrips (*Megalurothrips sjostedti* Trybom). *Int J Plant Breed Genet* 4(6):292–299.

Sodedji FA, Ryu D, Choi J, Agbahoungba S, Assogbadjo AE et al. (2022) Genetic diversity and association analysis for carotenoid content among sprouts of cowpea (*Vigna unguiculata* L. Walp). *Int J Mol Sci* 23(7):3696.

Tang JT, Vosman B, Voorrips RE, van der Linden GC, Leunissen JAM (2006) QualitySNP: A pipeline for detecting single nucleotide polymorphisms and insertions/deletions in EST data from diploid and polyploid species. *BMC Bioinform* 7:438–453.

Till BJ, Reynolds SH, Greene EA, Codomo CA, Enns LC et al. (2003) Large-scale discovery of induced point mutations with high-throughput TILLING. *Genome Res* 13(3):524–530.

Ubi BE, Mignouna H, Thottappilly G (2000) Construction of a genetic linkage map and QTL analysis using a recombinant inbred population derived from an intersubspecific cross of Cowpea: *Vignaunguicultа* (L.) Walp. *Breed Sci* 50(3):161–172.

Varshney RK, Nayak SN, May GD, Jackson SA (2009) Next-generation sequencing technologies and their implications for crop genetics and breeding. *Trends Biotechnol* 27(9):522–530.

Weckx S, Del Favero J, Rademakers R, Claes L, Cruts M (2005) novoSNP, a novel computational tool for sequence variation discovery. *Genome Res* 15(3):436–442.

Xu P, Wu X, Muñoz-Amatriaín M, Wang B, Wu X (2017) Genomic regions, cellular components and gene regulatory basis underlying pod length variations in cowpea (*V. unguiculata* L. Walp). *Plant Biotechnol J* 15:547–557.

Yu J, Buckler ES (2006) Genetic association mapping and genome organization of maize. *Curr Opin Biotech* 17:155–160.

Yu S, Liao F, Wang F, Wen W, Li J et al. (2012) Identification of rice transcription factors associated with drought tolerance using the Eco TILLING method. *PLoS One* 7:e30765.

Zeng YD, Sun JL, Bu SH, Deng KS, Tao T et al. (2016) EcoTILLING revealed SNPs in GhSus genes that are associated with fiber-and seed-related traits in upland cotton. *Sci Rep* 6(1):1–9.

5 Allele Mining in Cowpea (*Vigna Unguiculata*) Subspecies and Close Relatives

Matthew W. Blair, Max C. Miller II, Damba Yahaya, David Hickok, and Xingbo Wu

5.1 INTRODUCTION

The species *Vigna unguiculata* L. (Walp.), commonly referred to as the cowpea, is grown in a wide range of production environments across Africa, Asia, Europe, and North/South America, especially in the subtropics. Wild forms of cowpea, of the same species but varying in subspecies status, exist in various parts of sub-Saharan Africa, where the species is from. These include *V. u.* spp. *alba*, *dektiana*, *pubescens*, *stenophylla*, and *tenuis*, amongst others (Pasquet and Padulosi, 2014). The domesticated cowpea in its dry grain and fodder forms most likely originated in West Africa somewhere between the Volta and Niger Rivers. The largest number of wild subspecies occurs in Southern Africa (Singh, 2014). Overall, the species has a small genome with 11 pairs of chromosomes, typical of the genus/tribe (2n = 2x = 22). Cowpeas are one of the world's most important legumes for direct human consumption on a par with the temperate pulses of the Mediterranean or bean species of the Americas. The dry grain pulse has also become a worldwide commodity for export or local consumption. Production of dry cowpeas occurs mainly in West Africa (Nigeria, Niger, Burkina Faso, Senegal, Ghana, Mali, Cameroon, in order of production). Other regions of production include East Africa (e.g., Kenya, Sudan), Europe (e.g., Italy), North America (United States), South America (Brazil, Peru, and Venezuela) and South Asia (India, Myanmar, and Sri Lanka). In East Asia, China is a major producer with some exports. Peru exports to Europe, and the United States exports worldwide. Total worldwide production is currently almost 9 million (M) metric tons grown on over 15 M hectares (www.fao.org/faostat). Apart from the grain, a vegetable form has become very popular in East and South Asian cuisines, so much so that it has its own crop designation as yard-long bean and is a different subspecies (*V. u.* spp. *sesquipedilis*). This chapter covers allele mining in all the subspecies but emphasizes dry grain types, as they have been the best-studied and represent the subject most often published in major journals.

5.2 FIELD SCREENING OF TRAITS

Essential to any attempts at allele mining is the use of accurate phenotyping, and the most reliable and essential phenotyping efforts are conducted at the whole plant level in the field. Experimental trials are especially important for determining the adaptation of cowpeas to local conditions and their overall drought resistance and heat tolerance, given the climate and weather events occurring at each testing site. As examples of the projects we have sponsored, screening for the West African site of Ghana by the University of Development Studies (UDS) involved four locations near the regional capital of Tamale, while screening for Tennessee State University (TSU) in the United States involved two locations near the mid-state capital city of Nashville. Studies in the first case were conducted on a farm (Yahaya et al., 2019a), while for the second case, they were conducted on experiment stations. Any of these types of screenings make sense for local conditions but place

DOI: 10.1201/9781003385059-5

logistical limits at times on what sort of data are collected. Typical data for field drought experiments are drought stress index, water use efficiency, stomatal conductance, and yield components (grain weight per plant, hundred-seed weight, seeds per plant, seeds per pod), along with nitrogen fixation parameters in some cases (Yahaya et al., 2019b).

Accurate field phenotyping requires constant monitoring of environmental conditions that the cowpea crop is subjected to (Singh et al., 2003). Having this data is essential for the comparison of different studies since each year and each growing season have different weather patterns. Field conditions also vary greatly within a day, week, or month and from season to season. Examples of the Ghana–UDS study site weather conditions are shown in Figure 5.1, with evident pre–rainy season pattern of low rainfall but increasing humidity under stable average daily temperatures at or above 30°C (Yahaya et al., 2019a). That climate is typical of Guinea savanna ecology found in a zone directly below the Sahel in West Africa, a region of high cowpea production stretching from Senegal/Guinea in the west to Niger in the east and running through Mali and Burkina Faso along with northern portions of Ghana, Togo, Benin, and Nigeria.

Field trials in formal settings such as that of TSU in the United States (Figure 5.2) differ from those of experiments conducted by UDS in Ghana, which often used farmers' fields. When contrasting on-farm versus on-station trials, one must note plant spacing and other agronomic parameters differing due to constraints of hand versus mechanized planting. In northern Ghana, most work on farm is done with hand tools, while even in home gardens in the United States, tractors are commonplace. In Tennessee, the cowpea experiments are less likely to be planted with farmers who grow a few established varieties compared to the mixtures used in northern Ghana. Another major difference is that in West Africa, cowpeas are planted as an intercrop or under a savanna system of

(a)

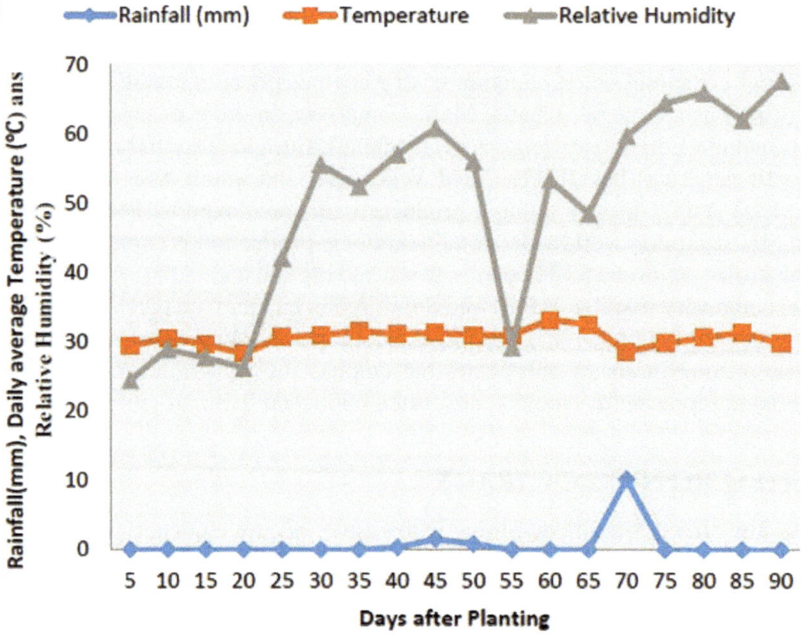

FIGURE 5.1 Experimental conditions of (a) rainfall, relative humidity, and average daily temperatures across the growing season for a UDS site at the Bontanga field station in the Kumbungu district within (b) semi-forested cowpea growing areas of the agroecology system of the Guinea savanna in northern Ghana.

(b)

FIGURE 5.1 (Continued)

dispersed fruit and nut trees. Meanwhile, in the United States, cowpeas are planted as a monoculture, but often following winter wheat or brassica vegetables.

Given differences between the two sites, it is appropriate that research be conducted with pure line varieties at the Agricultural and Extension Research Centres (AREC) of TSU, while for UDS, it is appropriate to do on farm trials. Management in each situation presents its own challenges. For example, labor requirements are higher in on-farm experiments, but these can be met by seasonal hiring, whereas on-station experiments are managed by graduate students and field assistants. The scientific knowledge of the latter group may be higher, but the former group excels at plant management. This is notable in the hillock design of on-farm experiments versus the row-to-row experiments on station.

FIGURE 5.2 Experiments grown in Tennessee State University, United States, showing (a) phenotyping of cowpea germplasm and (b) rotation with winter wheat.

5.3 GREENHOUSE SCREENING OF TRAITS

Biotic stress resistance is often cumbersome or inaccurate to evaluate for in the field. Too much inoculum or not enough can limit the accuracy of field screening for disease pathogens or insect pests. Therefore, greenhouse/screenhouses with more controlled conditions are used for screening of some pests and pathogens. The greenhouse is better for specialized screening, such as for inoculating with nematodes and evaluating fungal and bacterial pathogens. Resistance to some insects and parasitic weeds (e.g., *Striga*) is also best conducted in biosafety greenhouses to avoid the escape of potentially rapidly spreading plagues. Laboratory or specialized storage room tests are used for bruchids/weevils and should be separate from long-term seed storage rooms.

Abiotic stresses can also be evaluated in greenhouse and rainout shelter screening as another method to apply more controlled conditions to the evaluation of cowpea germplasm for heat tolerance (summer greenhouse) or drought resistance (drought box seedling assays or adult plant screening). Automated evaluation of physiological parameters is becoming possible in greenhouse settings with specialized pot units grown on arrays of analytical balances which constantly monitor water loss and carbon/nitrogen fixation in the soil–plant system.

While greenhouses may be more stable in temperature and watering, it is still very important to monitor humidity, irrigation, and temperature parameters within the growing area. This requires sensors to be installed above the plant canopy, and sometimes within the pots and soil, solution, or media used to grow the plants. Cowpea varieties vary in photoperiod response, so sensors for photosynthetically active radiation are necessary. Without enough duration and quality of light, certain cowpea plant genotypes can become very leggy with very few flowers. Different genotypes respond to light availability with varying plant architecture (erect to viny), which even in the greenhouse (Figure 5.3) should reflect the growth type in the field.

5.4 TOXICITY SCREENING EXAMPLE (ALUMINUM)

Laboratory or growth chamber analysis of traits is another way to mine for germplasm that contains novel phenotypic traits and which might therefore have novel alleles for a trait of interest. An

FIGURE 5.3 Greenhouse testing of cowpeas with varying growth habits at Tennessee State University.

example of specialized screening in the growth chamber with the use of hydroponics for evaluating aluminum toxicity tolerance in cowpea germplasm is shown (Figure 5.4a). When using hydroponics, root evaluation is automated by scanning and software analysis. An example is WinRHIZO scanning for root length, partitioning, and thickness (diameter) after aluminum stress treatment (Figure 5.4b). Seedling-stage plants must be carefully managed for hydroponics to function well.

FIGURE 5.4 Screening of cowpea with hydroponic tanks (A) and root scanning (B) for evaluation of aluminum toxicity stress tolerance among United States varieties at Tennessee State University.

Therefore, various germination media and techniques can be tried before deciding on the best approach. These can include treatments in the dark or under lights. Often, paper towel rolls are used upright in water trays, but we found this to lead to seed rots. Therefore, we used petri dishes with or without wicking disks of filter paper saturated with water at the beginning of the germination process. At Tennessee State University, researchers in our group also tried sand in petri dishes and sphagnum moss in canning jars (Figure 5.5).

As seen from the preceding growth characteristics, identification of cultivars with aluminum toxicity stress tolerance should be a main objective of cowpea improvement programs in the Southeastern USA, where aluminum toxic soils are common and where differential response is found in cultivars when grown on these soils. Two local cowpea genotypes (Dixie Lee and Queen Anne) were identified as suitable for germination in growth chamber transfer to hydroponics solution with 50 μM aluminum (Figure 5.4). In this case, seeds were germinated on petri dishes using autoclaved paper towels prior to transfer to hydroponics tanks.

For most assays, seedlings are carefully transferred and most often grown in a growth chamber for destructive sampling before flowering. Apart from these types of seedling assays, which are common when testing for high salt or low nutrient solutions, adult plants can also be used for screening. Intermediate one, two, or three trifoliate leafed plants can be used for drought boxes and for hydroponics in greenhouse (Figure 5.6).

(a)

FIGURE 5.5 Germination methods for cowpea seeds: (a) paper towel rolls in small tank; (b) petri dish with autoclaved wicking paper; and (c) sphagnum moss in canning jars.

FIGURE 5.5 (Continued)

FIGURE 5.6 (a) Growth and (b) screening of young cowpea seedlings from hydroponic tanks at TSU.

5.4.1 Specialized Screening of Other Traits

Rhizobia-inoculated pots can be used where large systems can be deployed. These adult plant systems are very necessary for nodulation traits, as nodules are specialized organs on legume root systems that require at least one month for full development. By that time, the plant is much larger than for most regular 4 or 6 in pots/hydroponic systems, and either greenhouse or field screening and soil washing from the root system are preferable. Another specialized method for phenotyping in a greenhouse was used with a Plant DiTech Inc. array by Wu et al. (2021a), who evaluated stomata-related drought-responsive traits across a long-period, well-tracked stress experiment based on constantly weighted pots with plants on top of load cells.

5.4.2 Gene Expression Goals of Specialized Screening

Going back to the aluminum trials, an experiment was set up to collect root tissues over a range of days (eight days in experiment 1, and five days in experiment 2). RNA was then extracted from lateral roots and analyzed for quality and quantity to later be used in RNA-seq for aluminum stress resistance candidate genes. The results from the two experiments can be used to guide further research and analysis of cowpea responses to aluminum in hydroponic solution. The ultimate goal is to correlate gene expression with the specific traits measured. These included seven root growth characteristics that are characteristic of root responses to aluminum: (1) root volume (VOL), (2) root surface area (SA), (3) number of crosses (NCROS), (4) number of forks (NFORKS), (5) root length (LEN), (6) average root diameter (AVGD), and (7) projected root area (PA). It was found that response differed between the two experiments. The first experiment resulted in root thickening in cowpea grown in the aluminum hydroponic solution, a result consistent with other plants. The second experiment found that root growth was not inhibited in the aluminum hydroponics solution. This experiment–genotype interaction is a warning even in hydroponic studies that environmental conditions can seriously affect gene and trait expression, confounding studies with RNA-seq or GWAS, both used in the search for candidate genes, as discussed in the next section.

5.5 CANDIDATE GENE AND GWAS STUDIES // PHENOTYPE TOGETHER WITH GENOTYPE

Just like for the activity of underground mining for coal, minerals, and stones, allele mining technologies vary in complexity and cost, especially between parts of the United States as a developed country and other parts of the world, primarily those considered middle-income or developing. Variability in technology used for genotyping is a characteristic of recent studies on cowpea species. Two major platforms have been used: one on genotyping-by-sequencing (GBS) and one based on SNP bead arrays read on a fluorescence detector (iScan/iSelect beads). Within the SNP array space technology, Illumina, based in San Diego, California, has been the major company involved in sequencing and this type of marker detection. They first developed a low-throughput system called GoldenGate about 18 years ago (Shen et al., 2005) and since then improved the same technique for the high-throughput system Infinium iSelect platform. This was used widely by University of California researchers on fully funded international projects and, to a lesser extent, in University of Virginia, due to its expense.

5.5.1 Genotyping with GoldenGate Beads

As stated, the GoldenGate Assay for cowpea was initially developed by Muchero et al. (2009) at the University of California–Riverside (UC-R). It consisted of 1,536 features and was tested on a total of 632 recombinant inbred lines from six separate mapping populations. The array was designed based on more than 10,000 high-confidence SNPs initially discovered in 183,118 expressed sequence tags

(ESTs) from 17 cDNA libraries. Using bioinformatics, the number of dependable markers was narrowed down to 1,375 loci.

Of these, 928 SNPs were polymorphic in more than one RIL population and incorporated into a consensus map. Lucas et al. (2011) further improved upon this consensus map by utilizing 1,293 individual RILs from 13 mapping populations to generate a consensus map containing 1,107 SNPs. The number of total bins was increased from 645 to 865 in the improved consensus map, showing greater saturation of the linkage groups.

Muchero et al. (2013) used one of these populations to perform a QTL study with the GoldenGate SNP dataset and then confirmed marker–trait associations with a GWAS panel. In this case, a population of 113 $F_{2:9}$ RILs from a drought tolerant × drought susceptible cross was evaluated for the three main traits: (1) seedling-stage delayed senescence, (2) grain and biomass yield, and (3) post-flowering senescence. A total of six QTLs for grain yield were detected, all of which were also associated with seedling-stage delayed senescence, and four of which were associated with biomass yield; all QTLs associated with delayed senescence were associated with grain yield or biomass yield.

The GWAS portion of this study evaluated 383 accessions, finding a total of 16 SNPs for seven QTL regions detected in at least one experiment – three collocated with delayed senescence from the RIL population; two loci associated with delayed senescence, biomass yield, and grain yield; five showing evidence of pleiotropy for at least two of the evaluated traits. Other QTL studies using the GoldenGate platform are described in Table 5.1.

In another set of studies with GoldenGate beads, Pottorff et al. (2012) and Pottorff et al. (2014) studied two traits: (1) leaf morphology with gene tagging, and (2) heat-induced brown discoloration of seed coats in cowpea with comparative mapping. The first study used one RIL population of 122 F_{10} lines and 416 polymorphic SNPs for QTL identification, followed by physical mapping and identification of candidate genes using synteny with other legumes species for the "hastate" and "sub-globose" leaf characteristics.

Meanwhile, the second study used two RIL populations from crosses between two lines resistant to heat-induced browning of seed coats (*Hbs*) and two lines susceptible to heat-induced browning of seed coats (*hbs*). These included 114 RILs from IT93K-503–1 (*Hbs*) × CB46 (*hbs*) and 136 RILs from IT84S-2246 (*Hbs*) × TVu14676 (*hbs*). After mapping with the 1,536 Illumina GoldenGate Assay, 374 SNP loci were polymorphic in the first population and 345 in the second. Two QTLs, one major and one minor, were discovered in the population derived from a cross between IT93K-503–1 and CB46. Two loci, one major and one minor, were discovered in the population derived from the cross IT84S-2246 × TVu14676. The major loci for heat-induced browning of seed coats were considered to be allelic with each other.

Lucas et al. (2013) from the same research group at UC-R also used the 1,536 Illumina GoldenGate Assay to study seed size (100-seed weight) in cowpeas compared to other legumes by synteny to reveal haplotypes for larger seeds. Two germplasm sets were used, consisting of (1) 804 RILs from eight biparental mapping populations (seven at the F_8 generation and one in F_4), plus (2) a subset of 171 accessions from the USDA core. The number of polymorphic SNPs depends on the population, with anywhere from 288 to 560 used to make maps. The GWAS study used 665 polymorphic SNPs.

A total of ten QTLs were associated with seed size across the eight mapping populations; most accounted for >30% of the phenotypic variance and contained multiple associated SNPs; 36 SNPs were significantly associated with seed size in the USDA core subset (six of the ten QTLs discovered in mapping were confirmed by GWAS). Xu et al. (2011, 2013) developed a separate GoldenGate marker consensus map for yard-long bean (also called asparagus bean), based on two populations of *V. u.* spp. *sesquipedialis* using the most dependable SNPs from Muchero et al. (2009) and Muchero et al. (2013).

Although 1,375 SNPs were screened by Xu et al. (2011), only 206 were polymorphic. This is likely due to the use of a narrow yard-long bean cross as opposed to the larger set for grain cowpea crosses and the limited molecular diversity of spp. *sesquipedialis* compared to spp. *unguiculata*. Xu

TABLE 5.1
Examples of GoldenGate Studies for Allele Mining in Cowpeas

Title	Authors	Journal	Number of Accessions and Origins	Genotyping Technology Used, Number of SNPs	Traits Evaluated
Leaf morphology in Cowpea [*Vigna unguiculata* (L.) Walp]:	Pottorff et al. (2012)	*BMC Genomics* 13:234	122 F_{10} RILs	1536 Illumina GoldenGate Assay 416 SNPs	Leaf morphology: classified as "hastate" or "sub-globose"
Genetic Architecture of Delayed Senescence, Biomass, and Grain Yield under Drought Stress in Cowpea	Muchero et al. (2013)	*PLOS One* 8(7) e70041	383 accessions and 113 $F_{2:9}$ RILs from a cross between drought-tolerant and drought-susceptible	1536 Illumina GoldenGate Assay 1,080 SNPs	Drought-related traits: (1) seedling-stage effects, (2) grain yield, (3) biomass yield, (4) post-flowering senescence
QTL mapping and epistatic interaction analysis in asparagus bean for several characterized and novel horticulturally important traits	Xu et al. (2013)	*BMC Genetics* 14:4	209 RILs derived from a cross between two spp. *sesquipedialis* cultivars (one landrace and one commercial cultivar)	1536 Illumina GoldenGate Assay	(1) Days to flowering, (2) nodes to first flower, (3) pod number per plant, (4) leaf senescence
Association studies and legume synteny reveal haplotypes determining seed size in *Vigna unguiculata*	Lucas et al. (2013)	*Frontiers in Plant Science* 4:95	804 RILs from eight biparental mapping populations; subset of 171 USDA core accessions	1536 Illumina GoldenGate Assay 665 SNPs for USDA core accessions	Seed size (100-seed weight)
Identification of candidate genes and molecular markers for heat-induced brown discoloration of seed coats in cowpea [*Vigna unguiculate* (L.) Walp]	Pottorff et al. (2014)	*BMC Genomics* 15:328	Two RIL populations from crosses (IT93K-503–1 (*Hbs*) × CB46 (*hbs*) and IT84S-2246 (*Hbs*) × TVu14676 (*bhs*))	1536 Illumina GoldenGate Assay 374 SNPs from T93K-503–1 × CB46; 345 from IT84S-2246 × TVu14676	Heat-induced seed browning based on two lines resistant to heat-induced browning of seed coats (*Hbs*) and two lines susceptible to heat-induced browning of seed coats (*hbs*)

et al. (2013) studied horticulturally important traits, such as nodes, to first flower and pods per plant for yard-long bean using QTL mapping and epistatic interaction analysis in a population of 209 RILs derived from a cross between two spp. *sesquipedialis* cultivars (one landrace and one popular commercial cultivar).

They also evaluated phenological traits of (1) days to flowering and (2) timing of leaf senescence. Two major QTLs (explaining up to 31.9% and 16% on average, respectively) and a potential minor QTL (explaining around 10% of phenotypic variance) were associated with days to flowering.

Two major QTLs (one explaining up to 22.1% phenotypic variation) were detected for nodes to first flower, and another two minor QTLs explained up to 11.7 and 13.2%. Two major QTLs (explaining up to 20.1 and 16.1% phenotypic variation, respectively) and another potential minor QTL explained 11.3% phenotypic variation.

5.5.2 Genotyping-by-Sequencing and Related Methods

GBS is an older technology (Elshire et al., 2011) that has been used by lower-budget projects in the Southern states of the USA (such as Arkansas, Tennessee, and Texas) and also in activities by national programs abroad. GBS has also been used by international projects in Africa, such as those based at the International Institute for Tropical Agriculture (IITA), or others in West African National Agricultural Research Institutes (NARIs) with financial support from USAID and DFID, which are the US and UK foreign aid agencies, respectively.

In these circumstances, where logistical concerns of DNA quality or amount were present, it was useful to employ less-expensive techniques, such as GBS. The method can be outsourced to other universities or to Diversity Arrays Technologies (DArT), a genotyping service company in Australia. DArT has run GBS with increasing resolution in recent cases (Ongom et al., 2022) compared to initial solid-state method (Jaccoud et al., 2001). Most of the biological parts of these studies have been related to basic plant characteristics, such as seed traits, as well as disease or insect resistance in grain-type dry cowpeas of the spp. *unguiculata* (Table 5.2).

One of the first research groups to use GBS technology in cowpeas was at the University of Arkansas (UofA)–Fayetteville, as shown by Shi et al. (2016), who evaluated cowpea bacterial blight (CoBB) resistance in the USDA core collection for the species consisting of 249 accessions from 42 countries. Their technique based on a single digestion with the enzyme *Ape*KI detected a total of 1,031 SNP loci that were not referenced to the genome, as the chromosomes of cowpea had not yet been sequenced. A total of eight loci were associated with CoBB resistance in their work.

In a follow-up study, two SNPs were found associated with aphid resistance, which were discovered using 338 of the core accessions (Qin et al., 2017). Finally, a total of six SNPs were found to be associated with cowpea mosaic virus (CPMV) resistance by the same UofA group and also using the USDA core collection, but in this case, a panel of 333 cowpea accessions from the USDA for which pre-existing genotypic data and CPMV phenotypic response data were available (Bhattarai et al., 2017).

In another study with a focus on biotic stress, Kpoviessi et al. (2022) used DArT technology combined with sequencing of fragments to identify 3,082 SNPs markers in 107 cowpea accessions, varying for bruchid resistance. The study found a total of 21 quantitative trait nucleotides (QTNs), including two each for percentage weight loss of the grain due to infestation/consumption and median development period of the bruchids; three for number of eggs laid on the grain; and five for percentage of bruchids emerged. These might represent antibiosis and anti-xenosis characteristics inherent in the cowpea germplasm, and the authors postulated 11 candidate genes for 14 of the QTNs found. Bruchids are the most serious source of post-harvest loss in cowpeas, and it is important to breed resistant materials, especially in sub-Saharan Africa, where storage conditions are not ideal (Mahesha et al., 2022).

Among the first to evaluate abiotic stress tolerance genes by GWAS was a study by Ravelombola et al. (2017) on low-phosphorous stress in a collection of 357 USDA accessions. The authors

TABLE 5.2
Examples of Allele Mining/Genotyping with Sequence-Based Technologies of Various Sorts (GBS/DArT/ddRADseq Technology) Used in Grain or Vegetable Cowpea-Type Genotyping Panels with the Number of Accessions, Traits Evaluated, and Restriction Enzymes Used, Along with the Number of Resulting Single-Nucleotide Polymorphisms (SNPs) Identified

Title	Authors	Journal	Number of Accessions and Origins	Restriction Enzyme(s) Used, Number of SNPs	Traits Evaluated
Association Analysis of cowpea bacterial blight resistance in USDA cowpea germplasm	Shi et al. (2016)	*Euphytica* 208:143–155	249 USDA accessions from 42 countries	Genotyping-by-sequencing/ *Ape*KI 1,031 SNP loci	Resistance to cowpea bacterial blight (CoBB); genetic diversity
Population structure analysis and association mapping of seed antioxidant content in USDA cowpea (*Vigna unguiculata* L. Walp.) core collection using SNPs	Qin et al. (2016)	*Canadian Journal of Plant Sciences* 96:1026–1036	369 USDA core accessions	Genotyping-by-sequencing 1,006	Seed coat color and antioxidant activity (AOA) using the free radical, 2,2-diphenyl-1-picrylhydrazyl (DPPH) method
Association mapping of aphid resistance in USDA cowpea (*Vigna unguic*ulata L. Walp.) core collection using SNPs	Qin et al. (2017)	*Euphytica* 213:36	338 USDA core accessions	Genotyping-by-sequencing 1,047	Aphid resistance mapped based on resistance and susceptible with intermediate not used in the analysis
Association analysis of cowpea mosaic virus (CPMV) resistance	Bhattarai et al. (2017)	*Euphytica* 213:230	333 USDA core accessions	Genotyping-by-sequencing 1,033	Cowpea mosaic virus (CPMV) resistance
Association mapping revealed SNP markers for adaptation to low phosphorus conditions and rock phosphate response in USDA cowpea (*Vigna unguiculata* (L.) Walp.) germplasm	Ravelombola et al. (2017)	*Euphytica* 213:183	357 USDA accessions	Genotyping-by-sequencing 1,018	Tolerance to low-phosphorus conditions and response to application of rock phosphorus
Association analysis of salt tolerance in cowpea (*Vigna unguiculata* (L.) Walp) at germination and seedling stages	Ravelombola et al. (2018)	*Theoretical and Applied Genetics* 131:79–91	155 USDA accessions	Genotyping-by-sequencing 1,049	Salt tolerance at germination and seedling stages using salt tolerance index
Quantitative trait loci influencing days to flowering and plant height in cowpea, *Vigna unguiculata* (L.) Walp	Angira et al. (2020)	*Molecular Genetics and Genomics* (2020) 295: 1187–1195	175 $F_{2:8}$ RILs	Double-digested restriction site–associated DNA sequencing/*Bam*HI and *Mlu*CI 4,154 SNP loci	Population: GoldenEye cream and IT98K-476–8 Days to flowering and plant height

Genetic dissection of yield associated traits in a cross between cowpea and yard-long bean (*Vigna unguiculata*) based on DArT markers	Garcia-Oliveira et al. (2020)	*Journal of Genetics* 99:57	204 $F_{2:3}$ progenies	Diversity arrays technology/*Bam*HI 1,627 SNP loci	Yield-associated traits: pod length, pedunce length and number; pods per plant; number of seeds per pod; 100-seed dry weight, seed length, seed width, seed height
Genetic dissection of iron deficiency chlorosis by QTL analysis in cowpea	Angira et al. (2022)	*Euphytica* 218:38	172 $F_{2:8}$ RILs	Double-digested restriction site–associated DNA sequencing/*Bam*HI and *Mlu*CI 4,154 SNP loci	Population: GoldenEye cream and IT98K-476–8 Iron deficiency chlorosis (IDC)
Construction of a single-nucleotide polymorphism linkage map and identification of quantitative trait loci controlling heat tolerance in cowpea, *Vigna unguiculata*.	Angira et al. (2022)	*Molecular Genetics and Genomics* (online)	175 $F_{2:8}$ RILs	Double-digested restriction site–associated DNA sequencing/*Bam*HI and *Mlu*CI 4,154 SNP loci	Population: GoldenEye cream and IT98K-476–8 Heat tolerance (flower abortion and other trait changes)
Application of multi-locus GWAS for detection of bruchid resistance loci in cowpea (*Vigna unguiculata*)	Kpoviessi et al. (2022)	*Plant Breeding* 141:439–450	107 cowpea accessions	Diversity Arrays Technology sequencing (DArT-Seq) markers (subset of 3,082 SNPs)	Bruchid resistance–related: (1) number of eggs lain, (2) mean number of holes, (3) number of emerged bruchids, (5) percent weight loss

discovered ten SNPs associated with tolerance to low-phosphorus conditions and eight SNPs associated with response to application of rock phosphorus. In a similar GWAS study, Ravelombola et al. (2018) used 155 cowpea accessions to find a total of ten SNPs, three during germination and seven at seedling stage, associated with salt tolerance.

Two recent studies used GBS technology to evaluate abiotic stresses in cowpea at Texas A&M University. A biparental mapping population was used rather than an association panel. This time the genetic analysis was based on a cross of GoldenEye cream and IT98K-476–8, which was used to create a population of over 170 $F_{2:8}$ recombinant inbred lines (RILs), a common breeding and genetics population type used for inbreeding legume crops that is developed by single seed descent (SSD).

The studies with this population evaluated different stresses, the first being heat tolerance, finding seven significant QTLs (Angira et al., 2022b), and the second being amount of iron deficiency chlorosis (IDC), finding eight significant QTLs (Angira et al., 2022a). The QTLs were found across a range of the chromosomes using a LOD threshold of 3.0.

The data of 4,154 SNPs for the genetic map construction used in these QTL studies were generated by double digestion of genomic DNA with the restriction enzymes *Bam*HI and *Mlu*CI, followed by low-pass sequencing of adaptor-ligated amplicons. The overall genetic resources described earlier (full set of 175RILs, SNP dataset of 4,154 loci) had been used earlier in an initial study to discover one major and one minor QTL for days to flowering, explaining up to 29.4% and two QTLs for plant height (Angira et al., 2020).

In a study by the UofA group, GBS was used to discover SNPs associated with seed quality traits in cowpeas. For example, a study published in the same year used a total of 369 accessions from the USDA core collection to discover two SNPs that were strongly associated with antioxidant activities. Two additional SNPs were associated with seed coat colors across three models for association, while other loci were associated with one or both traits, depending on the model used (Qin et al., 2016).

In a study from IITA, Garcia-Oliveira et al. (2020) used *Bam*HI digestion and sequencing by DArT to detect 1,627 polymorphic loci in 204 $F_{2:3}$ progenies and evaluate morphological characteristics and yield-associated traits (pod length, peduncle length, number of peduncles, pods per plant, number of seeds per pod, 100-seed dry weight, seed length, width, and thickness) in a cross between cowpea and yard-long bean. A total of 30 QTLs were found, 9 for pod and seed traits across ten chromosomes, 7 for pod length, and 1 each for peduncles per plant, seeds per pod, seed length, seed width, and seed height.

5.5.3 iSelect Arrays for Genotyping

Infinium SNP chips, a type of bead array for polymorphism detection made by Illumina Inc. (https://www.illumina.com/science/technology/microarray.html), have been used in multiple allele mining studies for dry grain cowpeas. An overall consortium-based iSelect array was made for cowpeas by Muñoz et al. (2017). It has 51,128 SNPs, but within any given population, the proportion of polymorphic loci can be lower.

Furthermore, the high cost of the Infinium assay both in terms of equipment where an iScan is needed and in terms of consumables where the iSelect beads are needed limited this type of work to high-budget projects, such as those sponsored by the Gates Foundation or National Science Foundation in the United States. Unlike the equipment needed for GBS, such as any brand sequencer, the Infinium is unique to the Illumina brand bead reader.

In a pioneering study, Lo et al. (2018) found QTLs controlling domestication-related traits in cowpea-evaluated 215 recombinant inbred lines derived from a wild × cultivated cross, and only 17,739 SNPs were useful (Table 5.3). However, this was enough to detect QTLs for nine domestication-related traits: (1) pod shattering, (2) peduncle length, (3) flower color, (4) days to flowering, (5) 100-seed weight, (6) pod length, (7) leaf length, (8) leaf width, (9) seed number per pod.

TABLE 5.3
GWAS Studies Using Infinium Technology from Illumina in Cowpea Evaluations

Title	Authors	Journal (Year)	Study Type/ Population(s) Used	Marker Type and # of Markers	Traits Evaluated
Genomic regions, cellular components and gene regulatory basis underlying pod length variations in cowpea	Xu et al. (2017)	*Plant Biotech Journal* 15:547–557	GWAS using 119 RILs and 299 diverse accessions	7,988 SNPs and 30,211 SNPs from the cowpea iSelect array	Pod length in yard-long bean × grain cowpea
A multi-parent advanced generation intercross (MAGIC) population for genetic analysis and improvement of cowpea	Huynh et al. (2018)	*The Plant Journal* 93:1129–1142	8-parent MAGIC population; 92 $F_{8:9}$ RILs	36,346 polymorphic SNPs (from the iSelect BeadArray), of which 11,848 were parent-unique	Drought-related traits: photoperiod sensitivity (under short- and long-day conditions); days to flowering; change in plant growth habit; days to maturity (95% dry pods); full yield; 100-seed rate
Identification of QTL controlling domestication-related traits in cowpea (*Vigna unguiculata* L. Walp)	Lo et al. (2018)	*Scientific Reports* 8:6261	215 recombinant inbred lines derived from a cross between cultivated and wild cowpeas	17,739 polymorphic SNPs from cowpea iSelect array (51,128 loci)	9 domestication-related traits: (1) pod shattering, (2) peduncle length, (3) flower color, (4) days to flowering, (5) 100-seed weight, (6) pod length, (7) leaf length, (8) leaf width, (9) seed number per pod
QTL mapping and transcriptome analysis of cowpea reveals candidate genes for root-rot nematode resistance	Pereira Santos et al. (2018)	*PLOS One* 13:e0189185	84 F_{10} recombinant inbred lines	14,202 polymorphic SNPs from the Illumina iSelect array	Average number of egg masses per RIL
QTL Mapping of Ineffective Nodulation and Nitrogen Utilization-related Traits in the IC-1 Mutant of Cowpea	Ohlson et al. (2018)	*Crop Science* 58:264	Two F_2 populations	6083 polymorphic SNPs from the iSelect consortium SNP array	Nitrogen use efficiency traits: (1) nodule color, (2) leaf color, (3) nitrogen concentration, (4) aboveground plant dry weight, (5) nodule number
Seed Coat Pattern QTL and Development in Cowpea (*Vigna unguiculata* L. Walp)	Herniter et al. (2019)	*Frontiers in Plant Science* 10:1346	8-parent MAGIC population, 4 RIL populations, 5 F_2 populations	SNPs from the Illumina iSelect consortium array	Seed coat patterns: (1) no color, (2) eye 1, (3) eye 2, (4) Holstein, (5) Watson, (6) full coat

(Continued)

TABLE 5.3 ***(Continued)***
GWAS Studies Using Infinium Technology from Illumina in Cowpea Evaluations

Title	Authors	Journal (Year)	Study Type/ Population(s) Used	Marker Type and # of Markers	Traits Evaluated
A genome-wide association and meta-analysis reveal regions associated with seed size in cowpea [*Vigna unguiculata* (L.) Walp]	Lo et al. (2019)	*Theoretical and Applied Genetics* 132:3079–3087	368 accessions from 51 countries	42,711 polymorphic SNPs from the iSelect consortium array	Seed length, width, and weight as component traits having to do with seed size
Loci discovery, network-guided approach, and genomic prediction for drought tolerance index in a multi-parent advanced generation intercross (MAGIC) cowpea population	Ravelombola et al. (2021)	*Horticulture Research* 8:24	MAGIC population (305 $F_{8:10}$) lines	32,058 high-quality SNPs of the Illumina cowpea consortium array)	Drought tolerance–related traits: (1) growth habit change, (2) tolerance indices for maturity, (3) flowering time, (4) 100-seed weight, (5) and grain yield
Genome-Wide Association Study Reveals Candidate Genes for Flowering Time in Cowpea (*Vigna unguiculata* [L.] Walp.)	Paudel et al. (2021)	*Frontiers in Genetics* 12:667038	Genome-wide association using 368 and 292 accessions	41,902 SNPS from the cowpea iSelect consortium array	Days to flowering at two locations (Colorado and Florida)

The findings showed clusters of QTLs associated with a domestication syndrome, especially with seed size. In a study of seed coat color patterns, the same group used the same array to find three strong QTLs all corresponding to candidate genes involved in flavonoid biosynthesis on three separate chromosomes (Herniter et al., 2019). The seed coat patterns studied were absence of color, two eye types, full coat color, versus Holstein and Watson patterning.

In more specialized assays (non-field ones), Pereira Santos et al. (2018) used 84 F_{10} recombinant inbred lines with the same iSelect array used fort the GWAS evaluations described earlier to find nematode resistance in cowpeas. In their case, 14,202 SNPs were polymorphic, and resistance affected an average number of egg masses, with one major gene hypothesized and three candidate genes confirmed by RNA studies of gene expression.

During studies with natural populations rather than biparental populations, the iSelect array was better at uncovering large numbers of polymorphic SNPs closer to the full potential of the 52K chip. As an example, a natural population or panel of landraces, advanced lines, and varieties for a total of 368 accessions from 51 countries was used in a series of genome-wide association studies (GWAS).

Lo et al. (2019) used this population along with 42,711 SNPs detected from the iSelect Consortium Array (83.5%) in a GWAS analysis of regions associated with seed size in cowpea. In that study, 17 loci were identified, with a hot spot identified on chromosome Vu03 associated with seed weight, width, and length, suggesting either pleiotropy or physical linkage of genes controlling seed size.

Paudel et al. (2021) used the same GWAS panel at two locations (Colorado and Florida, with the 368 and subset of 292 accessions grown, respectively) to evaluate candidate loci and regions for flowering time in cowpea. The common number of SNPs was 41,902, and 30 marker–trait associations corresponding to 20 SNPs explained 8–12% of phenotypic variation for days to flowering. These SNPs were distributed over seven chromosomes. A total of seven markers were considered reliable, as they were discovered by both GWAS studies across the two very different locations.

Drought was also the subject of another allele mining exercise by Wu et al. (2021a), who analyzed stomata-related drought responses using high-throughput plant physiological methods of phenotyping of a mini-core first used by Xu et al. (2017) and discussed later. In this case, a natural population of 106 cowpea accessions constituted the mini-core and were analyzed with 434 KASP markers derived from the list of Infinium SNPs.

QTLs (6 and 14, respectively) were found for two stomata-related drought traits: (1) critical soil water content (θcri) and (2) slope of transpiration rate decline (KTr). After retaining one representative locus per linkage disequilibrium block, no overlap in significant loci between QTLs for θcri and KTr was found, suggesting these two traits are independently controlled by different genes. Recently, a set of 50 KASP markers matching the most polymorphic Infinium SNPs was determined to be ideal in wider-ranging germplasm, with emphasis on yard-long beans (Wu et al., 2021b), by the same group in the Institute of Vegetables, Zhejiang Academy of Agricultural Sciences, Hangzhou, China.

As a legume, nitrogen fixation–related traits are of interest in both agronomic and vegetable types of cowpea. The first iSelect-based GWAS study to analyze nitrogen fixation was by Ohlson et al. (2018), who used two F_2 populations and QTL mapping to find SNPs associated with ineffective nodulation and nitrogen utilization traits.

From 6,083 polymorphic SNPs uncovered by the iSelect array, a single QTL controlling almost 50% of the phenotypic variance and based on the IC-1 mutant controlled leaf color, nodule color, nitrogen concentration, aboveground plant dry weight, and nodule number, with another minor QTL responsible for a smaller percent variance for nodule fresh weight. Overall, the study showed pleiotropy for multiple traits associated with nodulation and nitrogen utilization.

One aspect of cowpea germplasm studied with both the iSelect array and with DArT marker systems were flower and seed color, important traits for the plant breeder and cowpea consumer, respectively. Ongom et al. (2022) used GWAS in elite breeding lines and germplasm accessions from the Genetic Resources Center of IITA along with multi- and biparental RILs (376 genotypes in total) plus a subset of SNPs from the cowpea mid-density genotyping panel V1.0 (2,602 SNPs selected from the cowpea iSelect consortium array) to find 26 QTLs of interest with flower and/or seed color. Some of these SNPs were positioned in a cluster in a common genomic region on Vu07, where genes for basic helix-loop-helix DNA-binding superfamily proteins were found that could be involved in pigment regulation.

The first multi-parent advanced generation intercross (MAGIC) population created for cowpea was made by Huynh et al. (2018). It consisted of an eight-parent, 92-RIL population in the $F_{8:9}$ generation made by crossing the F_1s of two double crosses, each with four parents. The results of crossing so many lines made for a highly polymorphic population: in total, 36,346 SNPs from the iSelect BeadArray were readable, of which 11,848 were unique to one parent or the others. Drought-influenced traits were measured on the RILs, including days to flowering, change in plant growth habit, days to maturity (95% dry pods), full yield, and 100-seed rate. In addition, photoperiod sensitivity (under short- and long-day conditions) was evaluated.

Ravelombola et al. (2021) reported on the same MAGIC population together with a 305 genotype GWAS panel to evaluate additional drought tolerance–related traits, such as growth habit change, tolerance indices, maturity time, flowering time, 100-seed weight, and grain yield. The population consisted of $F_{8:10}$ lines, and the study was based on 32,058 high-quality SNPs and also used genomic prediction and a gene network approach to suggest a drought tolerance index. A total of 4 SNPs were associated in this index for plant growth habit change, 18 for maturity, and 35 for grain yield. No SNP was significant for either flowering time or 100-seed weight. Tolerance index for grain yield and a change in growth habit had common significantly associated loci according to network analysis.

5.6 PERSPECTIVES BEYOND DRY COWPEAS

Cowpea can be used not only as a dry grain crop, as considered in the preceding sections, but also as a fodder for animal feed and, in the case of yard-long bean, as a vegetable. An intermediate type is the "shelly" pea, a common form of eating cowpeas in the southeastern USA, where they are prized at a semi-mature stage shelled either by hand or with a shelling machine, first patented in Georgia in the nineteenth century. The direct consumption of immature seed or pods in the cases of the purple-hulled and yard-long bean varieties, respectively, usually requires direct-to-consumer fresh sales or processing into canned products.

Alternatively, the green tissues of easy-to-cook seeds and tender pods can be flash-frozen and packaged for the freezer aisles of supermarkets. Yard-long bean is also known as asparagus bean and is grouped as a separate subspecies from grain types. In this case, ssp. *sesquipedalis* is unique and famous for its long pods (up to 100 cm) and climbing growth habit. The pods are narrow and light green.

Yard-long bean is a popular vegetable in much of East, Southeast, and South Asia, with growing importance in Africa and the Americas among Asian immigrants. An advantage of the vegetable crop compared to the grain crop is that the long pods are harvested on a continuous basis throughout the season. They are immature pods and can be cooked in stir fry or boiling water in a short amount of time.

Pods can also be pickled or consumed fresh and serve as an important source of vitamins and minerals, along with digestible fiber, dietary proteins, health-promoting secondary metabolites, as well as special amino acids and starches. Overall, yard-long bean is a high-yield legume which can be produced year-round, and the green tender pods can be harvested up to four times after 50 planting days.

As yard-long bean and purple-hulled peas are of the same species, *Vigna unguiculata* L. Walp, they both can share genomic studies and allele mining. An example of this is the work of Xu et al. (2017), who used the iSelect Infinium SNP assay to evaluate 119 RILs from a cross of yard-long bean and grain cowpea compared to 299 diverse accessions from China to discover genomic regions, cellular components, and gene regulatory bases underlying pod length variations. Although only one-sixth of the SNPs (7,988 loci) were polymorphic in the cross, a total of 30,211 SNPs were useful to confirm QTLs found for pod length for the 299 accessions.

Garcia-Oliveira et al. (2020) experimented with a cross between the spp. *sesquipedelis* and spp. *unguiculata* but used DArT technology and a population of 204 $F_{2:3}$ progenies in early generation for QTL detection. They found below 500 markers but still had enough loci to evaluate yield-associated traits for the vegetable types, such as (1) pod length, (2) peduncle length, (3) number of peduncles, (4) pods per plant, (5) number of seeds per pod, (6) 100-seed dry weight, (7) seed length, and (8) seed width. In total, 30 putative QTLs were found to be associated with nine pod and seed traits across ten chromosomes; only seven QTLs (two for pod length and one each for peduncles per plant, seeds per pod, seed length, width, and thickness) were QTLs responsible for more than 10% phenotypic variance.

Xu et al. (2017) analyzed genomic regions, cellular components, and gene regulatory basis underlying pod length variations in a 299 accession GWAS panel and with a yard-long bean × grain cowpea cross with 119 RILs (with 30,211 and 7,988 SNPs polymorphic from the iSelect array). A total of 55 SNPs with known map positions were associated with pod length. Many of these SNPs overlapped with domestication-related trait QTLs; transcriptome analysis implicated genes involved in sugar signaling, cell division, and hormonal and nutritional signaling.

A mystery exists as to where the yard-long bean came from. It was presumed to have been derived from African germplasm of spp. *unguiculata* perhaps from East Africa. Then the *sesquipedalis* group was believed to have been domesticated in South or Southeast Asia for its long pods and vegetable use (Steele and Mehra, 1980). The genetic bases of subspecies *unguiculata* and subspecies *sesquipedalis* are very similar (Xu et al., 2010), but Wu et al. (2022) were able to distinguish the two groups based on GBS markers, as were Fatokun et al. (2018).

The yard-long bean is considered to share an unknown progenitor with cowpea, as this specific type of cowpea is never found in Africa but diversely exists in Asia (Xu et al., 2010). However, the extent of genome conservation or diversification between the two subspecies is not well understood (Gillaspie et al., 2005).

Returning to the use of GBS for genotyping, Wu et al. (2022) predicted that sufficient diversity existed between most accessions of yard-long bean and dry grain cowpeas to make crosses of interest for genetic analysis, although segregants would vary in growth habit. The distinct morphological traits in the two subspecies reflect the intense breeding pressure between adaptation and selection in Asia.

In Africa, cowpea grows in hot and droughty environments, where abiotic stress is the main issue (Veeranagappa et al., 2022), compared to the humid, rainy season in Asia, where biotics stresses are primary concerns (Mahesha et al., 2022). As the dry grain and fodder are the main objectives of cowpea production in Africa, high yield, nutritional seeds, and fast maturity are preferred traits in dry grain types (Singh et al., 2003).

Meanwhile, for yard-long beans in Asia, it is the tenderness of pods, their high production, and their pod length that are favorable characteristics for vegetable use (Wu et al., 2022), while the spp. *unguiculata* is mainly grown as a grain crop; the seed has high lysine and tryptophan amino acid concentrations even in immature stages so can be harvested earlier than at full maturity.

This characteristic leads to cowpea being used in all growth stages. Indeed, the tender green leaves are prepared as porridge like spinach in Africa (Bressani, 1985). The green shelly beans can be steam-cooked or boiled and then mixed with rice or corn. The matured seeds can be processed into flour, protein concentrates and isolates, starch, and extruded products. The only disadvantage is that cowpea seeds are deficient in cysteine and methionine

5.7 USING THE FULL DIVERSITY OF COWPEA GENOTYPES

Dry cowpeas and yard-long beans, respectively, as pulse and vegetable subspecies divisions of *V. unguiculata*, along with a group of wild relatives in related *Vigna* species, represent a diverse gene pool with moderate amounts of population structure and easy crossability, making for a primary gene pool that is ample and complex at the same time. The wild relatives have been less well studied than the agronomic and horticultural subspecies. These latter types are represented in gene banks to a greater extent since they are collectable from farmers and markets. The wild relatives must be collected from natural places or places where mixtures of cultivars and wild seed co-exist in the centers of origin for *V. unguiculata* and related species of *Vigna* in sub-Saharan Africa.

Among wild cowpeas as a whole, there are two recognized groups of subspecies (Pasquet and Padulosi, 2014; Pasquet et al., 2021). These are *dekindtiana* (savanna ecotypes) and *menensis* (forest ecotypes). Within the *dekindtiana* group, there are those close to the agronomically and horticulturally important subspecies, *unguiculata* and *sesquipedelis*, compared to others that are more distant. The wild ecotypes *of dekindtiana* include subspecies (1) *alba*, (2) *dekindtiana*, (3) *pubescens*, (4) *stenophylla*, and (5) *tenuis*. These are distinguished on a morphological basis more readily than they can be separated even by molecular diversity. Wu et al. (2022) found a greater differentiation of the two domesticated subspecies compared to *V. u.* spp. *cylindrica* and *unguiculata*, which were not clearly distinct with 11,083 GBS markers evaluated on USDA accessions despite having very different seed size, shape, and colorations. The *menensis* group appears to be more primitive than those of the *dekindtiana* group. The *menensis* group includes the three subspecies (1) *baoulensis*, (2) *letouzeyi*, and (3) *pawekiae*. Overall, the group shows high molecular diversity but little morphological variation. All these wild relatives are an untapped resource for improvement of the cultivated types of cowpeas and yard-long beans.

Various markers have been used to identify the relationship between wild and cultivated germplasm. With the multi-locus systems of interest to this chapter being GBS and SNP arrays, we concentrate on results of diversity assessment with these technologies. Few studies have used these methods for wild accessions. For example, Zuluaga et al. (2021) used *Ape*KI-based GBS to evaluate allogamous wild perennials, perennial outbred/inbreds, and annual inbreds (subsp. *unguiculata*) with controls of Italian landraces, separated by seed color. Nkhoma et al. (2020) used the same GBS technique and genome reduction enzyme to look at diversity in 100 genotypes sourced from different locations across sub-Saharan Africa. With 14,116 SNP loci, they found that the relatedness of the cowpeas based on phenotypic and genotypic results did not necessarily cluster based on geographic origin.

The importance of evaluating diversity in specific national germplasm collections cannot be over-emphasized, given their utility for local adaptation. Two examples are genotyping studies from countries in the African continent, as shown by Ketema et al. (2020) and Gbedevi et al. (2021), respectively, who evaluated cowpeas found in Ethiopia, East Africa (357 accessions), and Togo, West Africa (255 genotypes).

Both studies used DArT genotyping with 6,498 and 4,600 markers, respectively. Different levels of population structure were found to be ideal in each country, depending on the inclusion of improved varieties, which tended to cluster together, and clusters based on geographic origin. High variance existed within populations and individuals.

5.8 PERSPECTIVES AND CONCLUSIONS

Allele mining is well-established in cowpea but can benefit from new technologies, such as whole-genome sequencing, an alternative for high-density SNP maps even better than the Infinium assays. We have already shown the evolution from GoldenGate bead arrays to iSelect with displacement of the former technology for use in allele mining. GBS technologies are evolving and have potential to segue into low-pass "skim" sequencing without genomic reduction. That is beneficial for reduction in costs and evaluation of wider arrays of germplasm in more locations.

The future is also likely to hold evaluation of more populations, including new types of crosses and various germplasm sets, and more traits of interest. For example, aluminum toxicity tolerance is an abiotic stress that has not yet been evaluated with GWAS methods. The diversity of cowpea landraces is still difficult to compartmentalize; therefore, more national collections should be evaluated to help breeders make cowpeas with local adaptation incorporating the genes and QTLs discovered through broader studies of the genetic structure of cowpeas.

Cowpea is a crop that is widely grown in areas that are having issues with food scarcity, such as sub-Saharan Africa, Southern Asia, South America, and in the southeastern United States. This makes research on cowpea improvement a likely way to supply the increased demand for food products in these areas. This is the process we are also seeing in yard-long bean, which as a different subspecies from grain types is being analyzed separately by some groups in Southern China. The diversity of grain cowpeas in the southeastern United States is poorly understood, and GWAS in this germplasm could lead to discovery of crop improvement genes for advanced breeding lines for this region, where cowpeas once covered almost 4 million acres.

Apart from adaptation to the varied landscape these regions represent, GWAS studies in the southern USA could be relevant to adaptation to climate change, as combinations of heat waves with early, intermittent, and terminal drought are occurring at a high rate in the Northern Hemisphere. As climate change is the most pressing issue of our times, expansion of cowpea production in the Indo-Gangetic plain, the Deccan highlands of India, and parts of Southeast Asia is likely, as heat-tolerant legumes are needed in these highly populous regions (Singh, 2014).

Intensification in parts of Central and Western Africa are also needed as the border of the Sahelian savanna regions fluctuate with climate change (Hall et al., 1997). The few remaining agricultural frontiers in the tropics around the Amazon basin, the Congo basin, and areas of Southern Africa with open savanna could also benefit from heat- and drought-tolerant cowpeas adapted to other abiotic and biotic stresses developed by allele mining and intensified breeding efforts. Allele mining studies for local adaptation should continue in all these new environments and in new settings outside of academia. Industry participation in cowpea seed will be key too.

We have shown throughout this review that GWAS results can have a large impact on agricultural adaptation and crop improvement both in the high-tech studies of developed countries (USA and parts of Europe) as well as more and more in many parts of Africa and Asia, leaders, respectively, in the phenotyping and genotyping of cowpea and yard-long bean.

ACKNOWLEDGMENTS

Evans Allen funding (TEN-X07) to Tennessee State University (TSU) from the United States Department of Agriculture (USDA), for funding PhD studies of MCMII. This work was also supported by the Norman Borlaug PhD graduate fellowship to YD under the BHEARD program of the United States Agency for International Development. Further USDA funding via the AFRI 1401 Plant Breeding section is acknowledged (TENX-2020–03568) for establishing aluminum tolerance screening at TSU.

REFERENCES

Angira B, Zhang Y, Scheuring CF, Zhang Y, Masor L et al. (2020) Quantitative trait loci influencing days to flowering and plant height in cowpea, *Vigna unguiculata* (L.) Walp. *Mol Genet Genom* 295:1187–1195.

Angira B, Zhang Y, Scheuring CF, Zhang Y, Masor L et al. (2022a) Construction of a single nucleotide polymorphism linkage map and identification of quantitative trait loci controlling heat tolerance in cowpea, *Vigna unguiculata* (L.) Walp. *Mol Genet Genom* 297:1481–1493.

Angira B, Zhang Y, Zhang Y, Scheuring CF, Masor L et al. (2022b) Genetic dissection of iron deficiency chlorosis by QTL analysis in cowpea. *Euphytica* 218:38.

Bhattarai G, Shi A, Qin J, Weng Y, Morris JB et al. (2017) Association analysis of cowpea mosaic virus (CPMV) resitance in the USDA cowpea germplasm collection. *Euphytica* 213:230.

Bressani R (1985) Nutritive value of cowpea. In: Singh SR, Rachie RO (eds) *Cowpea Research, Production and Utilization*. Cambridge University Press, pp. 353–359.

Elshire RJ, Glaubitz JC, Sun Q, Poland JA, Kawamoto K et al. (2011) A robust, simple genotyping-by-sequencing (GBS) approach for high diversity species. *PLoS One* 6(5):e19379.

Fatokun C, Girma G, Abberton M, Gedil M, Unachukwu N et al. (2018) Genetic diversity and population structure of a mini-core subset from the world cowpea (*Vigna unguiculata* (L.) Walp.) germplasm collection. *Sci Rep* 8(5):e19379.

Garcia-Oliveira AL, Zate ZZ, Olasanmi B, Boukar O, Gedil M et al. (2020) Genetic dissection of yield associated traits in a cross between cowpea and yard-long bean (*Vigna unguiculata* (L.) Walp.) based on DArT markers. *J Genet* 99:57.

Gbedevi KM, Boukar O, Ishikawa H, Abe A, Ongom PO et al. (2021) Genetic diversity and population structure of cowpea [*Vigna unguiculata* (L.) Walp.] germplasm collected from Togo based on DArT markers. *MDPI Genes* 12(9):1451.

Gillaspie A, Hopkins M, Dean R. (2005) Determining genetic diversity between lines of *Vigna unguiculata* subspecies by AFLP and SSR markers. *Genet Resour Crop Evol* 52(3):245–247.

Hall AE, Singh BB, Ehlers JD (1997) Cowpea breeding. In: Janick J (ed) *Plant Breeding Reviews 15*. John Wiley & Sons, pp. 215–274.

Herniter IA, Lo R, Muñoz-Amatriaín M, Lo S, Guo Y-N et al. (2019) Seed coat pattern QTL and development in cowpea (*Vigna unguiculata* L. Walp). *Front Plant Sci* 10:1346.

Huynh B-L, Ehlers JD, Huang BE, Muñoz-Amatriaín M, Lonardi S et al. (2018) A multi-parent advanced generation inter-cross (MAGIC) population for genetic analysis and improvement of cowpea. *Plant J* 93(6):1129–1142.

Jaccoud D, Peng K, Feinstein D, Kilian A (2001) Diversity arrays: A solid state technology for sequence information independent genotyping. *Nucleic Acids Res* 29(4):e25.

Ketema S, Tesfaye B, Keneni G, Fenta BA, Assefa E et al. (2020) DArTSeq SNP-based markers revealed high genetic diversity and structured population in Ethiopian cowpea [*Vigna unguiculata* (L.) Walp] germplasms. *PLoS One* 15(10):e0239122.

Kpoviessi AD, Agbahoungba S, Agoyi EE, Chougourou DC, Adoukonou-Sagbadja H (2022) Application of multi-locus GWAS for detection of bruchid resistance loci in cowpea (*Vigna unguiculata*). *Plant Breed* 141:439–450.

Lo S, Muñoz-Amatriaín M, Boukar O, Herniter I, Cisse N et al. (2018) Identification of QTL controlling domestication-related traits in cowpea (*Vigna unguiculata* L. Walp). *Sci Rep* 8:6261.

Lo S, Muñoz-Amatriaín M, Hokin SA, Cisse N, Roberts PA et al. (2019) A genome-wide association and meta-analysis reveal regions associated with seed size in cowpea [*Vigna unguiculata* (L.) Walp]. *Theor Appl Genet* 132:3079–3087.

Lucas MR, Diop N-N, Wanamaker S, Ehlers JD, Roberts PA et al. (2011) Cowpea-soybean synteny clarified through an improved genetic map. *Plant Genome* 4(3):218–225.

Lucas MR, Huynh B-L, Vinholes PDS, Cisse N, Drabo I et al. (2013) Association studies and legume synteny reveal haplotypes determining seed size in *Vigna unguiculata*. *Front Plant Sci* 4:95.

Mahesha HS, Keerthi MC, Shivakumar KV, Bhargavi HA, Saini RP et al. (2022) Development of biotic stress resistant cowpea. In: Kole C (ed) *Genomic Designing for Biotic Stress Resistant Pulse Crops*. Springer, Cham, pp. 213–251.

Muchero W, Diop NN, Bhat PR, Fenton RD, Wanamaker S et al. (2009) A consensus genetic map of cowpea [*Vigna unguiculata* (L.) Walp.] and synteny based on EST-derived SNPs. *Proc Natl Acad Sci USA* 106(43):18159–18164.

Muchero W, Roberts PA, Diop NN, Drabo I, Cisse N et al. (2013) Genetic architecture of delayed senescence, biomass, and grain yield under drought stress in cowpea. *PLoS One* 8(7):e70041.

Muñoz N, Liu A, Kan L, Li M-W, Lam H-M (2017) Potential uses of wild germplasms of grain legumes for crop improvement. *Int J Mol Sci* 18:328–328.

Nkhoma N, Shimelis H, Laing MD, Shayanowako A, Mathew I (2020) Assessing the genetic diversity of cowpea [*Vigna unguiculata* (L.) Walp.] germplasm collection using phenotypic traits and SNP markers. *BMC Genet* 21:110.

Ohlson EW, Seido SL, Mohammed S, Santos CAF, Timko MP (2018) QTL mapping of ineffective nodulation and nitrogen utilization-related traits in the IC-2 mutant of cowpea. *Crop Sci* 58(1):264–272.

Ongom PO, Fatokun C, Togola A, Garcia-Oliveira AL, Ng EH et al. (2022) A medium density DArTag single nucleotide polymorphism panel for genetic dissections and deployment in cowpea improvement. *ESS Open Archive*. https://doi.org/10.1002/essoar.10512138.1.

Pasquet RS, Feleke Y, Gepts P (2021) Cowpea [*Vigna unguiculata* (L.) Walp.] maternal lineages, chloroplast captures, and wild cowpea evolution. *Genet Resour Crop Evol* 68:2799–2812.

Pasquet RS, Padulosi S (2014) Genus Vigna and cowpea (*Vigna unguiculata* [L.] Walp.) taxonomy: Current status and prospects. In: Boukar O, Coulibaly O, Fatokun CA, Lopez K, Tamo M (eds) *Innovative Research Along the Cowpea Value Chain*. IITA, Ibadan, pp. 66–87.

Paudel D, Dareus R, Rosenwald J, Muñoz-Amatriaín M, Rios EF (2021) Genome-wide association study reveals candidate genes for flowering time in cowpea (*Vigna unguiculata* [L.] Walp.). *Front Genet* 12:667038.

Pereira Santos JR, Ndeve AD, Huynh B-L, Matthews WC, Roberts PA (2018) QTL mapping and transcriptome analysis of cowpea reveals candidate genes for root-rot nematode resistance. *PLoS One* 13(1):e0189185.

Pottorff M, Ehlers JD, Fatokun C, Roberts PA, Close TJ (2012) Leaf morphology in cowpea [*Vigna unguiculata* (L.) Walp]: QTL analysis, physical mapping and identifying a candidate gene using synteny with model legume species. *BMC Genomics* 13:234.

Pottorff M, Roberts PA, Close TJ, Lonardi S, Wanamaker S et al. (2014) Identification of candidate genes and molecular markers for heat-induced brown discoloration of seed coats in cowpea [*Vigna unguiculata* (L.) Walp]. *BMC Genomics* 15:328.

Qin J, Shi A, Mou B, Bhattarai G, Yang W et al. (2017) Association mapping of aphid resistance in USDA cowpea (*Vigna unguiculata* L. Walp.) core collection using SNPs. *Euphytica* 213:36.

Qin J, Shi A, Xiong H, Mou B, Motes D et al. (2016) Population structure analysis and association mapping of seed antioxidant content in USDA cowpea (*Vigna unguiculata* L. Walp.) core collection using SNPs. *Can J Plant Sci* 96(6):1026–1036.

Ravelombola W, Qin J, Shi A, Lu W, Weng Y et al. (2017) Association mapping revealed SNP markers for adaptation to low phosphorus conditions and rock phosphate response in USDA cowpea (*Vigna unguiculata* (L.) Walp.) germplasm. *Euphytica* 213:183.

Ravelombola W, Shi A, Huynh B-L (2021) Loci discovery, network-guided approach, and genomic prediction for drought tolerance index in a multi-parent advanced generation intercross (MAGIC) cowpea population. *Hortic Res* 8:24.

Ravelombola W, Shi A, Weng Y, Mou B, Motes D et al. (2018) Association analysis of salt tolerance in cowpea (*Vigna unguiculata* (L.) Walp) at germination and seedling stages. *Theor Appl Genet* 131:79–91.

Shen R, Fan J-B, Campbell D, Chang W, Chen J et al. (2005) High-throughput SNP genotyping on universal bead arrays. *Mutat Res* 573(1–2):70–82.

Shi A, Buckley B, Mou B, Motes D, Morris JB et al. (2016) Association analysis of cowpea bacterial blight resistance in USDA cowpea germplasm. *Euphytica* 208:143–155.

Singh BB (2014) *Cowpea: The Food Legume of the 21st Century*. Crop Science Society of America, Madison, WI, USA.

Singh BB, Ajeigbe HA, Tarawali SA, Fernandez-Rivera S, Abubakar M (2003) Improving the production and utilization of cowpea as food and fodder. *Field Crops Res* 84(1–2):169–177.

Steele WM, Mehra K (1980) Structure, evolution, and adaptation to farming systems and environments in Vigna. In: Summerfield RJ, Bunting AH (eds) *Advances in Legume Science*. Royal Botanic Gardens, UK, pp. 393–404.

Veeranagappa P, Manu B, Prasad G, Blair MW, Hickok D et al. (2022) Advanced breeding strategies for abiotic stress tolerance in cowpea. In: Kole C (ed) *Genomic Designing for Abiotic Stress Resistant Pulse Crops*. Springer, Cham, pp. 115–144.

Wu X, Cortes AJ, Blair MW (2022) Genetic differentiation of grain, fodder and pod vegetable type cowpeas (*Vigna unguiculata* L.) identified through single nucleotide polymorphisms from genotyping by sequencing. *Mol Hort* 2:8.

Wu X, Sun T, Xu W, Sun Y, Wang B et al. (2021a) Unraveling the genetic architecture of two complex, stomata-related drought-responsive traits by high-throughput physiological phenotyping and gwas in cowpea (*Vigna unguiculata* L. Walp). *Front Genet* 12:743758.

Wu X, Wang B, Wu S, Li S, Zhang Y et al. (2021b) Development of a core set of single nucleotide polymorphism markers for genetic diversity analysis and cultivar fingerprinting in cowpea. *Legume Sci* 3(3):e93.

Xu P, Wu X, Muñoz-Amatriaín M, Wang B, Wu X et al. (2017) Genomic regions, cellular components and gene regulatory basis underlying pod length variations in cowpea. *Plant Biotechnol J* 15(5):547–557.

Xu P, Wu X, Wang B, Hu T, Lu Z et al. (2013) QTL mapping and epistatic interaction analysis in asparagus bean for several characterized and novel horticulturally important traits. *BMC Genet* 14:4.

Xu P, Wu X, Wang B, Liu Y, Ehlers JD et al. (2011) A SNP and SSR based genetic map of asparagus bean (*Vigna unguiculata* ssp. *sesquipedalis*) and comparison with the broader species. *PLoS One* 6(1):e15952.

Xu P, Wu X, Wang B, Liu Y, Qin D et al. (2010) Development and polymorphism of *Vigna unguiculata* ssp. *unguiculata* microsatellite markers used for phylogenetic analysis in asparagus bean (*Vigna unguiculata* ssp. *sesquipedalis* (L.) Verdc.). *Mol Breed* 25(4):675–684.

Yahaya D, Denwar N, Blair MW (2019a) Effects of moisture deficit on the yield of cowpea genotypes in the Guinea savannah of northern Ghana. *Agric Sci* 10(4):577–595.

Yahaya D, Denwar N, Mohammed M, Blair MW (2019b) Screening cowpea (*Vigna unguiculata* (L.) Walp.) genotypes for enhanced N_2 fixation and water use efficiency under field conditions in Ghana. *Am J Plant Sci* 10(4):640–658.

Zuluaga DL, Lioi L, Delvento C, Pavan S, Sonnante G (2021) Genotyping-by-sequencing in *Vigna unguiculata* landraces and its utility for assessing taxonomic relationships. *MDPI Plants* 10(3):509.

6 Allele Mining in Pea (*Pisum sativum* L.)

Current Status and Future Perspectives

A. K. Parihar, Arpita Das, Soumyayan Roy, Mousumi Murmu, Kiran Vishwakarma, and G. P. Dixit

6.1 INTRODUCTION

Pea (*Pisum sativum* L.) is a cool-season legume crop that is being grown all over the world to be consumed as food, feed, and fodder (Parihar et al., 2022a, 2022b, 2022c, 2023). The fresh green pods and seeds are used as vegetables, while the dry round seeds are utilized in different forms, particularly intact or split seeds and flour (Bastida-Garcia et al., 2011). In addition, pea is used to prepare various cuisine, in confectioneries, and in several by-products-based food industries (Parihar et al., 2020, 2022a, 2022b). Pea is a nutritionally enriched crop that offers 15–32% protein along with vitamins, minerals, and fiber. Pea intake positively affects the cardio health for being a rich source of potassium, folate, and dietary fibers, which also support the gut health and lower the risk of several malignancies (Chung et al., 2009; Tosh and Yada, 2010). Being a legume crop, it offers sufficient nitrogen, increasing the magnitude and activities of soil microorganisms, which subsequently enhance soil properties (Biederbeck et al., 2005; Clark, 2007). Pea is the fourth most important legume crop globally, with 9.80 mha acreage and 34 mt productions. The regional dry pea production scenario demonstrates that the USA and Europe together contribute more than 77% and Asia contributes about 16.9%. Quite the reverse, Asia is having more than 87% of green pea's production, whereas Europe and America account for 9.0% only. Canada and China are the leading producers of dry pea and green peas, respectively (FAOSTAT, 2020).

The available genetic resources, such as germplasm, landraces, mutants, and obsolete varieties, are constantly embraced by breeders to create variability and develop new crop varieties with desirable traits, such as high yields and nutritionally dense and climate-resilient cultivars to cater to the needs of farmers and other stakeholders (Premkumar et al., 2020). The sufficient genetic diversity in the relevant crop species is the basic requirement for further genetic improvement, since, without diversity, there is no chance for selection or the development of new varieties. Crop gene pools include a large amount of genetic variation that can be utilized to identify novel alleles of agronomic value and their further utilization to develop/design improved cultivars. Allele mining is an excellent approach to identify superior and/or novel alleles and their further validation for a specific characteristic. The natural mutation because of transitions, transversions, insertion, and deletion is the main driving force for the evolution of new alleles (Kumar et al., 2010). During the post-genomics era, allele mining in several crops can contribute to genomic exploration for desired genetic enhancement (Varshney et al., 2005). The understandings of genetic variability in general and allelic diversity at candidate genes of interest in particular are very important to judiciously exploit available diversity (Kaur, 2008). For the use of genetic resources in crop development, the availability of sequence and sequence variation that impact the plant phenotype is of

 DOI: 10.1201/9781003385059-6

the highest relevance (Graner, 2006). Single-nucleotide polymorphisms (SNPs) or insertion and deletion (InDels) mutations in the particular genomic region are crucial for producing new alleles or changing the current alleles and allelic combinations. The structure and function of the encoded proteins can be changed by mutations in the coding regions and/or regulatory regions, which can significantly impact the phenotype of crops (Ashkani et al., 2015). Generally, there are several approaches available for allele mining and/or identification of sequence polymorphisms for a given gene, including (1) targeting induced local lesions in genomes (TILLING); (2) ecotype-TILLING (EcoTILLING) or modified TILLING; (3) re-sequencing or sequencing-based allele mining; (4) haplotype-based allele mining; and (5) association mapping–based allele mining. TILLING uses artificial chemical treatments to identify single base pair allelic variation to identify polymorphisms (more specifically point mutations) in a target gene by performing heteroduplex analysis. EcoTILLING represents a means to determine the extent of natural variation in selected genes in crops. The method is essentially the same as TILLING, except that the mutations are not induced artificially and are detected from naturally occurring alleles in the primary and secondary crop gene pools (Till et al., 2006; Comai et al., 2004; Comai and Henikoff, 2006; Huang et al., 2009). Initially, both TILLING and EcoTILLING were proposed as cost-effective approaches for haplotyping and SNP discovery; however, these techniques require more sophistication and involve several steps. Later, a rapid and cost-effective method for detecting novel allelic variants of known candidate on agarose gels and its utility in candidate gene mapping has been described (Raghavan et al., 2007). The sequencing-based allele mining technique includes amplification of alleles in diverse genotypes using polymerase chain reaction (PCR) and subsequent identification of nucleotide variation by DNA sequencing. This technique is less sophisticated and more effective as well as efficient (Kumar et al., 2010), but the sequencing cost is the vital concern for this type of allele mining program. However, with the advancement of sequencing methods and techniques like massively parallel sequencer using supported oligonucleotide ligation and detection system (SOLiD), the sequencing cost has been reduced (Ashkani et al., 2015). High-throughput (HT) sequencing, also termed as next-generation sequencing (NGS), can generate DNA sequence data inexpensively and in a faster way as compared with traditional technologies. NGS-based approach has increased the sequence capacity and made the whole-genome re-sequencing cheaper (Gupta, 2008; Mardis, 2008). Now it has been possible to re-sequence the candidate genes for thousands of individuals, entire transcriptomes, or entire plant genomes more efficiently, in greater depth and more economically than ever before (Kilian and Graner, 2012). Identification of genetic variations has become more precise, and thousands of candidate genes can be traced from large gene bank collections (Varshney et al., 2009). Large-scale screening of germplasm collections using NGS with high accuracy is now possible by employing several modern sequencing platforms, that is, Illumina/GA, Roche/GS FLX, Applied Biosystems/SOLiD, and cPAL sequencing. Genotyping through sequencing is being used to obtain information on SNP and haplotype patterns at large scale. In this chapter, we have briefly covered the current status of the allele mining approaches in pea and its future prospective to develop climate-resilient varieties of pea.

6.2 SIGNIFICANCE OF PEA REFERENCE GENOME

The pea genome, with its considerable size of 4.45 Gb, poses challenges for sequencing and identifying genes of interest, mainly owing to the presence of an array of transposable elements (Praça-Fontes et al., 2014; Tayeh et al., 2015; Gali et al., 2018). Previous researchers proved that pea genome is constituted by around 75–97% of heterogeneous repetitive DNAs mostly from the Tyl-copia/gypsy family (Murray et al., 1981; Macas et al., 2015). Meager similarity was reflected concerning repetitive DNAs between pea and soybean, but close synteny was observed with model legume *Medicago truncatula* (Macas et al., 2015). Several schools of thought reported that the pea genome has undergone faster evolution compared to other studied Leguminosae genomes. This accelerated evolution is likely driven by processes such as transposon-mediated non-homologous recombination, leading to the acquisition or loss of genes.

In the later phase, the fully annotated genome sequencing of the inbred pea cultivar "Caméor" was accomplished by the international consortium led by Institut national de la recherche agronomique (INRA), France (Kreplak et al., 2019). Over the past decade, extensive genomic resources have been developed with cv. Caméor. With this fully annotated assembly, a new vista has opened for enriching genomics-assisted pea improvement program. The assembly of the pea nuclear genome resulted in the organization of its genetic material into seven pseudomolecules using Illumina sequencing platform, where short reads were first transformed into contigs and further converted to scaffolds. They published 3.92 Gb length of Caméor genome assembly with 3.16 Gb of contigs, which represented approximately 82.5% of the estimated pea genome. The size gap generated was due to the presence of highly repetitive DNAs and transposable elements which reflected as unassembled reads. *De novo* gene prediction was accomplished by deploying several bioinformatic tools, like AUGUSTUS, and potential gene functions were determined by identifying the best matches to the SwissProt and TrEMBL databases. Out of 44,756 genes detected, 30,687 genes were annotated in this pea genome program (Kreplak et al., 2019).

Over the last few decades, the advent of NGS technology has significantly advanced crop genomics research, providing valuable insights into genome architecture and complexity. Previous researchers have demonstrated that relying solely on a single reference genome for a particular crop is inadequate to encompass the genetic diversity within a species, explicitly in the context of crop domestication, where genetic structures undergo diversification and alterations during the course of domestication and evolution. As a solution to this limitation, pan-genomes constructed from a diverse set of cultivated species and crop wild relatives are becoming increasingly popular. These pan-genomes offer a powerful tool to capture and represent the full extent of genetic diversity present within the gene pool of a particular species (Liu et al., 2020). Additionally, the earlier available pea reference genome lacks the complete annotation of the genes with improved reference genome assembly to gain a deeper understanding of the phenotypic variation and genome evolution of this crop.

A high-quality fully annotated reference genome of pea has been documented, offering essential tools for population genomics and molecular genetics research (Yang et al., 2022). This valuable resource enhances our understanding of pea domestication, synteny, and expedites molecular breeding efforts. They reported a *de novo* genome assembly of the Chinese pea cultivar "ZW6," which was created by integrating an array of sequencing platform and workflow along with Illumina NGS technologies. It was observed that the assembled anchored contigs accounted for 97.96% of pea cv. "ZW6" genome, with a total size of 3.71 Gb, in comparison to 82.5% coverage of pea cv. "Caméor." The numbers of annotated genes in the pseudomolecules were enhanced than "Caméor" with 46,607 number of annotated genes, representing 1.19 Gb size in "ZW6" (Yang et al., 2022). So it was evidenced that the pea genome cv. "ZW6" has undergone a high-quality 243-fold increase in contig length, demonstrating significant enhancements in the continuity and quality of complex repetitive regions and mobile elements. These improvements effectively address the gaps that were present in the previous reference genome. Additionally, this recent pea genome project also highlighted the pea genomic polymorphism and duplicity detection, as well as genetic basis of important agronomic traits for better elucidation of evolutionary significance of the genes underlying the traits. Further, to gain a comprehensive understanding of pea genome diversity, a pan-genome analysis was conducted by individual *de novo* assembly consisting of the genomes of 118 cultivated and wild pea accessions, which will facilitate the enrichment of pea genomic resources for further exploration in the genomics-assisted pea improvement program (Yang et al., 2022).

6.3 ALLELE MINING AND EVOLUTION OF ALLELIC VARIATION IN THE PERSPECTIVE OF PEA

Alleles represent alternative forms of a gene and are situated at the same locus on homologous chromosomes, thus conferring alternate expression of a trait. Allele mining involves the exploration and discovery of new allelic variation for the targeted traits within the gene pool, thus specifically seeking out superior alleles from naturally variable populations, followed by introgression into the cultivated background.

In the natural population, mutation serves as a fundamental driving force in evolution, contributing to the diverse array of alleles present in any crop species. The creation of new alleles or variations in existing genes and allelic combinations relies on mutations occurring within the genic regions of the genome, either in the form of SNPs or as InDels. These mutations play a crucial role in remediating genetic diversity of a crop species. Mutations that occur in exon and/or regulatory regions can have a significant impact on the phenotype by modifying the structure and/or function of the encoded protein. Conversely, it was envisaged that mutations that take place in non-coding regions of a gene may often be silent, having no discernible effect on the phenotype. However, recent reports have highlighted that allelic variation occurring in non-coding regions, such as the 5' UTR (untranslated region), promoter, introns, and 3' UTR region, can also exert significant effects on transcript synthesis and accumulation. These alterations, in turn, have an impact on the expression of specific traits (Sullivan and Gray, 2000; Kumar et al., 2010; Moreau et al., 2022).

Pea has a wealth of morphological mutations, with hundreds of them well documented, and a considerable number have been precisely located on a genetic map. These mutations serve as valuable resources for investigating and understanding regulatory pathways related to essential aspects, such as seed development (*def*: development funiculus), floral morphology (*det*: determinate; *sm*: stem fasciation; *deh*: determinate habit), compound leaves (*af*: afila; *unitac*: tendrilled acacia), and nodulation for biological nitrogen fixation (Sym; Nod) (Makasheva et al., 1987; Goldmann et al., 1992; Kof et al., 2006; Ayeh et al., 2009; Tsyganov and Tsyganov, 2020; Lamichaney et al., 2021; Sinjushin et al., 2022). Thus, molecular identification of certain valuable mutants is very pertinent for their successful introgression and elucidation of genetics behind them. Several examples of mutation in pea are available, which can shed light on the impact of mutation and subsequent generation of allelic variation both in the coding and non-coding region with significant alteration in the phenotype. An interesting example of pea lip1 (light-independent photomorphogenesis1) mutant characterized by small deletions and point mutations within the introns subsequently led to variable expression of light-induced development instead of continuous exposure to darkness (Sullivan and Gray, 2000). In pea, two homologs of the terminal flower1 (*TFL1*) gene exist: *PsTFL1a*, which corresponds to the DETERMINATE (*DET*) gene, and *PsTFL1c*, which corresponds to the LATE FLOWERING (*LF*) gene. These genes play crucial roles in controlling flowering time and reproductive phenology in pea. To investigate the relationship between *PsTFL1a* and *DET*, *PsTFL1a* was sequenced from three different det mutants and their corresponding wild-type lines, namely, "Paloma." The sequencing analysis revealed that the *det-1* mutant retained the first intron, leading to the generation of a stop codon and resulting in the production of a truncated PsTFL1a protein (Foucher et al., 2003). Mutations occurring within intronic regions are characterized by mostly substitution, insertions of mobile genetic elements, or variations in the numbers of repeats in microsatellite sequences. The Is-1 mutant of pea, responsive to gibberellic acid (GA), demonstrated reduced biosynthesis of GA in a manner specific to both developmental stages and organs when compared to the wild type. This reduction was attributed to an intronic G-to-A substitution, which led to impaired RNA splicing in the mutant (Ait-Ali et al., 1997). Another example of pea with the rugosus mutant (R), found in the starch-branching enzyme I gene (*SBEI*), occurs naturally and has a remarkable impact on starch structure by determining the ratio of branched and unbranched starch in pea seed. This mutation leads to significantly increased amylose content, ranging from 50% to 70% amylose (Bhattacharyya et al., 1990). The molecular explanation for the r mutation has been elucidated, revealing an insertion event within the genic region that encodes the carboxy-terminal part of *SBEI* and generated the truncated version of the gene (Bhattacharyya et al., 1990). Later, the near-isogenic lines developed through backcrossing the series of r mutants known as seed-induced mutant (SIM) with the round seeded parental cultivar were subjected to detection of the position of mutations. It revealed an array of mutation events like missense mutations, point mutation, InDels, and cryptic splicing in the intron region. These series of mutants having variable expression in terms of starch content thus validated the significance of mutation in both exon as well as intronic region towards changing the trait expression (Moreau et al., 2022).

6.4 TRUE ALLELE MINING *VIS-À-VIS* PROMOTER MINING

Comprehensive allele mining involves not only analyzing sequence variations in the coding regions of the important candidate genes but also encompassing the examination of non-coding and regulatory regions of these candidate genes. This approach ensures that functional variations of relevance within the genes are thoroughly investigated (Foucher et al., 2003; Kumar et al., 2010; Gokidi et al., 2017; Moreau et al., 2022). In recent decades, the concept of promoter mining is also considered instrumental toward elucidating gene expression studies. Promoter region is the binding site of RNA polymerase and transcription factors to initiate the transcription process of genes. Differential gene expression is commonly attributed to detect sequence variation in the *cis*-acting or *trans*-acting regulatory elements, which are binding sites of the transcription factors situated in the upstream region of the target gene. Research on unveiling the allelic variation in the promoter region of major genes is scanty in pea. However, many examples in other food legumes represented allelic variation in the promoter region generated due to either spontaneous or induced mutation, significantly impacting the phenotypic expression of the targeted gene. In pea as well as other food legumes, indeterminate growth habit is governed by *Dt1* or *Det*, an ortholog of *AtTFL1* gene (Foucher et al., 2003; Liu et al., 2010; Repinski et al., 2012), encodes a signaling protein which is usually expressed in shoot as well as apical meristem. In mung bean, it was detected that, during the course of domestication from wild to cultivated form, the shift from an indeterminate to a determinate stem growth habit was accomplished through the selection of two closely linked point mutations in two potential *cis*-regulatory elements in the promoter region of *Vigna radiata det* (*Vrdet*) gene. This led to a noteworthy reduction in gene expression (Li et al., 2018). In soybean, significant allelic variation observed in the promoter region of the flowering time regulation gene (*E1*) resulted in a drastic alteration in the expression level of the *E1* gene. Consequently, this change in gene expression led to a modified phenotype in terms of flowering time (Tsubokura et al., 2014; Zhai et al., 2015). Strategies for promoter mining involved *in silico* analysis to detect nonsynonymous SNPs and InDels within the transcription factor binding sites as well as changes in the number and frequency of regulatory element binding sites within the promoter regions, which subsequently transform the corresponding transcription factors associated with the specific expression patterns in the form of constitutive or tissue/development-specific manner (Heidari et al., 2015; Piya et al., 2023). Promoter mining offers insights into the molecular mechanisms of gene expression and the knowledge on the conserved regulatory motifs that are specific to different growth stages and tissues. With the rapid expansion of global gene expression experiments aimed at identifying gene sets with similar expression patterns, it has become imperative to integrate approaches such as *in vitro* and *in silico* promoter mining into routine allele mining programs. Thus, incorporating promoter mining with allele mining would enable us to efficiently identify and classify cis-regulatory elements, leading to a deeper comprehension of their function in gene regulation.

6.5 DEVELOPMENT OF ALLELE MINING SET

Strategies for allele mining are usually carried out in core as well as mini-core collection of the specific crop species (Figure 6.1). Creating core collections has been proposed as a strategy to optimize the utilization of genetic diversity in crop improvement programs, especially in the field of allele mining (Upadhyaya et al., 2009). The concept of core collection was proposed by Frankel (1984) as expressing the genetic diversity of a crop species and its wild relatives precisely without redundancy. The fundamental requirement for the creation of core collection is that it fully represents the entire diversity of the collection from which it was derived. The steps involved in captivating diversity through core collection are presented elaborately. At the outset, characterization of the accession

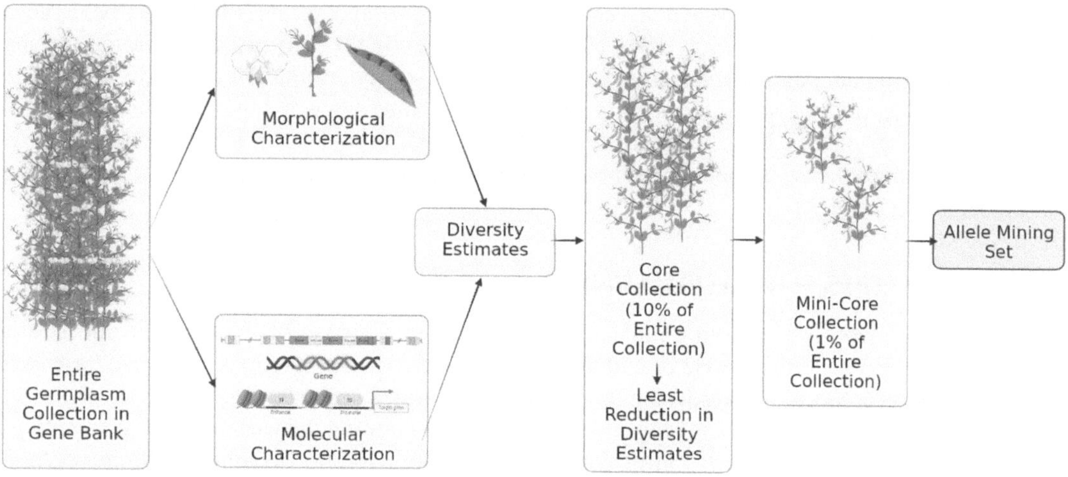

FIGURE 6.1 Steps involved in the development of allele mining set.

through passport data/descriptors and morphological and molecular markers has been carried out by taking into account the total available genetic resources of any crop species maintained in the gene banks (Brown and Spillane, 1999; Balfourier et al., 2007; Odong et al., 2013). Subsequently, based on the diversity index and genetic distance between the entries/accessions, hierarchical clusters have been formed, considering the morphological and molecular diversity as well as integrating the taxonomic and geographic origin of the accessions. As per the result of diversity analysis and following the sampling theory of Brown (1989), approximately 10% entries of the total collection, with a maximum limit of 3,000 per species, should be selected. This strategy can effectively capture 70% of the allelic variation of the full collection. For judging the quality of the core set in terms of maximum representation of the range and pattern of variation present in the entire gene bank collection, several summary statistics have been proposed based on the nature of variables. In case of continuous variable, mean, range, variance, coefficient of variation, and other summary statistics of the core set should be compared with the whole gene bank collection. Additionally, computation of quantile–quantile (q-q) plot (Gnanadesikan and Wilk, 1968) is also suggested for comparing the distribution of the traits in the entire collection with the core through enumerating Kullback–Leibler distance (Kullback and Leibler, 1951) between the core set and the entire collection of the particular crop. In case of categorical variables, chi-square (χ^2) goodness of fit test is usually followed (Dasgupta and Alldredge, 1998). After analyzing the summary statistics for each trait separately, which include t-tests for comparing population means and F-tests for comparing variances, several evaluation criteria have been proposed to test the significant difference between core collection and the entire collection, as well as the property of the core collection. These criteria usually involve mean difference percentage (MD%), variance difference percentage (VD%), coincidence rate (CR%), variable rate (VR%), and the sign test (Diwan et al., 1995; Basigalup et al., 1995; Hu et al., 2000; Tai and Miller, 2001). Numerous statistical programs and algorithm options are presently accessible, such as MSTRAT (Gouesnard et al., 2001), genetic distance sampling (Jansen and Van Hintum, 2007), PowerCore (Kim et al., 2007), CoreHunter (Thachuk et al., 2009), and Core Hunter3 (De Beukelaer et al., 2018), which are helpful in the selection of core collections that achieve maximum representation of the entire collection. In the next step, the core collection set should be evaluated at multi-locational trials to judge the stability of the accessions prior to their utilization in the breeding program. Further, the entries within the core collection can be subdivided into a mini-core collection when the size of the core collection becomes impractical to manage. The entries in the mini-core collection represent either 10% of the core collection or 1% of the entire collection. The Shannon and Weaver diversity index (Shannon and Weaver, 1963) can be deployed to measure the phenotypic diversity for each trait independently for both core and mini-core collection to determine the quality of the mini-core set in terms of representation of the diversity of the core set. Finally, the core set of the specific crop species can be subjected to elucidate allelic variation for traits of interest. In pea, presently, no international organization is conducting extensive breeding activities; however, pea core set has been generated and maintained by some organization at global level (Table 5.1).

6.6 STRATEGIES FOR ALLELE MINING

Allele mining involves identifying molecular alterations in the gene of interest through multiple strategies that differ in their mode of action. Initially, all the methods require a diverse base set of germplasm to proceed. But considering the entire germplasm is arduous, the process involves sampling strategies to narrow it to a manageable size while maintaining the variability. The sampling is undertaken by developing core and mini-core from the entire collection to simplify the conservation of germplasm resources while properly utilizing the existing variation in gene banks. Following this, several studies have utilized various techniques to utilize the inherent variation in the germplasm or induce variations to identify the varying alleles. The techniques are elaborated hereafter.

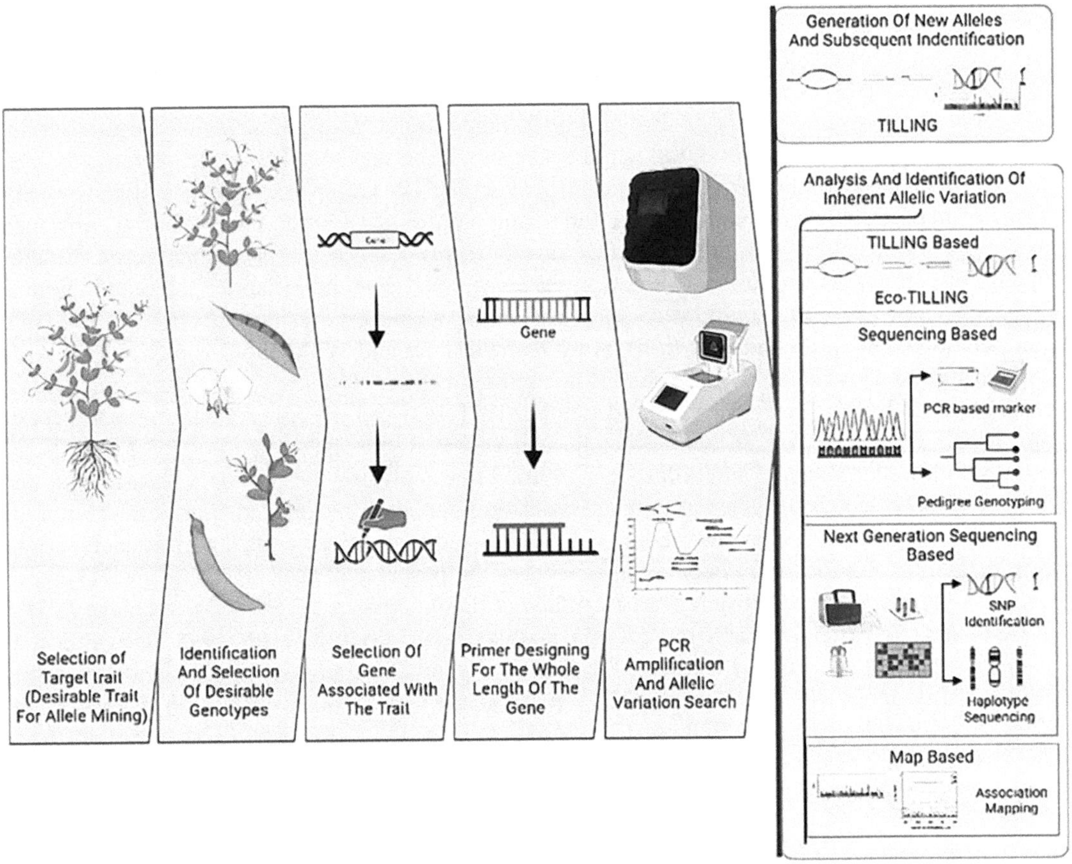

FIGURE 6.2 Steps involved in allele mining in pea for detection of inherent and induced alleles.

6.6.1 Generation of New Alleles and Subsequent Identification of Allelic Variation

Mutation is a powerful tool for generation of allelic variation in an otherwise non-variable population. Induced mutations lead to variations in the traits of interest and their identification can lead to identification of alleles of the gene generated through mutation. TILLING is a powerful reverse genetics tool for functional genomics for identification of these allelic variations of target gene in a high-throughput manner.

6.6.1.1 TILLING

TILLING method includes the generation of a high number of arbitrary mutations with mutagens and subsequent mutational screening to detect induced mutations within the stretch of DNA targets (McCallum et al., 2000a). Pea has been a model system in plant biology since the work of Sir Gregor Johann Mendel to current genomic era; in spite of a large genome and the difficulty to transformation, it is still considered a good model system. Compared with other legume crops, pea presents low sequence redundancy, and it is likely to stand among crop plants as an important model system for reverse genetics approach. This crop has a large account of mutant collections for single or combined characters, first established by De Vilmorin (1913) and further developed by Blixt (1972). Several attempts have been made to create variability through induced mutagenesis by employing chemical mutagens like ethyl methanesulfonate (EMS) in pea. Seeds of the pea variety "Rondo" were treated by EMS and, subsequently, a chlorate-resistant mutant identified with decreased *in vitro* activity of the enzyme nitrate reductase. The mutation was monogenic and recessive (Feenstra and Jacobsen, 1980). From the same population, a mutant efficiently nodulating in the presence of 15 mM KNO3 was selected which was also expressed in monogenic and recessive pattern, and the gene was designated as *nod3* (Jacobsen and Feenstra, 1984). A stable nodulation-resistant pea mutant, sym-5, was detected by screening of the M2 progeny of EMS-treated pea cv. Sparkle (Kneen and LaRue, 1984). On a similar way, mutants for nodulation were identified from pea variety "Frisson," and all mutations were under the control of single recessive genes (Duc and Messager, 1989). The mutants' population of pea variety "Sparkle" also produced a stable mutant having ineffective nodulation which was controlled by a recessive gene *sym 13* located on chromosome 2. Another gene, *sym 14*, was also detected, conditioning non-nodulation and mapped to a locus on a different part of chromosome 2 (Kneen et al., 1990). Seeds of pea cv. Frisson were treated by EMS to create mutations affecting nodulation and identified two genes, namely, *Sym28* and *Sym29*, involved in regulation of nodulation (Sagan and Duc, 1996).

An EMS-mutagenized library derived from pea cultivar "Terese" (Rameau et al., 1997) was generated. Later on, attempt was also made to develop TILLING platform from an EMS-mutagenized library of 8,000 M2 families derived from cultivar Terese and observe the relevance of the population for carrying out large-scale reverse genetic project (Triques et al., 2007). In this study, specific mutants were screened from the *LE* gene (conferring stem length) encoding for gibberellin 3β-hydroxylase and were further characterized to determine the effect on internode length. Induced mutagenesis is currently under revival state and being used to create novel genetic variations for selection using molecular techniques. Several pea EMS mutant populations already existed, but they were unsuitable for a genomic approach as they were not prepared or maintained under rigorously controlled conditions and suffer from cross-contamination. Hence, the need was felt for a high-quality pea genetic mutant reference collection which could be extended for both forward and reverse genetics studies. Consequently, genomics tools such as fast neutron and TILLING mutant populations have been developed for reverse genetics approaches (Dalmais et al., 2008; Wang et al., 2008). Sufficiently large TILLING population made in the French variety "Cameor" is available for pea, and the generated data were available in an online database (UTILLdb) that contains phenotypic as well as sequence information on mutant genes (Dalmais et al., 2008). Currently, it has 4,817 lines, of which 1,840 are with phenotyping data, and 464 identified mutations adhered with genotype by sequencing. In addition, fast neutron–generated deletion mutant resources (around

3,000 lines) are also available for pea. The major pea mutant collections include John Innes Collection, Norwich, UK (575 accessions); Plovdiv, Bulgaria (122 accessions); and a TILLING population of 4,817 lines at INRA, Dijon, France. In addition, fast neutron–derived deletion mutant resources (around 3,000 lines) are available in pea and used to identify several developmental genes (Wang et al., 2008; Le Signor et al., 2009). A fast neutron (FN)–mutagenized population was developed in pea using recombinant inbred line JI 2822 to identify and isolate genes underlying the trait (Hofer et al., 2009; Domoney et al., 2013). Hofer et al. (2009) used pea TILLING population to support the identification of the tendril-less gene, while D'erfurth et al. (2012) used mutant lines of EMS population generated from the cultivar "Cameor" by TILLING developed by Dalmais et al. (2008) and reported a subtilase gene *SBT1.1* controlling seed size by providing molecules that can act as signals to control cell division within the embryo. Zhuang et al. (2012) used FN population of JI 2822 and characterized *lathyroides* (*LATH*), a classical locus in pea, whose mutation displays pleiotropic alteration of lateral growth of organs and predominant effects on tendril and dorsal petal development. This population so far has been exploited by several research groups and is mainly focused on the genetics of phenotypic traits and indicated that individual mutant lineages can segregate for several mutations, necessitating the backcrossing of lines for rigorous analysis of mutations. Several of these studies have been successful in identifying the genes underlying phenotypic mutations (Sainsbury et al., 2006; Hellens et al., 2010; Chen et al., 2012; Moreau et al., 2012; Domoney et al., 2013; Moreau et al., 2018).

6.6.2 Analysis and Identification of Inherent Allelic Variations of Genes

Over the years, natural evolution has resulted in mutation of genes to incorporate variation either as SNPs or as InDels in the gene pool of pea. Such natural variation in the genes leads to allelic variation in the genes and provides an array of DNA polymorphisms in the different alleles of the gene of interest. These polymorphisms necessitate allele mining for identification of the variation in their sequence through modified TILLING approach (EcoTILLING) or analysis of sequence through PCR, NGS, or association maps.

6.6.2.1 TILLING-Based Identification

TILLING in general requires mutagenesis for initial induction of mutation, but heteroduplex-based mismatch identification for allele identification can also be used for the inherent natural mutations in a population using a modification of the TILLING approach proposed by Comai et al. (2004) called EcoTILLING.

6.6.2.1.1 EcoTILLING

This is an extension of the TILLING approach wherein natural variants/alleles at a locus are identified without induced mutations. It is a cost-effective, more efficient, and rapid technique as compared to TILLING (Comai et al., 2004; Rashid et al., 2011). It can also be used to estimate phylogenetic diversity, selection, and linkage disequilibrium (Gilchrist et al., 2006). EcoTILLING can identify DNA polymorphisms in highly genetically similar populations and to mine for SNPs in collections of plant germplasm. The EcoTILLING platform was first exploited in *Arabidopsis thaliana* to mine natural polymorphisms (Comai et al., 2004). Another important application is to find out allelic variation in resistance genes against various diseases, for example, *Mla* (powdery mildew) resistance gene of *Hordeum vulgare* (barley) (Mejlhede et al., 2006). Among pulse crop, it has been employed to identify polymorphisms in mung bean (Barkley et al., 2008). Overall, EcoTILLING has shown great potential toward identification of allelic variation in disease resistance genes, which can be utilized in crop improvement.

6.6.2.2 Sequence-Based Allele Mining

Sequence-based allele mining entails identifying new alleles among the diverse cultivars by multiplying and sequencing a certain region of nucleotides. The method is more cost-effective than

TILLING approaches and can be used for regions that contain insertion and deletions (InDels) in the gene of interest and have show variation in PCR-based amplification.

6.6.2.2.1 PCR-Based Allele Mining

The process of allele mining utilizing PCR involves simple sequence repeat (SSR) markers for identifying diversity among the genotypes. The subsequent variation obtained is checked for the phenotypically differing genotypes, and the marker is associated with the trait of interest through statistical analyses. Over time, several researches have reported SSR markers in pea. Gong et al. (2010) reported 49 SSRs in pea and identified nine novel expressed sequence tag (EST)–SSRs for their transferability and polymorphism. Similarly, Mishra et al. (2012) identified 40 polymorphic EST-SSRs for use in molecular mapping. Subsequently, Yang et al. (2015) reported 1,644 novel SSR markers, and 33 markers were anchored on the genetic linkage map. Recently, Smýkal et al. (2018) used SSR markers to evaluate pea accessions' genetic diversity and population structure, highlighting the importance of preserving and utilizing diverse germplasm. But allelic variation for a specific trait utilizing SSR marker was not undertaken until recently. Iglesias-Garcia et al. (2015) identified SSR markers associated with drought tolerance, providing insights into genetic resources for improving drought resilience in the crop. Similar studies for powdery mildew resistance was undertaken by Gupta et al. (2023), and they reported two polymorphic SSRs associated with the trait. These trait-linked SSRs can be utilized to identify the alleles for the associated trait in the phenotypically diverse population and help transfer promising alleles in breeding programs with molecular confirmation.

6.6.2.2.2 Pedigree Genotyping–Based Allele Mining

Pedigree approach in crops is generally used for understanding the heritability and genome-wide co-ancestries for specific characters through field-based genetic analysis without incorporating molecular techniques. Pedigree genotyping is a modification of the pedigree approach by incorporating marker data (SSRs or SNPs) to find the alleles of the genes and subsequent marker–trait association studies.

Pedigree genotyping can be initially applied to identify loci related to the trait in consideration, followed by deployment of SSR- or SNP-based markers linked or associated with the locus. The process requires a set of multi-allelic markers for the genome segment of the loci. Subsequently, the markers are validated for a diverse set of genotypes and their subsequent generation to identify the marker–trait association and the alleles available for the trait at the desired locus. The process is cost-effective, as SSRs are easily implemented at multiplex PCRs. This can identify InDels through PCR and subsequent sequencing to identify allelic variations that differ in their sequence.

In the case of pea, as the whole genome sequence is known, many SSR-based markers for the yield-related traits and the biotic and abiotic stress-related traits are mostly identified. This poses a unique position for pea in the case of pedigree genotyping, as advanced lines can be checked for allelic variation with the related genotypes, and trait-associated alleles can be identified easily and identify allelic interactions. Pedigree genotyping will also expedite the analysis of mapping population when many loci are known from the mapping population.

6.6.2.3 Next-Generation Sequencing–Based Allele Mining

Sequencing helps unravel the exact arrangement of nucleotides in the sequenced section of the DNA. High-throughput sequencing techniques, termed as next-generation sequencing techniques, involve "massively parallel" approaches for sequencing, leading to a minimized cost of sequencing, more precise and accurate results, and increased sequencing capacity to a great extent. These approaches make re-sequencing, aligning sequence data, and comparing sequence data to a data reference genome much more feasible. The sequencing of large populations of individuals for a candidate gene is possible to identify genetic variations. This information can be harnessed to link the variation with the traits leading to the identification of alleles related to the candidate gene. Modern sequencing platforms, including Illumina/GA (Illumina; http://www.illumina.com), Roche/GS FLX

(Roche Applied Science; http://454.com/products/index.asp), Applied Biosystems/SOLiD (Applied Biosystems; https://products.appliedbiosystems.com), and cPAL sequencing (Shendure, 2008; Metzker, 2010) and the HeliScope Single Molecule Sequencer (Helicos; http://www.helicosbio.com), facilitate identification of SNPs and subsequent identification of haplotypes for efficient detection of alleles related to phenotypically varying characters and associate them in a precise manner.

6.6.2.3.1 *Allele Mining through SNP Identification*

Single-nucleotide polymorphism is a variation in the nucleotide sequence at the same loci of a gene for two different individuals. The variation generally permeates to change in the gene's phenotypic expression and leads to allelic variation for the target trait. Crop EST project for pea identified 9,377 ESTs through BLAST from two cDNA libraries of pea (http://pgrc.ipk-gatersleben.de/crest) (Künne et al., 2005). In the following years, gene-specific studies for economic characters have identified SNP markers representing the diversity and relationships among pea genotypes (Jing et al., 2007; Aubert et al., 2006). Subsequently, many genes were sequenced using high-throughput sequencing methods to identify SNPs, leading to allelic changes in the gene. Deulvot et al. (2010) identified map positions for 37 new gene markers through genotyping of mapping populations. Bordat et al. (2011) identified map positions for 5,460 pea Unigenes identified through transcriptome on the consensus map after confirmation with model plant *Medicago truncatula*. Duarte et al. (2014) and Sindhu et al. (2014) validated 1,538 SNPs and 1,536 SNPs for structure analysis through the GoldenGate assay. Among them, 1,340 SNPs were genetically mapped by Duarte et al. (2014) to build a new consensus map covering all seven linkage groups using 2,070 markers in pea. Ma et al. (2017) further improved the marker development status in pea by identifying 1,608 SNPs and generating a linkage map of size 1,310.1 cM. The study also identified 46 seed mineral concentration QTLs, 37 seed mineral content–related QTLs, and six seed weight QTLs. In 2020, Beji et al. analyzed 11,366 SNP markers using GWAS and identified 62 SNPs distributed over six linkage groups of pea and significantly associated with frost tolerance. Recently, Bari et al. (2023) associated 260 SNPs with Aphanomyces resistance in pea, while Delvento et al. (2023) associated three QTLs with broomrape resistance using SNP-based genotyping-by-sequencing (GBS) for a RIL population of 148 F_7 lines. Recent studies also implement SNPs for genetic diversity analysis in pea for leaf type characters by Tran et al. (2023). Overall, SNP variation is well documented in pea and has been at the forefront in identifying alleles and QTLs for phenotypic traits. This provides a basis for other analyses like haplotype and association map development for a better understanding of the genetic variation related to the trait of interest.

6.6.2.3.2 *Haplotype Sequencing–Based Allele Mining*

Haplotype refers to a group of genetic variations inherited on the same chromosome. High-throughput genotyping using NGS techniques has paved the way for haplotype construction based on SNP data. Genome-wide association studies (GWAS) identify SNPs in the genome for a specific trait, and subsequently, the SNPs inherited together and linked on the same chromosome are used to determine the haplotype of the character for the identified SNPs. Haplotype genotyping proceeds to the next step in SNP utilization by the subsequent association of haplotypes and traits of interest. These haplotypes can then be sequenced in phenotypically differing populations to identify the SNPs and alleles associated with the trait of interest present within the haplotype.

The process of generation of haplotype and subsequent haplotype sequencing proceeds by initial identification of SNPs in the genome for the trait of interest, followed by GWAS studies to detect the haplotype associated with the SNPs identified. Then, haplotype is linked with the trait of interest. The haplotype is then sequenced in the population phenotypically diverse for the trait of interest, and the haplotypes are used as markers for the trait, and allelic variation is identified (Ogawa et al., 2018).

Haplotypes for several biotic and abiotic stress resistance–related characters are identified in pea (Desgroux et al., 2016a; Boutet et al., 2023; Leprévost et al., 2023). Still, identification of alleles by haplotype sequencing has been undertaken only recently by Tayeh et al. (2023) for afila (*af*) mutant in pea that results in leaves that are composed of a basal pair of stipules, a petiole, and a branched mass of tendrils. The study identified eight haplotypes in the genomic region of chromosome 2 that harbors *af*. It associated the haplotypes with the deletion region change in the genotypes for the *PsPALM1a-b* and close genes. Representative cultivars were assigned to the mutations, resulting in the phenotype being characterized with the specific haplotype-based allele. This process holds potential for pea, as unexplored haplotypes can be associated with the alleles of the phenotypically diverse traits of interest, leading to a more concise understanding of the alleles for the genes and their underlying allelic variations.

6.6.2.4 Map-Based Approaches for Allele Mining

Map-based approach involves the identification of allelic variation and its molecular justification through analysis of sequence variation and association of those variations with molecular markers through identification of the physical location of the gene and the associated marker in a fine map of the genome. The most significant approach for map-based allele mining involves association mapping of the genes implementing sequencing-based tools for construction of association map like GWAS.

6.6.2.4.1 Association Mapping–Mediated Allele Mining

Association mapping identifies molecular markers closely linked to quantitative traits using linkage disequilibrium through population-based surveys (Zhu et al., 2008). The method is helpful in the estimation of the natural variation in the genetic resources for the specific genes in the crop. Also, it helps establish areas of the genome related to the trait of interest. Genetic analysis of pea has provided a well-documented precise annotation of morphological mutations and diverse alleles on genetic map. This provides the basis for association mapping, as the pea's reference genome expedites the association of traits of interest with the molecular markers (Yang et al., 2022).

The association mapping approach necessitates initially identifying core germplasm with enough diversity to construct association panels. The panel generally includes landraces, cultivars, breeding lines, exotic cultivars, and wild relatives. The subsequent step is the comprehensive and precise phenotyping of the panel for the traits of interest. This must be conducted in varying environments and replicated to eliminate environmental effects. Phenotyping is followed by genotyping with suitable unlinked molecular markers (SSR markers are preferred) spread uniformly and separated by a minimum of 20 cM in the genome. The markers are used to determine the population structure using STRUCTURE software (Pritchard et al., 2000) and kinship analysis using TASSEL software (Thornsberry et al., 2001; Yu et al., 2006) to remove the false positives. Following this, the extent of linkage disequilibrium (LD) is calculated between the markers to determine the number of markers required to cover the entire genome. Finally, the model-based analysis of the genotypic and phenotypic data with LD value is done to identify the proximal location of markers to alleles of the gene of interest to tag the marker to the trait.

The implementation of association mapping was first undertaken by Murray et al. (2009) for genes concerning yield-related traits like ADP glucose pyrophosphorylase, APETALA2 (*AP2*), cytokinin receptor family proteins, cytokinin oxidase family proteins, and GA 3-oxidase (*GA3ox*). They identified three candidate gene homologs: PsAP2, PsGA3ox (*Le*), and a cytokinin receptor family gene. Subsequently, Carpenter et al. (2017) identified polymorphisms in genes involved in the substrate supply, chain elongation, and branching stages of the pea carbohydrate and starch metabolic pathways through population structure and kinship analysis. In the following years, SNP-based association mapping and genome-wide association studies were undertaken for association mapping of genes and related markers. Deulvot et al. (2010) implemented SNPs for genetic mapping and identified 37 new gene markers. Cheng et al. (2015) subsequently utilized SNPs to demonstrate the importance of association mapping for allele mining in pea. They identified 71 marker–trait associations with 1–34 markers per trait in 25 valuable traits, including disease resistance, seed type/color, flower color, seed low-molecular-weight carbohydrate concentration, and seed mineral nutrient concentration. In the following years, Jiang et al. (2017) identified 60 SNP markers associated with characters for reproductive development in pea, out of which 33 SNPs were associated with the onset of flowering, 8 SNPs with pod development, and 19 SNPs with the number of reproductive nodes. After that, Desgroux et al. (2018) utilized GWAS to identify QTLs associated with resistance to *Aphanomyces euteiches* and elucidate novel resistance loci. Following this, Gali et al. (2019) undertook a similar approach for the identification of loci related to agronomic and seed quality traits and documented two loci each for plant height and grain yield, one each for lodging resistance and seed protein concentration and four loci for seed starch concentration. Recently, similar study by Tafesse et al. (2020) unveiled six loci for chlorophyll concentration, two each for canopy temperature and photochemical reflectance index, seven loci for reproductive stem length, six loci for internode length, and nine loci for pod number.

Similar studies for yield-related genes and resistance genes for major diseases and pests of pea and abiotic stress–related genes can be undertaken to identify the associated markers and utilize them in resistance breeding for biotic and abiotic stress tolerance in pea.

6.7 BIOINFORMATICS TOOLS FOR ALLELE MINING

Allele mining is a breeding approach involving allelic modification analysis of a gene by identifying polymorphisms in the gene of interest. The process consists of scrutinizing gene sequences of various genotypes through various software tools. The tools revamp the ease in the access of large sets of nucleotide data and their subsequent analysis for the identification of polymorphisms, the modifications in the transcripts for the prediction of putative alterations in the functional and structural components of complexes, and finally, the probing of these modifications for identification of morphological changes. The tools also discern the over-expression of genes and identify the variation in regulatory elements corresponding to the gene's expression changes. The various bioinformatics tools involved in the multiple steps of allele mining are mentioned hereafter as per the steps in which they are incorporated (Table 7.1).

6.8 CURRENT STATUS AND APPLICATION OF ALLELE MINING IN PEA BREEDING PROGRAM

The main objective of allele mining is to detect unique sequence polymorphism and favorable haplotypes associated with the trait of interest and their further validation in a large panel of germplasm for subsequent utilization toward the development of markers to facilitate marker-assisted breeding programs. The following sections discuss the application of allele mining in the pea breeding program.

6.8.1 Development of Allele-Specific Markers for Enriching Genomic Resources in Pea

Basic genomic resources are required for conducting a comprehensive molecular breeding program. Molecular markers, genetic maps, and sequence information availability are the prerequisites for initiating improvement in any crop deploying molecular breeding approaches. Development and application of different technologies started with the genetic study of model plants. Molecular markers play a crucial role in QTL identification, preparing genetic map, genetic diversity assessment, identification and fingerprinting of genome, identification of useful candidate gene. A worldwide online database provides full and free access to pea markers, named Pea Marker Database (PMD) (www.peamarker.arriam.ru). This provides information about 2,484 genic markers and their locations in linkage groups. Another version of PMD comprises of information about 15,944 pea markers in the same format with several advanced features. NGS proteomic analysis facilitates polymorphic studies and genome map construction. Pea consensus linkage mapping was done using different segregating populations (JI15/JI1194, JI15/JI399 and JI281/JI399) (Weeden, 1998).

A composite genetic map was obtained from three populations (Terese/K586, Champagne/Terese, Shawnee/Bohatyr) comprising 239 SSR markers (Loridon et al., 2005). These markers were useful to localize numerous QTLs for disease resistance as well as quality- and morphology-related traits. Another consensus map was built using six RIL populations – Terese/K586, Champagne/Terese, VavD265/Cameor, Ballet/Cameor, VavD265/Ballet, China/Cameor – that included 214 functional markers linked to developmental biology, carbohydrate metabolism, amino acid metabolism, mineral transport, and transcriptional regulation (Bordat et al., 2011).

Identification of candidate genes by expressed sequence tag (EST)–based approach is frequently followed in pea breeding programs. EST analysis proves valuable in rapidly and efficiently providing preliminary information on expression profiles, especially in specific tissues under different physiological conditions. It serves to identify regulatory genes involved in these processes. "CROP-EST," an online database (http://pgrc.ipk-gatersleben.de/cr-est/) developed for crop EST projects, includes two cDNA libraries for pea and a collection of 9,377 ESTs (Kunne et al., 2005).

With the advancement of low-cost NGS-based sequencing platforms, SNPs are currently in use in pea crop. There is a possibility of unlimited discovery of SNPs than retrotransposon and microsatellite markers. Gene amplicon studies in different structured populations of pea provided data on identification of a total of 669 SNPs and 84 indels (Aubert et al., 2006; Bordat et al., 2011). NGS and gene capture arrays have the potential to identify gene-based SNP markers, thus can be extended to pea breeding program. Recent study of QTL identification using SNP markers discovered different agronomic and qualitative traits–related QTLs and also reported the highest level of synteny with *Medicago* (Ma et al., 2017). A report by Klein et al. (2020) provided information on QTL analysis carried out in 1,213 RIL population using SNP-based markers that resulted in the identification of 17 QTLs for seed weight per plant, 16 QTLs for seed number per plant, 35 QTLs for thousand-seed weight, and 21 QTLs for seed protein content in pea. QTLs associated with pea seed protein content were identified using SNP markers (Zhou et al., 2022). Linkage map was constructed using SNP and SSR marker linked to *Aphanomyces* resistance gene in pea (Wu et al., 2021). Similarly, high-density linkage mapping was done using 4,489 SNP marker for broomrape resistance (Delvento et al., 2023). Enrichment of efficient marker system in pea thus facilitates marker-aided introgression of trait of interest.

6.8.2 Detection of Haplotype

Construction of the haplotypes from the sequence data through haplotype estimation is known as *phasing*. Development of haplotypes can be done by using whole-genome sequencing approach. High-density SNP data are obtained by NGS-based or array-based approaches. The integration of various phasing tools and bioinformatics tools along with long-read sequencing technologies has revealed the potential of full exploitation of haplotype construction. There is availability of an array of phasing tools, like reference-based phasing, *de novo* genome assembly, and strain-resolved metagenome assembly, as well as bioinformatics tools like WhatsHap, HapCut2, HapTree, WhatsHap polyphase, Falcon phase, Hifiasm, SDip, POLYTE, DESMAN, MetaMaps, and ProxiMeta, which have made possibilities of efficient and precise identification of *de novo* and rare variants from the long-read sequencing data. Sun et al. (2022) deployed the PacBio HiFi reads (30× per haplotype) and Hifiasm to generate the assembly of the autotetraploid genome of potato. This was the first study demonstrating the haplotype-resolved assembly of potato crop. There is another method of haplotype discovery, which is "Haplotagging," where molecular barcoding of long DNA molecules is carried out primarily before sequencing. However, its practical application is limited or not extensively utilized (Amini et al., 2014; Wang et al., 2019). Haplotypes are useful for conducting GWAS, identification of epistatic interactions between SNPs at a locus, providing information about evolutionary history. Haplotype-based GWAS analysis has been done in different economically important crops, like rice for grain size and agronomic trait (Hamazaki et al., 2020; Ogawa et al., 2021); wheat for heading date and grain yield (Luján Basile et al., 2019; Sehgal et al., 2020); maize for agronomic and reproductive traits (Maldonado et al., 2019; Mayer et al., 2020); oat for heading date (Bekele et al., 2020); barley for deoxynivalenol content in kernel, date of maturity, heading time, yield, and quality trait (Gawenda et al., 2015; Abed et al., 2019); and rapeseed for seed glucosinolate content and days to flowering (Jan et al., 2019). Further, the development of haplotype-specific markers is useful in the identification of the most desirable combination of haplotype concerning the different important phenotypes.

Information on important phenotype-controlling haplotypes is limited in the genetic studies in pea. In the face of climatic change, adaptation of flowering duration is an important attribute in crop plant. Pea can only flower under long-day condition. A study was conducted to unveil the genetics behind flowering time regulation and photoperiodic adaptation in wild and cultivated accessions of pea concerning the major genes' *HIGH RESPONSE TO PHOTOPERIOD* (*HR*), which is an ortholog of *EARLY FLOWERING 3* (*ELF3*) gene involved in circadian clock regulation in Arabidopsis (Weller et al., 2012). Sequence analysis of *ELF3* gene, as well as examining the TILLING population for the said gene, detected several SNP variations, 15 haplotypes with four discrete haplotype groups. The prevalence of the mutation in the *hr/elf3* and its presence within a closely related but separate group of haplotypes indicated that the mutation likely emerged quite recently within a distinct lineage and has quickly spread throughout the population with possibility of summer cultivation of pea. A similar study was carried out using 15 Swedish pea lines. In this study, data on reproductive phenology was collected, and the allelic diversity of genes such as *HR*, *LATE FLOWERING* (*LF*), which regulate different flowering phases in peas, and the photoperiod responsive gene *STERILE NODES* (*SN*), an ortholog of the *LUX ARRHYTHMO* gene family from *Arabidopsis thaliana*, were elucidated (Vanhala et al., 2016). Sequence variation of all three genes detected several InDels and SNPs, while more specifically for *LF* gene, six haplotypes, and for *SN* genes nine haplotypes, were detected with associated significant differences in flowering time.

In the context of winter pea, frost poses a significant abiotic stress limitation. Consequently, the identification of genetic factors responsible for frost tolerance is a crucial challenge in breeding varieties suitable for cold northern regions. Therefore, detection of the genomic region concerning this trait is very crucial for further implementation in a marker-assisted breeding program in pea. A GWAS study was performed using a set of 365 pea accessions at INRAE, France, to detect the genomic region controlling frost tolerance. Result revealed detection of 62 SNPs widely distributed within six linkage groups, along with favorable haplotypes associated with frost tolerance in pea (Beji et al., 2020).

Biotic stresses are a major bottleneck associated with pea growing systems globally. Pea is frequently infected with soilborne pathogen *Aphanomyces euteiches*, causing Aphanomyces root rot in USA, pea growing areas of Europe, and Canada (Chatterton et al., 2015). GWAS was conducted deploying a diverse collection of 175 pea accession with varied disease reaction against *Aphanomyces* at INRA, France (Desgroux et al., 2016a). GWAS analysis integrated with multi-locus mixed model has detected 52 QTLs with small intervals associated with resistance against *A. euteiches*. Furthermore, it offered marker haplotypes for 14 consistent QTL regions linked to resistance, emphasizing the concentration of favorable haplotypes in the most resistant lines. Powdery mildew incited by *Erysiphe pisi* is a detrimental disease in pea. The disease resistance against this fungal disease is conferred by three distinct genes, including two single recessive genes, namely, er1 and er2, as well as one dominant gene, viz., er3 (Fondevilla et al., 2007; Humphry et al., 2011). Screening of 310 pea accessions was carried out at CSK Himachal Pradesh Krishi Vishvavidyalaya, Palampur, India, to unveil the host–pathogen interaction concerning the pea–erysiphe system (Banyal et al., 2022). Molecular studies detected 11 distinct haplotype groups with a frequency of the haplotypes varied from 1 to 24 among the resistant cultivars, where Hap-1 was the most frequent one. Most of the resistant cultivars were detected with having er1 gene sharing a common haplotype, implying that these lines likely descended from a common ancestor. GWAS analysis rendered with 240 pea accessions to elucidate the molecular mechanism behind pea aphid resistance (*Acyrthosiphon pisum*) involving one adapted and one non-adapted biotype detected major

QTL "ApRVII" spanning 55 candidate genes on chromosome 7. Additionally, distinct haplotypes were detected associated with resistance and susceptible reaction against pea aphid, thus facilitating marker-assisted selection of resistant locus for introgression of the trait of interest (Ollivier et al., 2022). Therefore, haplotype detected through various approaches thus sets the ground to aid in marker-assisted selection for development of climate-resilient pea cultivars.

6.8.3 Synteny Analysis toward Elucidating Evolutionary Relationship

Synteny of pea can be identified after availability of full-genome sequence of other legume crops. Identification and searching of homologous gene sequences mapped on pea consensus map in other legume established the syntenic relationship among pea, *M. truncatula*, *L. japonicus*, and soybean (Bordat et al., 2011). Comparative mapping showed the gene order conservation and correspondence among pea, *M. truncatula*, *L. Japonicas*. There are reports of two comparative mapping studies informing a good conservation of marker order between pea and *M. truncatula* for a set of 57 genes (Choi et al., 2004), and between pea and *M. sativa*, a crop species closely related to *M. truncatula*, for a set of 103 genes (Kalo et al., 2004). Translational genomics also facilitate in identifying the candidate gene in pea, like genomic information from *L. Japonicus*, which helped identify two floral zygomorphy mutant loci keeled wings (K) and lobed standard 1 (LST1) (Wang et al., 2008). Additionally, in another report of comparative study by Kalo et al. (2004), it was revealed that pea gene *SYM19* is an ortholog of the Medicago *NODULATION RECEPTOR KINASE* (*NORK*) gene.

The high level of synteny between the two genomes observed in between *Pisum* and *Medicago* predicted that there was five- to ten-fold difference in genome size (Arumuganathan et al., 1991). It did not support the idea of pea genome as a polyploid one and deriving from a simple ancient Medicago-like genome. It was suggested that pea genome evolved by multiple transposition events that permitted the larger genome size (Vershinin et al., 2003). The comparison of the maps indicated that the differences in chromosome number in the two genera can be explained by a few segmental genetic rearrangements (resulting from chromosomal aberrations, including breakage and reunion). However, the order and temporal sequence of these rearrangements cannot be deduced from the data available.

6.8.4 Traits Improvement in Pea Breeding Program

The traits of priority in pea breeding programs include high yield, standing ability, biotic and abiotic cues, early maturity, seed and pod quality, cooking quality, flavor, and enhanced nutritional value for diverse market use.

With it being an important legume crop, different breeding programs are followed in different countries focusing on sustainable development in this crop. These programs mainly followed targeted approaches for enhancing yield and other quality traits. There are many constraints associated with pea breeding under different agroecological limitations, where climate-changing factors are prevailing. The primary focus of the pea breeding program revolves around enhancing traits such as biotic stress tolerance; resistance against abiotic factors like frost, salinity, and drought; along with other agronomically important characteristics, including flowering, growth habit, high photosynthetic efficiency, and nodulation. A climate smart pea breeding program looks after strategies like characterization and conservation of germplasm accessions that capture the prevailing diversity, transferring the target traits, development of molecular markers that aid the search of diversity, as well as facilitate MAS, construction of genetic map, linkage map, association mapping for detection of genes and QTLs, as well as mining desirable allelic variations and haplotypes associated with trait of interest.

There is a strong emphasis on developing gene-specific markers to support diversity studies, construct linkage maps, and carry out QTL analysis, among other research endeavors. Additionally, there have been reports of advanced omics tools developed for pea, including transcriptomic, proteomic, and phenomic atlases, fully annotated reference genome database. The advancement of NGS technology and the availability of high-throughput marker systems have facilitated the development of high-density genetic linkage maps and functional maps using EST-SSRs and SNPs, which have been documented in various studies (Duarte et al., 2014; Sindhu et al., 2014; Sun et al., 2014; Tayeh et al., 2015; Boutte et al., 2016; Ma et al., 2017a, 2017b).

QTL mapping for agronomic trait allowed the localization of traits like stem length and number of nodes (Irzykowska et al., 2002), seed weight (Timmerman-Vaughan et al., 1996), seed yield, and seed protein content (Tar'an et al., 2004; Timmerman-Vaughan et al., 2005). GWAS was conducted for several yield attributing as well as seed quality traits in 135 pea accessions to detect SNP markers associated with the trait of interest (Gali et al., 2019). Ferrari et al. (2016) reported consensus mapping of traits using 384 SNP markers for grain yield, seed protein content, seed weight, onset of flowering, plant height, and lodging susceptibility in three biparental recombinant inbred line (RIL) populations and detected one QTL for grain yield, one QTL for flowering in LGII, three QTLs for seed weight in LG-III, VI, VII.

Biotic and abiotic stresses have a devastating impact on pea yield, causing significant losses. Several biotic stresses associated with pea cultivation include fungal, bacterial, and viral diseases, like fusarium wilt (*Fusarium oxysporum.* f. sp. pisi), common root rot (*Aphanomyces euteiches*), pseudomonas (*Pseudomonas syringae* pv. pisi), and Ascochyta blight (*Peyronellaea pinodes*), and insects like pea weevil (*Bruchus pisorum*) and pea aphid (*Acyrthosiphon pisum*) have been reported. Different molecular studies related to biotic stress resistance provided valuable insights against host–pathogen interaction in pea system. QTLs associated with Ascochyta blight resistance were identified, and their localization was confirmed on all the seven linkage groups of pea (Timmerman-Vaughan et al., 2002; Tar'an et al., 2004; Timmerman-Vaughan et al., 2004; Prioul et al., 2004; Fondevilla et al., 2008; Carrillo et al., 2014; Timmerman-Vaughan et al., 2016; Gali and Shunmugam, 2018). Recently, Jha et al. (2015) reported genotyping of diverse pea accessions using 15 SNPs and found that SNP loci PsDof1p308 and RGA-G3Ap103 were significantly associated with Ascochyta blight resistance. Likewise, complete or partial resistance found for rust resistance trait was conferred by single partially dominant gene *Ruf*, and two other QTLs were further mapped on LG VII (Rai et al., 2011). In case of fusarium wilt resistance, nine QTLs and markers linked to these QTLs were identified (Weeden and Porter, 2007; Feng et al., 2011; Coyne et al., 2015). In a similar way, QTL analysis for Aphanomyces root rot resistance detected 27 meta-QTLs associated to resistance gene (Pilet-Nayel et al., 2002, 2005; Hamon et al., 2011) and seven QTLs located on six linkage groups in pea (Hamon et al., 2013). Additionally, for Aphanomyces root rot, six QTLs were identified using GWAS analysis by Desgroux et al. (2016a, 2016b). Bacterial blight resistance in pea was associated with eight QTLs that were detected through a study utilizing three RIL populations. These QTLs were localized on LG II, III, VI, and VII (Fondevilla and Rubiales, 2012; Sudheesh et al., 2015a, 2015b). Besides these, Pereira et al. (2010) isolated different mutants of powdery mildew resistance genes (er1mut1 and er1mut2) using ethyl nitrosourea mutagenesis. Additionally, diethyl sulphate mutagen treatment allowed to identify a novel er1 resistant allele and the development of cleaved amplified polymorphic sequence (CAPS) marker and high-resolution melting (HRM)–based marker in this allelic region (Pavan et al., 2011, 2013). QTL analysis has been carried out for elucidating the genetics behind *Sclerotinia sclerotiorum* resistance in pea using 1196 SNP markers in two RIL populations (Ashtari et al., 2020).

Insects are also responsible for ultimate yield loss in pea. A recent study for weevil resistance by Aznar-Fernández et al. (2020) explained his discovery about six markers linked to four resistant genes associated with LG- I, II, IV. Barilli et al. (2018) identified three QTLs significantly associated with aphid tolerance on linkage group II, IV in 84 RIL population between the cross involving susceptible cultivar "IFPI3260" and the resistant cultivar "IFPI3251" using 12,058 SNP and SSR markers. Similarly, pea aphid resistance involved six markers linked to six genes and associated with LG-I, II, III, IV, V (Barili et al., 2020).

The important abiotic stresses having significant impact on pea cultivation are cold, drought, salinity, etc. Drought-tolerant QTLs were identified using 98 F7-derived RILs developed from parental lines P665 and cv. Messire, and ten QTLs were found to be associated with the trait (Iglesias-García et al., 2015). Salt tolerance traits were in focus of the study of Koyama et al. (2001) and Leonforte et al. (2013), and they have reported molecular markers associated with two QTLs conferring salinity tolerance identified on linkage groups Ps III and Ps VII. Marker–trait association studies were done for frost tolerance by using seven SSR markers in 672 pea collections and detected seven markers with significant association with frost tolerance (Liu et al., 2017). The information generated through these studies shed light on the physiological mechanism involving stress tolerance as well as detection of genomic region and marker system for further exploration in pea breeding program.

Seed qualities hold significant importance in determining consumer preference. Nowadays, reports on breeding for improving quality traits in pea are gaining priorities. Protein-quality loci were found and located on LGIIa using proteomic and QTL mapping approach (Bourgeois et al., 2011). Another report detected the gene for phytic acid–phosphorus concentration which was co-localized with the QTL for iron (Fe) bioavailability, and both were mapped together on LGV using 1,536 SNPs in 163 RIL populations (Shunmugam et al., 2015). Diapari et al. (2015) carried out GWAS using 1,233 EST-based SNPs in eight subpopulation of pea and reported nine SNPs associated with Fe content, and two SNPs with zinc (Zn) content in pea. Another study with the aim of detection of genome-wide SNPs and QTLs associated with seed minerals content through genotyping-by-sequencing (GBS) reported 1,609 polymorphic SNPs between the contrasting parents with varying levels of seed minerals (Ma et al., 2017a). A GWAS study resulted in marker–trait association of traits like seed starch concentration and seed protein concentration on LGVII, LG5 using 16,877 SNP markers. A GWAS was conducted using 85 pea accessions to detect SNP markers associated with the folate profile of pea and detected five SNPs with significant association with folate content (Jha et al., 2020). A similar study was reported with 135 diverse pea accessions for detection of SNPs associated with grain Fe, Zn, and selenium (Se) through deploying 16,877 SNP markers for GWAS study (Dissanayaka et al., 2020). Their study reported five SNPs each were significantly associated with Fe and Zn concentrations. A report from Zhou et al. (2023) suggested the marker–QTL association for different amino acid concentrations in seed of pea. These studies enriched marker resources for further deployment for improving quality traits in pea.

Some useful mutants were discovered in pea for important traits. The mutants identified were related to seed development (*def*), growth habit (*det*: determinate; *sm*: stem fascination; *deh*: determinate habit), and nodulation for nitrogen fixation (*Sym*, *Nod*), *lip1* (light independent photomorphogenesis 1) (Sullivan and Grey, 2000); homologs for terminate flower 1 gene (*TFL1*), *PsTFL1* a corresponding to *DETERMINATE* (*DET*) gene, and *PsTFL1* corresponding to late flowering (*LF*), *Is-1* mutant for gibberellic acid (Ait-Ali et al., 1997); *rigosus* mutant (R) in starch-branching enzyme gene (*SBE-1*) (Bhattachryya et al., 1990). Development of molecular markers linked with these traits and genome mapping of these mutants are lagging. Further improvement can be done by utilizing the developed markers in MAS and MABC breeding program.

6.9 CHALLENGES TOWARD ALLELE MINING IN PEA

Allele mining in pea is a cumbersome task due to several constraints posed by the processes involved. The initial process involves choosing diverse genetic resources for the identification of various alleles. The germplasm collections in various centers around the globe provide a viable solution to the diversity required for identification of alleles. Now the challenge is to efficiently identify the allelic variation using suitable techniques. The multitude of challenges that impede the progress of allele mining is discussed in brief hereafter, with suggestions to overcome the challenges for more efficient allele mining.

6.9.1 Selection of Genotypes

The initial criterion for allele mining requires sufficient variability in the genotypes selected for allele mining to identify varied new alleles as well as rare alleles. But the analysis of an entire collection is an inefficient and unwieldy task. The construction of core and mini-core sets of germplasm, followed by precise and concise phenotyping and use of updated computational tools, can facilitate and expedite the task. Core germplasm for pea is available at several global germplasm collection centers, though the number is meager (Smýkal et al., 2008). Most of the international organizations with large germplasm collections were unable to generate core and mini-core sets for pea. Stable tools for identification of genotypes by Kim et al. (2007) (PowerCore) and De Beukelaer et al. (2018) (Core Hunter 3) are helpful to characterize the core germplasm with all the essential characteristics for utilization of more varied alleles from gene banks.

6.9.2 Handling of Enormous Genomic Data

The analysis of the germplasm generates molecular data that needs precise handling to identify all the different alleles recognized and documented over time. The functional nucleotide identification and assigning the variations to the changes in expression need to be incorporated in the data. All these tasks need multi-disciplinary involvement to concisely analyze and understand the data and implement the data successfully in the allele mining set for future development of the pea.

6.9.3 Characterization of Promoter and Regulatory Regions

Promoter regions are also included in true allele mining. But demarcation of regulatory regions poses a challenge, as discerning the regions is not possible even in fully sequenced genomes. Promoters and cis-acting regulatory elements are generally present upstream, but the regions of their presence vary from −100 bp to −3,000 bp (Park et al., 2002; Veerla and Hoglund, 2006). Thus, functional annotation of promoter through over-expression analysis is a necessary task for true allele mining and needs the development of computational tools for more precise identification of regulatory regions. Moreover, in regards to transcriptional factors, there is a chance that more than one gene is controlled by the same factor (Hehl and Wingender, 2001). The presence of trans-regulatory elements (Dai et al., 2007) also adds to the complexity. These can be ameliorated to some extent by experimental analysis and examination by using the bioinformatics tools mentioned in this chapter.

6.9.4 Exorbitant Sequencing Costs

The sequence changes in the loci of interest need sequencing of the diverse panel for the gene of interest as well as information related to promoter regions from genome data. The sequencing incurs high cost and is also time-consuming. The development of NGS tools can remediate the process and provide for high-throughput detection of alleles.

6.10 WAY FORWARD

The paradox of pea breeding program lies in the significant disparity between its well-established status as a model organism in genetics study and its scantiness in genomics-enabled breeding program. The outcomes of international genome sequencing projects for pea (Kreplak et al., 2019; Yan et al., 2022) have created exciting opportunities to enhance genomic resources. These projects have facilitated to gather information from NGS-based whole-genome sequencing, genome-wide SNP mining, gene mapping, and their subsequent association with phenotypic variations. This integration opens new possibilities for their application in improved variety development programs in pea.

The allele mining approach holds significant promise for uncovering and utilizing allelic variations present in genetic resources associated with resistance against biotic and abiotic cues, as well as yield attributing and quality traits. This creates a crucial link between genetic and genomic resources, enabling the efficient incorporation of these findings into current pea breeding programs through extending genomics-driven breeding. To keep up with the continuously growing sequence data in GenBank and the expanding pea core collection maintained in the various global gene banks, it is of utmost priority to develop innovative and efficient strategies and computational tools toward formulation of core and mini-core sets integrating phenotyping and genotyping approaches to capture maximum allelic diversity of the gene pool. Therefore, the development of a trait-specific subset is a vital endeavor for digging out novel alleles and desirable allelic variations associated with trait of interest. Fortunately, pea with limited species enrichment has a wealth of vast germplasm collection with exciting natural mutant populations. Pressing need is felt to integrate cutting-edge tools that leverage pea genetic resources and mutant collections for identifying functional nucleotide polymorphisms in superior alleles linked to candidate genes and their further introgression in the desired genetic background through marker-assisted breeding programs. There is tremendous potential to expand the application of allele mining to unveil the evolutionary significance of favorable alleles as well as in elucidating synteny with related species. The unprecedented development of NGS-based technology enabled the development of "GenoPea 13.2K SNP Array" (Tayeh et al.,

TABLE 6.1
List of Institutes in Different Countries Maintaining Pea Core Collection Set

S. N.	Country	Code	Website	Institute	Accession No.	Activity
1.	USA	USDA	https://www.ars.usda.gov/	Western Regional Plant Introduction Station, Washington State University, Pullman	6,904	Online cataloguing, partial genotyping, phenotyping, SNP calling, genomic selection, core set
2.	Australia	ATFCC	https://www.genesys-pgr.org/wiews/AUS039	Australian Temperate Field Crops Collection, Department of Primary Industries (DPI), Horsham, Victoria	7,432	Genotyping, phenotyping, core set generation
3.	United Kingdom	JIC	https://www.jic.ac.uk/	John Innes Centre, Norwich	3567	Online cataloguing, genotyping, phenotyping, core set generation
4.	France	INRA	https://urgi.versailles.inra.fr/siregal/siregal/grc.do	Institut National pour la Recherche Agronomique, CRG Protéagineux, INRA, 17 rue de Sully, 21065 Dijon	8,839	Online cataloguing, partial genotyping, phenotyping, core set generation
5.	Czech Republic	CZE	https://dms.vurv.cz/	Agritec Research, Breeding & Services Ltd., Šumperk	1,326	Online cataloguing, genotyping, phenotyping, core set generation
6.	Canada	CAN	https://pgrc-rpc.agr.gc.ca/gringlobal/landing	Plant Gene Resources of Canada, Saskatchewan	616	Genotyping, phenotyping, core set generation

Source: Adopted from Smýkal et al. (2013); Warkentin et al. (2015).

TABLE 6.2
List of Bioinformatics Tools Useful for Allele Mining

S. N.	Name	Utility	URL	Reference
Development of the Core Set of Germplasm				
1.	PowerCore	Preparation of core set based on characterization data on descriptor traits	http://genebank.rda.go.kr/powercore/	Kim et al. (2007)
2.	Core Hunter 3	Core set preparation using distance-based measures to account for diversity and representativeness	https://github.com/corehunter/corehunter3	De Beukelaer et al. (2018)
Sequence Analysis				
1.	MEGA	Integrated tool for multiple sequence alignment, construction of phylogenetic trees, divergence times estimation, online database mining, and testing evolutionary hypotheses by calculating rate of molecular evolution through ancestral sequence identification	http://www.megasoftware.net/	Tamura et al. (2013)
2.	MUSCLE	Nucleotide sequence analysis	https://www.ebi.ac.uk/Tools/msa/muscle/	Madeira et al. (2019)
3.	BioEdit	Nucleotide sequence analysis and further manipulation	www.mbio.ncsu.edu/BioEdit/BioEdit.html	Hall (1999)
4.	ClustalW	Multiple sequence alignment	www.ebi.ac.uk/Tools/clustralw/	-
5.	ClustalOMEGA	Multiple sequence alignment	https://www.ebi.ac.uk/Tools/msa/clustalo/	Madeira et al. (2019)
6.	DnaSP	Identifies SNPs and InDels along with phylogeny analysis based on the dissimilarity of DNA sequences	http://www.ub.edu/dnasp	Librado and Rozas (2009)
Primer Design				
1.	FastPCR	PCR primers or probes design, along with in silico PCR, oligonucleotide assembly and analyses, alignment and repeat searching for a single sequence or for comparisons of two sequences	http://en.bio-soft.net/pcr/FastPCR.html http://www.fastpcr.software.informer.com/ www.biocenter.helsinki.fi/bi/Programs/fastpcr.htm	Kalendar et al. (2009)
2.	Primer3	Suggests PCR primers for a variety of applications, for example, to create STSs (sequence-tagged sites) for radiation hybrid mapping, or to amplify sequences for single-nucleotide polymorphism discovery; Primer3 can also select single primers for sequencing reactions and can design oligonucleotide hybridization probes	http://frodo.wi.mit.edu/primer3/	Rozen and Skaletsky (2000)
3.	PrimerBLAST	Primer design and binding site analysis tool	https://www.ncbi.nlm.nih.gov/tools/primer-blast/	Ye et al. (2012)

Promoter Databases				
1.	Plantprom DB	Annotated, non-redundant plant proximal promoter for RNA polymerase II and transcription start site(s) (TSS) database	http://mendel.cs.rhul.ac.uk/mendel.php?topic=plantprom http://linux1.softberry.com/berry.phtml?topic=plantprom&group=data&subgroup=plantprom	Shahmuradov et al. (2003)
2.	PLACE (plant cis acting regulatory DNA elements)	Transcription factor and cis-acting regulatory element motif database	http://www.dna.affrc.go.jp/PLACE/index.html	Higo et al. (1999)
3.	GRAMENE	Data source for functional genomics research comparing crops and model plant species	https://www.gramene.org/	TelloRuiz et al. (2018)
4.	PlantTFDB	Plant transcription factor database	http://plantregmap.cbi.pku.edu.cn	Tian et al. (2020)
5.	EPD	Database of eukaryotic transcription start site (TSS) for POL II	http://www.epd.isb-sib.ch/	Schmid et al. (2004)
6.	TRRD	Database for information regarding localization and functions of the cis-acting regulatory elements and transcription factor binding site (TFBS) as well as gene expression patterns of the genes	http://wwwmgs.bionet.nsc.ru/mgs/gnw/trrd/	Kolchanov et al. (2000)
7.	JASPAR	Database of transcription factor binding site (TFBS)	http://jaspar.genereg.net/	Bryne et al. (2008)
8.	DCPD	Drosophila core promoter database for upstream as well as downstream promoter elements	http://www.biology.ucsd.edu/labs/Kadonaga/DCPD.html	Kutach and Kadonaga (2000)
9.	SCPD	*Saccharomyces cerevisiae* promoter database; SCPD also provides tools for promoter sequence analysis	http://rulai.cshl.edu/SCPD/	Zhu and Zhang (1999)
10.	AGRIS	Database of *Arabidopsis* cis- promoter sequences, transcription factors (TF), and their target genes	http://arabidopsis.med.ohio-state.edu	Davuluri et al. (2003)
11.	PlantCARE	Plant cis-acting regulatory elements and transcription factor database	http://bioinformatics.psb.ugent.be/webtools/plantcare/html	Lescot et al. (2002)
12.	TRANSFAC	Database of transcription factors and their regulatory sites along with target genes	http://www.gene-regulation.com/pub/programs.html	Matys et al. (2003)
13.	TRED	Mammalian regulatory elements database	http://rulai.cshl.edu/cgi-in/TRED/tred.cgi?process=home	Jiang et al. (2007)
14.	ooTFD	Object-oriented transcription factors database	http://www.ifti.org/ootfd/	Ghosh (1998)

(Continued)

TABLE 6.2 ***(Continued)***
List of Bioinformatics Tools Useful for Allele Mining

S. N.	Name	Utility	URL	Reference
Analysis Tools				
1.	W-AlignACE	Effective tool for discovering TF binding sites from gene expression or ChIP-chip data; motif discovery tool	http://www1.spms.ntu.edu.sg/~chenxin/W-AlignACE/	Chen et al. (2008)
2.	MEME (Multiple EM for Motif Elicitation)	Analyzing DNA and protein sequence motifs for similarities among them; motif discovery tool	http://meme.nbcr.net/meme/cgi-in/meme.cgi	Bailey et al. (2006)
3.	MATinspector	To predict TFBS and for promoter analysis	http://www.genomatix.de/products/MatInspector/	Cartharius et al. (2005)
4.	ModuleFinder and CoReg	Alternative tools for linking gene expression modules with promoter sequences motifs to uncover gene regulation mechanisms in plants; ModuleFinder provided a subset of co-expressed gene modules that are more logically related to biological functions than did subsets derived from traditional hierarchical clustering techniques	http://www.plantmethods.com/content/2/1/8/	Holt et al. (2006)
5.	DiAlign TF	DiAlign TF displays transcription factor (TF) binding site matches within a multiple alignment; database for promoter analysis	http://www.genomatix.de/online_help/help_dialign/dialign_TF.html	Cartharius et al. (2005)
6.	FrameWorker	A complex software tool that allows users to extract a common framework of elements from a set of DNA sequences; motif discovery tool	http://www.genomatix.de/online_help/help_gems/FrameWorker.html	Cartharius et al. (2005)
7.	AlleleCatalog	A web-based interactive tool for allele discovery and analysis	https://kbcommons.org/system/tools/AlleleCatalogTool/Zmays (Maize) https://kbcommons.org/system/tools/AlleleCatalogTool/Athaliana (*Arabidopsis*) https://soykb.org/SoybeanAlleleCatalogTool/ (Soybean)	Chan et al. (2023)

2015; Siol et al., 2017; Tran et al., 2023), which facilitated easy and cost-effective detection of genome-wide SNPs and characterization of novel allelic variations at the targeted trait. In the realm of efficient pea breeding programs, the current imperative lies in utilizing NGS-based allele mining to unearth natural variations, identify novel and desirable alleles, and subsequently validate them in diverse genetic backgrounds to facilitate trait improvement. Unfortunately, the lack of international support has severely impacted pea breeding programs. The availability of extensive genomic resources and efficient data mining tools enables the effective integration of genetic and genomic resources, thereby supporting the development of improved pea cultivars to ensure food security.

REFERENCES

Abed A, Belzile F (2019) Comparing single-SNP, multi-SNP, and haplotype-based approaches in association studies for major traits in barley. *Plant Genome* 12(3):190036.

Ait-Ali T, Swain S M, Reid JB, Sun TP, Kamiya Y (1997) The LS locus of pea encodes the gibberellin biosynthesis enzyme ent-kaurene synthase A. *Plant J* 11(3):443–454.

Amini S, Pushkarev D, Christiansen L, Kostem E, Royce T et al. (2014) Haplotype-resolved whole-genome sequencing by contiguity-preserving transposition and combinatorial indexing. *Nat Genet* 46(12):1343–1349.

Arumuganathan K, Earle ED (1991) Nuclear DNA content of some important plant species. *Plant Mol Biol Rep* 9:208–218.

Ashkani S, Yusop MR, Shabanimofrad M, Azadi A, Ghasemzadeh A et al. (2015) Allele mining strategies: Principles and utilisation for blast resistance genes in rice (*Oryza sativa* L.). *Curr Iss Mol Biol* 17(1):57–74.

Ashtari Mahini R, Kumar A, Elias EM, Fiedler JD, Porter LD et al. (2020) Analysis and identification of QTL for resistance to *Sclerotinia sclerotiorum* in pea (*Pisum sativum* L.). *Front Genet* 11:587968.

Aubert G, Morin J, Jacquin F, Loridon K, Quillet MC et al. (2006) Functional mapping in pea, as an aid to the candidate gene selection and for investigating synteny with the model legume *Medicago truncatula*. *Theor Appl Genet* 112:1024–1041.

Ayeh KO, Lee Y, Ambrose MJ, Hvoslef-Eide, AK (2009) Characterization and structural analysis of wild type and a non-abscission mutant at the development funiculus (*Def*) locus in *Pisum sativum* L. *BMC Plant Biol* 9, 1–7.

Aznar-Fernández T, Barilli E, Cobos MJ, Kilian A, Carling J et al. (2020) Identification of quantitative trait loci (QTL) controlling resistance to pea weevil (*Bruchus pisorum*) in a high-density integrated DArTseq SNP-based genetic map of pea. *Sci Rep* 10(1):33.

Bailey TL, Williams N, Misleh C, Li WW (2006) MEME: Discovering and analyzing DNA and protein sequence motifs. *Nucleic Acids Res* 34(suppl-2):W369–W373.

Balfourier F, Roussel V, Strelchenko P, Exbrayat-Vinson F, Sourdille P et al. (2007) A worldwide bread wheat core collection arrayed in a 384-well plate. *Theor Appl Genet* 114(7):1265–1275.

Banyal DK, Dixit H, Chaudhary J, Malannavar AB, Thakur N (2022) Deciphering diversity at *er* loci for diversification of powdery mildew resistance in pea. *Sci Rep* 12(1):16037.

Bari MAA, Fonseka D, Stenger J, Zitnick-Anderson K, Atanda SA et al. (2023) A greenhouse-based high-throughput phenotyping platform for identification and genetic dissection of resistance to Aphanomyces root rot in field pea. *Plant Phenome J* 6(1):e20063.

Barilli E, Carrillo-Perdomo E, Cobos MJ, Kilian A, Carling J et al. (2020) Identification of potential candidate genes controlling pea aphid tolerance in a *Pisum fulvum* high-density integrated DArTseq SNP-based genetic map. *Pest Manag Sci* 76(5):1731–1742.

Barilli E, Cobos MJ, Carrillo E, Kilian A, Carling J et al. (2018) A high-density integrated DArTseq SNP-based genetic map of *Pisum fulvum* and identification of QTLs controlling rust resistance. *Front Plant Sci* 9:167.

Barkley NA, Wang ML, Gillaspie AG, Dean RE, Pederson GA, Jenkins TM (2008) Discovering and verifying DNA polymorphisms in a mung bean [V. radiata (L.) R. Wilczek] collection by EcoTILLING and sequencing. *BMC Res Notes* 1:1–7.

Basigalup DH, Barnes DK, Stucker RE (1995) Development of a core collection for perennial Medicago plant introductions. *Crop Sci* 35(4):1163–1168.

Bastida Garcia JL, González-Ronquillo M, Dominguez Vara IA, Romero-Bernal J, Castelan Ortega O (2011) Effect of field pea (*Pisum sativum* L.) level on intake, digestion, ruminal fermentation and in vitro gas production in sheep fed maintenance diets. *Anim Sci J* 82(5):654–662.

Beji S, Fontaine V, Devaux R, Thomas M, Negro SS et al. (2020) Genome-wide association study identifies favorable SNP alleles and candidate genes for frost tolerance in pea. *BMC Genomics* 21(1):1–21.

Bekele WA, Itaya A, Boyle B, Yan W, Mitchell Fetch J et al. (2020) A targeted genotyping-by-sequencing tool (rapture) for genomics-assisted breeding in oat. *Theor Appl Genet* 133:653–664.

Bhattacharyya MK, Smith AM, Ellis TN, Hedley C, Martin C (1990) The wrinkled-seed character of pea described by Mendel is caused by a transposon-like insertion in a gene encoding starch-branching enzyme. *Cell* 60(1):115–122.

Blixt S (1972) Mutation genetics in *Pisum. Agri Hort Genet* 30.

Biederbeck VO, Zenter RP, Campbell CA (2005) Soil microbial populations and activities as influenced by legume green fallow in a semiarid climate. *Soil Biol Biochem* 37:1775–1784.

Bordat A, Savois V, Nicolas M, Salse J, Chauveau A et al. (2011) Translational genomics in legumes allowed placing *in silico* 5460 unigenes on the pea functional map and identified candidate genes in *Pisum sativum* L. *G3 Genes Genomes Genet* 1(2):93–103.

Bourgeois M, Jacquin F, Cassecuelle F, Savois V, Belghazi M et al. (2011) A *PQL* (protein quantity loci) analysis of mature pea seed proteins identifies loci determining seed protein composition. *Proteomics* 11(9):1581–1594.

Boutet G, Lavaud C, Lesné A, Miteul H, Pilet-Nayel ML et al. (2023) Five regions of the pea genome co-control partial resistance to *D. pinodes*, tolerance to frost, and some architectural or phenological traits. *Genes* 14(7):1399.

Boutte J, Aliaga B, Lima O, Ferreira de Carvalho J, Ainouche A et al. (2016) Haplotype detection from next-generation sequencing in high-ploidy-level species: 45S rDNA gene copies in the hexaploid Spartina maritima. *G3: Genes Genom Genet* 6(1):29–40.

Brown AHD (1989) Core collections: A practical approach to genetic resources management. *Genome* 31(2):818–824.

Brown AHD, Spillane C (1999) Implementing core collections—principles, procedures, progress, problems and promise. In: Johnson RC, Hodgkin T (eds) *Core Collections for Today and Tomorrow*. Crop Science Society of America, Madison, WI, pp. 1–10.

Bryne JC, Valen E, Tang MHE, Marstrand T, Winther O et al. (2007) JASPAR, the open access database of transcription factor-binding profiles: New content and tools in the 2008 update. *Nucleic Acids Res* 36(suppl_1):D102–D106.

Carpenter MA, Shaw M, Cooper RD, Frew TJ, Butler RC et al. (2017) Association mapping of starch chain length distribution and amylose content in pea (*Pisum sativum* L.) using carbohydrate metabolism candidate genes. *BMC Plant Biol* 17:1–23.

Carrillo E, Satovic Z, Aubert G, Boucherot K, Rubiales D et al. (2014) Identification of quantitative trait loci and candidate genes for specific cellular resistance responses against *Didymella pinodes* in pea. *Plant Cell Rep* 33:1133–1145.

Cartharius K, Frech K, Grote K, Klocke B, Haltmeier M et al. (2005) MatInspector and beyond: Promoter analysis based on transcription factor binding sites. *Bioinformatics* 21(13):2933–2942.

Chan YO, Dietz N, Zeng S, Wang J, Flint-Garcia S et al. (2023) The Allele Catalog Tool: A web-based interactive tool for allele discovery and analysis. *BMC Genomics* 24(1):107.

Chatterton S, Bowness R, Harding MW (2015) First report of root rot of field pea caused by *Aphanomyces euteiches* in Alberta, Canada. *Plant Dis* 99(2):288.

Chen J, Moreau C, Liu Y, Kawaguchic M, Hofer J et al. (2012) Conserved genetic determinant of motor organ identity in *Medicago truncatula* and related legumes. *Proc Natl Acad Sci USA* 109(11):723–728. https://doi.org/10.1073/pnas.1204566109

Chen X, Guo L, Fan Z, Jiang T (2008) W-AlignACE: An improved Gibbs sampling algorithm based on more accurate position weight matrices learned from sequence and gene expression/ChIP-chip data. *Bioinformatics* 24(9):1121–1128.

Cheng P, Holdsworth W, Ma Y, Coyne CJ, Mazourek, M et al. (2015) Association mapping of agronomic and quality traits in USDA pea single-plant collection. *Mol Breed* 35:1–13.

Choi HK, Mun JH, Kim DJ, Zhu H et al. (2004) Estimating genome conservation between crop and model legume species. *Proc Natl Acad Sci USA* 101(43):15289–15294.

Chung HJ, Liu Q, Hoover R (2009) Impact of annealing and heat-moisture treatment on rapidly digestible, slowly digestible and resistant starch levels in native and gelatinized corn, pea and lentil starches. *Carbohydr Polym* 75(3):436–447.

Clark A (2007) *Managing Cover Crops Profitably*. Sustainable Agriculture Research and Education Program Handbook Series-9, 3rd edn. Sustainable Agriculture Research and Education, College Park, MD.

Comai L, Young K, Till BJ, Reynolds SH, Greene EA et al. (2004) Efficient discovery of DNA polymorphisms in natural populations by Ecotilling. *Plant J* 37(5):778–786.

Comai L, Henikoff S (2006) TILLING: Practical single-nucleotide mutation discovery. *Plant J* 45(4):684–694.

Coyne CJ, Pilet-Nayel ML, McGee RJ, Porter LD, Smýkal P et al. (2015) Identification of QTL controlling high levels of partial resistance to *Fusarium solani* f. sp. *pisi* in pea. *Plant Breed* 134(4):446–453.

Dai LY, Liu XL, Xiao YH, Wang GL (2007) Recent advances in cloning and characterization of disease resistance genes in rice. *J Integr Plant Biol* 49(1):112–119.

Dasgupta N, Alldredge JR (1998) A multivariate chi-square analysis of resource selection data. *J Agric Biol Environ Stat* 22:583–589.

Davuluri RV, Sun H, Palaniswamy SK, Matthews N, Molina C et al. (2003) AGRIS: Arabidopsis gene regulatory information server, an information resource of Arabidopsis cis-regulatory elements and transcription factors. *BMC Bioinformatics* 4:1–11.

De Beukelaer H, Davenport GF, Fack V (2018) Core Hunter 3: Flexible core subset selection. *BMC Bioinformatics* 19:203. https://doi.org/10.1186/s12859-018-2209-z

Delvento C, Arcieri F, Marcotrigiano AR, Guerriero M, Fanelli V et al. (2023) High-density linkage mapping and genetic dissection of resistance to broomrape (*Orobanche crenata* Forsk.) in pea (*Pisum sativum* L.). *Front Plant Sci* 14. https://doi.org/10.3389/fpls.2023.1216297

Desgroux A, Baudais VN, Aubert V, Le Roy G, De Larambergue H et al. (2018) Comparative genome-wide-association mapping identifies common loci controlling root system architecture and resistance to *Aphanomyces euteiches* in pea. *Front Plant Sci* 8:2195.

Desgroux A, L'anthoëne V, Roux-Duparque M, Rivière JP, Aubert G et al. (2016a) Genome-wide association mapping of partial resistance to *Aphanomyces euteiches* in pea. *BMC Genomics* 17:1–21.

Desgroux A, L'anthoëne V, Roux-Duparque M, Riviere JP, Aubert, G et al. (2016b) Effects of seed priming with growth regulator biolan on salt tolerance of *Pisum sativum* L. *Biologiqnictydi* 10(3–4):107–118.

Dalmais M, Schmidt J, Le Signor C, Moussy F, Burstin J et al. (2008) UTILLdb, a Pisum sativum in silico forward and reverse genetics tool. *Genome Biol* 9:1–12.

D'erfurth I, Le Signor C, Aubert G, Sanchez M, Vernoud V et al. (2012) A role for an endosperm-localized subtilase in the control of seed size in legumes. *New Phytol* 196(3):738–751.

Deulvot C, Charrel H, Marty A, Jacquin F, Donnadieu C et al. (2010) Highly-multiplexed SNP genotyping for genetic mapping and germplasm diversity studies in pea. *BMC Genomics* 11:1–10.

De Vilmorin P (1913) Sur une race de blé nain infixable. *J Genet* 3(1):67–76.

Diapari M, Sindhu A, Warkentin TD, Bett K, Tar'an B (2015) Population structure and marker-trait association studies of iron, zinc and selenium concentrations in seed of field pea (*Pisum sativum* L.). *Mol Breed* 35:1–14.

Dissanayaka DN, Gali KK, Jha AB, Lachagari VR, Warkentin TD (2020) Genome-wide association study to identify single nucleotide polymorphisms associated with Fe, Zn, and Se concentration in field pea. *Crop Sci* 60(4):2070–2084.

Diwan N, McIntosh MS, Bauchan GR (1995) Methods of developing a core collection of annual Medicago species. *Theor Appl Genet* 90:755–761.

Domoney C, Knox M, Moreau C, Ambrose M, Palmer S et al. (2013) Exploiting a fast neutron mutant genetic resource in Pisum sativum (pea) for functional genomics. *Funct Plant Biol* 40:1261–1270.

Duarte J, Rivière N, Baranger A, Aubert G, Burstin J et al. (2014) Transcriptome sequencing for high throughput SNP development and genetic mapping in Pea. *BMC Genomics* 15:1–15.

Duc G, Messager A (1989) Mutagenesis of pea (Pisum sativum L.) and the isolation of mutants for nodulation and nitrogen fixation. *Plant Sci* 60(2):207–213.

FAO (2020) *Food and agriculture organization statistics*. https://www.fao.org/faostat/en/#data/QCL

Feenstra WJ, Jacobsen E (1980) A Chlorate resistant nitrate reductase-deficient pea mutant. *Mutat Res* 51:29–35.

Feng J, Hwang R, Chang KF, Conner RL, Hwang SF et al. (2011) Identification of microsatellite markers linked to quantitative trait loci controlling resistance to Fusarium root rot in field pea. *Can J Plant Sci* 91(1):199–204.

Ferrari B, Romani M, Aubert G, Boucherot K, Burstin, J et al. (2016) Association of SNP markers with agronomic and quality traits of field pea in Italy. *Czech J Genet Plant Breed* 52(3):83–93.

Fondevilla S, Rubiales D (2012) Powdery mildew control in pea. A review. *Agron Sustain Dev* 32:401–409.

Fondevilla S, Rubiales D, Moreno MT, Torres AM (2008) Identification and validation of RAPD and SCAR markers linked to the gene Er3 conferring resistance to Erysiphe pisi DC in pea. *Mol Breed* 22:193–200.

Fondevilla S, Torres AM, Moreno MT, Rubiales D (2007) Identification of a new gene for resistance to powdery mildew in *Pisum fulvum*, a wild relative of pea. *Breed Sci* 57(2):181–184.

Foucher F, Morin J, Courtiade J, Cadioux S et al. (2003) Determinate and late flowering are two terminal flower1/centroradialis homologs that control two distinct phases of flowering initiation and development in pea. *Plant Cell* 15(11):2742–2754.

Frankel OH (1984) Genetic perspectives of germplasm conservation. In: Arber WK, Limensee K, Peacock WJ, Stranlinger P (eds) *Genetic Manipulation: Impact on Man and Society*. Cambridge University Press, Cambridge, England, pp. 161–170.

Gali KK, Liu Y, Sindhu A, Diapari M, Shunmugam AS et al. (2018) Construction of high-density linkage maps for mapping quantitative trait loci for multiple traits in field pea (*Pisum sativum* L.). *BMC Plant Biol* 18(1):1–25.

Gali KK, Sackville A, Tafesse EG, Lachagari VR et al. (2019) Genome-wide association mapping for agronomic and seed quality traits of field pea (*Pisum sativum* L.). *Front Plant Sci* 10:1538.

Gawenda I, Thorwarth P, Günther T, Ordon F et al. (2015) Genome-wide association studies in elite varieties of German winter barley using single-marker and haplotype-based methods. *Plant Breed* 134(1):28–39.

Ghosh D (1998) OOTFD (Object-Oriented Transcription Factors Database): An object-oriented successor to TFD. *Nucleic Acids Res* 26(1):360–361.

Gilchrist EJ, Haughn GW, Ying CC, Otto SP, Zhuang J et al. (2006) Use of Ecotilling as an efficient SNP discovery tool to survey genetic variation in wild populations of *Populus trichocarpa*. *Mol Ecol* 15(5):1367–1378.

Gnanadesikan R, Wilk MB (1968) Probability plotting methods for the analysis of data. *Biometrika* 55(1):1–17.

Gokidi Y, Bhanu AN, Chandra K, Singh MN, Hemantaranjan A (2017) Allele mining – An approach to discover allelic variation in crops. *J Plant Sci Res* 33(2):167–180.

Goldmann IL, Gritton ET (1992) Evaluation of the Afila-Tendriled Acacia (*afaf-tactac*) pea foliage type under minimal composition. *Crop Sci* 32:851–855.

Gong YM, Xu SC, Mao WH, Hu QZ, Zhang, GW et al. (2010) Developing new SSR markers from ESTs of pea (*Pisum sativum* L.). *J Zhejiang Univ Sci B* 11:702–707.

Gouesnard B, Bataillon TM, Decoux G, Rozale C, Schoen DJ et al. (2001) MSTRAT: An algorithm for building germplasm core collections by maximizing allelic or phenotypic richness. *J Hered* 92(1):93–94.

Graner A (2006) *Barley Research at IPK: Molecular Markers for Allele Mining*. IPGRI, Rome, p. 25.

Gupta P (2008) Ultrafast and low-cost DNA sequencing methods for applied genomics research. *Proc Indian Natl Sci Acad Part B Biol Sci* 78:91–102.

Gupta A, Singh B, Kumar M, Sirohi U, Yadav SK et al. (2023) Identification of SSR Markers Linked to Powdery Mildew Resistance in Table Pea (*Pisum sativum* var. *hortense* L.). *Legume Res* 1:7. https://doi.org/10.18805/LR5085

Hall TA (1999) BioEdit: A user-friendly biological sequence alignment editor and analysis program for Windows 95/98/NT. *Nucleic Acids Symposium Series* 41:95–98.

Hamazaki K, Iwata H (2020) RAINBOW: Haplotype-based genome-wide association study using a novel SNP-set method. *PLoS Comput Biol* 16(2):e1007663.

Hamon C, Baranger A, Coyne CJ, McGee RJ, Le Goff I et al. (2011) New consistent QTL in pea associated with partial resistance to *Aphanomyces euteiches* in multiple French and American environments. *Theor Appl Genet* 123:261–281.

Hamon C, Coyne CJ, McGee RJ, Lesné A, Esnault R et al. (2013) QTL meta-analysis provides a comprehensive view of loci controlling partial resistance to *Aphanomyces euteichesin* four sources of resistance in pea. *BMC Plant Biol* 13(1):1–19.

Hehl R, Wingender E (2001) Database-assisted promoter analysis. *Trends Plant Sci* 6(6):251–255.

Heidari P, Ahmadizadeh M, Najafi-Zarrini H (2015) In silico analysis of Cis-regulatory elements on co-expressed genes. *J Biol Environ Sci* 9(25):1–9.

Hellens RP, Moreau C, Lin-Wang K, Schwinn KE, Thomson SJ et al. (2010) Identification of Mendel's white flower character. *PLoS One* 5:e13230. https://doi.org/10.1371/journal.pone.0013230

Higo K, Ugawa Y, Iwamoto M, Korenaga T (1999) Plant cis-acting regulatory DNA elements (PLACE) database. *Nucleic Acids Res* 27(1):297–300.

Hofer J, Turner L, Moreau C, Ambrose M, Isaac P et al. (2009) Tendril-less regulates tendril formation in pea leaves. *Plant Cell* 21(2):420–428.

Holt KE, Harvey Millar A, Whelan J (2006) ModuleFinder and CoReg: Alternative tools for linking gene expression modules with promoter sequences motifs to uncover gene regulation mechanisms in plants. *Plant Methods* 2(1):1–15.

Hu J, Zhu J, Xu HM (2000) Methods of constructing core collections by stepwise clustering with three sampling strategies based on the genotypic values of crops. *Theor Appl Genet* 101:264–268.

Huang X, Feng Q, Qian Q, Zhao Q, Wang L et al. (2009) High-throughput genotyping by whole-genome resequencing. *Genome Res* 19:1068–1076.

Humphry M, Reinstaedler A, Ivanov S, Bisseling TON, Panstruga R (2011) Durable broad-spectrum powdery mildew resistance in pea er1 plants is conferred by natural loss-of-function mutations in PsMLO1. *Mol Plant Pathol* 12(9):866–878.

Iglesias-García R, Prats E, Fondevilla S, Satovic Z, Rubiales D (2015) Quantitative trait loci associated to drought adaptation in pea (*Pisum sativum* L.). *Plant Mol Biol Rep* 33:1768–1778.

Irzykowska L, Wolko B, Swiecicki WK (2002) Interval mapping of QTLs controlling some morphological traits in pea. *Cell Mol Biol Lett* 7(2A):417–422.

Jacobsen E, Feenstra WJ (1984) A new pea mutant with efficient nodulation in the presence of nitrate. *Plant Sci Lett* 33(3):337–344.

Jan HU, Guan M, Yao M, Liu W, Wei D et al. (2019) Genome-wide haplotype analysis improves trait predictions in Brassica napus hybrids. *Plant Sci* 283:157–164.

Jansen J, Van Hintum TH (2007) Genetic distance sampling: A novel sampling method for obtaining core collections using genetic distances with an application to cultivated lettuce. *Theor Appl Genet* 114:421–428.

Jha AB, Gali KK, Zhang H, Purves RW, Tar'an B et al. (2020) Folate profile diversity and associated SNPs using genome wide association study in pea. *Euphytica* 216:1–16.

Jha AB, Tar'an B, Diapari M, Sindhu A, Shunmugam A et al. (2015) Allele diversity analysis to identify SNPs associated with ascochyta blight resistance in pea. *Euphytica* 202:189–197.

Jiang C, Xuan Z, Zhao F, Zhang MQ (2007) TRED: A transcriptional regulatory element database, new entries and other development. *Nucleic Acids Res.* 35(suppl 1):D137–D140.

Jiang Y, Diapari M, Bueckert RA, Tar'an B, Warkentin TD (2017) Population structure and association mapping of traits related to reproductive development in field pea. *Euphytica* 213:1–20.

Jing R, Johnson R, Seres A, Kiss G, Ambrose MJ et al. (2007) Gene based sequence diversity analysis of field pea (*Pisum*). *Genetics* 177:2263–2275.

Kalendar R, Lee D, Schulman AH (2009) FastPCR software for PCR primer and probe design and repeat search. *G3 Genes Genomes Genom* 3(1):1–14.

Kalo P, Seres A, Taylor SA, Jakab J, Kevei Z et al. (2004) Comparative mapping between Medicago sativa and Pisum sativum. *Mol Genet Genom* 272:235–246.

Kaur N (2008) Allele Mining and Sequence Diversity at the Wheat Powdery Mildew Resistance Locus Pm3. Doctoral dissertation, University of Zurich, Zurich.

Kilian B, Graner A (2012) NGS technologies for analyzing germplasm diversity in genebanks. *Brief Funct Genom* 11(1):38–50.

Kim KW, Chung HK, Cho GT, Ma KH, Chandrabalan D et al. (2007) PowerCore: A program applying the advanced M strategy with a heuristic search for establishing core sets. *Bioinformatics* 23(16):2155–2162.

Klein A, Houtin H, Rond-Coissieux C, Naudet-Huart M, Touratier M et al. (2020) Meta-analysis of QTL reveals the genetic control of yield-related traits and seed protein content in pea. *Sci Rep* 10(1):15925.

Kneen BE, LaRue TA (1984) Nodulation resistant mutant of *Pisum sativum* (L.). *J Hered* 75(3):238–240.

Kneen BE, LaRue TA, Welch RM, Weeden NF (1990) Pleiotropic effects of brz. A mutation in Pisum sativum (L.) cv 'Sparkle' conditioning decreased nodulation and increased iron uptake and leaf necrosis. *Plant Physiol* 93:717–722.

Kof EM, Vinogradova IA, Oorzhak AS, Kalibernaya ZV (2006) The rates of shoot and root growth in intact plants of pea mutants in leaf morphology. *Russ J Plant Physiol* 53:116–125.

Kolchanov NA, Podkolodnaya OA, Ananko EA, Ignatieva EV, Stepanenko IL et al. (2000) Transcription regulatory regions database (TRRD): Its status in 2000. *Nucleic Acid Res* 28:298–301.

Koyama ML, Levesley A, Koebner RM, Flowers TJ, Yeo AR (2001) Quantitative trait loci for component physiological traits determining salt tolerance in rice. *Plant Physiol* 125(1):406–422.

Kreplak J, Madoui MA, Cápal P, Novák P, Labadie K et al. (2019) A reference genome for pea provides insight into legume genome evolution. *Nat Genet* 51(9):1411–1422.

Kullback S, Leibler RA (1951) On information and sufficiency. *Ann Math Stat* 22(1):79–86.

Kumar GR, Sakthivel K, Sundaram RM, Neeraja CN, Balachandran SM et al. (2010) Allele mining in crops: Prospects and potentials. *Biotechnol Adv* 28(4):451–461.

Künne C, Lange M, Funke T, Miehe H, Thiel T et al. (2005) CR-EST: A resource for crop ESTs. *Nucleic Acids Res* 33:619–621.

Kutach AK, Kadonaga JT (2000) The downstream promoter element DPE appears to be as widely used as the TATA box in Drosophila core promoters. *Mol Cell Biol* 20(13):4754–4764.

Lamichaney A, Parihar AK, Dixit GP, Singh AK, Kumar N, Revanasidda, Singh NP (2021) Intact funiculus in mature harvested seeds of field pea (*Pisum sativum* L.): Preliminary investigation and possible implications. *Crop Sci* 61(5):2863–2871.

Leonforte A, Sudheesh S, Cogan NO, Salisbury PA, Nicolas ME et al. (2013) SNP marker discovery, linkage map construction and identification of QTLs for enhanced salinity tolerance in field pea (*Pisum sativum* L.). *BMC Plant Biol* 13(1):1–14.

Leprévost T, Boutet G, Lesné A, Rivière JP, Vetel P et al. (2023) Advanced backcross QTL analysis and comparative mapping with RIL QTL studies and GWAS provide an overview of QTL and marker haplotype diversity for resistance to Aphanomyces root rot in pea (*Pisum sativum*). *Front Plant Sci* 14:1189289.

Lescot M, Déhais P, Thijs G, Marchal K, Moreau Y et al. (2002) PlantCARE, a database of plant cis-acting regulatory elements and a portal to tools for in silico analysis of promoter sequences. *Nucleic Acids Res* 30(1):325–327.

Le Signor C, Savois V, Aubert G, Verdier J, Nicolas M et al. (2009b) Optimizing TILLING populations for reverse genetics in *Medicago truncatula*. *Plant Biotechnol J* 7:430–441.

Li S, Ding Y, Zhang D, Wang X, Tang X, Dai D et al. (2018) Parallel domestication with a broad mutational spectrum of determinate stem growth habit in leguminous crops. *Plant J* 96(4):761–771.

Librado P, Rozas J (2009) DnaSP v5: A software for comprehensive analysis of DNA polymorphism data. *Bioinformatics* 25(11):1451–1452.

Liu B, Watanabe S, Uchiyama T, Kong F, Kanazawa A et al. (2010) The soybean stem growth habit gene *Dt1* is an ortholog of Arabidopsis *TERMINAL FLOWER1*. *Plant Physiol* 153(1):198–210.

Liu R, Fang L, Yang T, Zhang X, Hu J et al. (2017) Marker-trait association analysis of frost tolerance of 672 worldwide pea (*Pisum sativum* L.) collections. *Sci Rep* 7(1):5919.

Liu Y, Du H, Li P, Shen Y, Peng H et al. (2020) Pan-genome of wild and cultivated soybeans. *Cell* 182(1):162–176.

Loridon K, McPhee K, Morin J, Dubreuil P, Pilet-Nayel ML et al. (2005) Microsatellite marker polymorphism and mapping in pea (*Pisum sativum* L.). *Theor Appl Genet* 111:1022–1031.

Luján Basile SM, Ramírez IA, Crescente JM, Conde MB, Demichelis M et al. (2019) Haplotype block analysis of an Argentinean hexaploid wheat collection and GWAS for yield components and adaptation. *BMC Plant Biol* 19:1–16.

Ma Y, Coyne CJ, Grusak MA, Mazourek M, Cheng P et al. (2017a) Genome-wide SNP identification, linkage map construction and QTL mapping for seed mineral concentrations and contents in pea (*Pisum sativum* L.). *BMC Plant Biol* 17(1):1–17.

Ma Y, Coyne CJ, Main D, Pavan S, Sun S et al. (2017b) Development and validation of breeder-friendly KASPar markers for er1, a powdery mildew resistance gene in pea (*Pisum sativum* L.). *Mol Breed* 37:1–7.

Macas J, Novák P, Pellicer J, Čížková J, Koblížková A et al. (2015) In depth characterization of repetitive DNA in 23 plant genomes reveals sources of genome size variation in the legume tribe Fabeae. *PLoS One* 10(11):e0143424.

Madeira F, Park YM, Lee J, Buso N, Gur T et al. (2019) The EMBL-EBI search and sequence analysis tools APIs in 2019. *Nucleic Acids Res* 47(W1):W636–W641.

Makasheva RK, Drozd AM (1987) Determinate growth habit (*det*) in peas: Isolation, symbolization and linkage. *Pisum Newsl* 19:31–32.

Maldonado C, Mora F, Scapim CA, Coan M (2019) Genome-wide haplotype-based association analysis of key traits of plant lodging and architecture of maize identifies major determinants for leaf angle: Hap LA4. *PLoS One* 14(3):e0212925.

Mardis ER (2008) The impact of next-generation sequencing technology on genetics. *Trends Genet* 24:133–141.

Matys V, Fricke E, Geffers R, Gößling E, Haubrock M et al. (2003) TRANSFAC®: Transcriptional regulation, from patterns to profiles. *Nucleic Acids Res* 31(1):374–378.

Mayer M, Hölker AC, González-Segovia E, Bauer E, Presterl T et al. (2020) Discovery of beneficial haplotypes for complex traits in maize landraces. *Nat Commun* 11(1):4954.

McCallum CM, Comai L, Greene EA, Henikoff S (2000a) Targeted screening for induced mutations. *Nat Biotechnol* 18:455–457.

Mejlhede N, Kyjovska Z, Backes G, Burhenne K, Rasmussen SK et al. (2006) Eco TILLING for the identification of allelic variation in the powdery mildew resistance genes *mlo* and *Mla* of barley. *Plant Breed* 125:461–467.

Metzker ML (2010) Sequencing technologies – the next generation. *Nat Rev Genet* 11(1):31–46.

Mishra RK, Gangadhar BH, Nookaraju A, Kumar S, Park SW (2012) Development of EST-derived SSR markers in pea (*Pisum sativum*) and their potential utility for genetic mapping and transferability. *Plant Breed* 131(1):118–124.

Moreau C, Ambrose MJ, Turner L, Hill L, Ellis THN et al. (2012) The b gene of pea encodes a defective flavonoid 30, 50-hydroxylase, and confers pink flower color. *Plant Physiol* 159:759–768. https://doi.org/10.1104/pp.112.197517

Moreau C, Hofer JM, Eléouët M, Sinjushin A, Ambrose M et al. (2018) Identification of Stipules reduced, a leaf morphology gene in pea (*Pisum sativum*). *New Phytol* 220(1):288–299.

Moreau C, Warren FJ, Rayner T, Perez-Moral N, Lawson DM, Wang TL et al. (2022) An allelic series of starch-branching enzyme mutants in pea (*Pisum sativum* L.) reveals complex relationships with seed starch phenotypes. *Carbohydr Polym* 288:119386.

Murray MG, Peters DL, Thompson WF (1981) Ancient repeated sequences in the pea and mung bean genomes and implications for genome evolution. *J Mol Evol* 17:31–42.

Murray S, Falloon L, Frew T, Coyne CJ, Butler R et al. (2009) Association mapping of yield candidate gene homologs in a diverse collection of pea (Pisum sativum L.) lines. In: *Meeting Proceedings, 19th Annual Queenstown Molecular Biology Meetings*, Queenstown, New Zealand, vol. 30.

Odong TL, Jansen J, Van Eeuwijk FA, van Hintum TJ (2013) Quality of core collections for effective utilisation of genetic resources review, discussion and interpretation. *Theor Appl Genet* 126:289–305.

Ogawa D, Sakamoto T, Tsunematsu H, Kanno N, Nonoue Y et al. (2021) Haplotype analysis from unmanned aerial vehicle imagery of rice MAGIC population for the trait dissection of biomass and plant architecture. *J Exp Bot* 72(7):2371–2382.

Ogawa D, Yamamoto E, Ohtani T, Kanno N, Tsunematsu H et al. (2018) Haplotype-based allele mining in the Japan-MAGIC rice population. *Sci Rep* 8(1):4379.

Ollivier R, Glory I, Cloteau R, Le Gallic JF, Denis G et al. (2022) A major-effect genetic locus, ApRVII, controlling resistance against both adapted and non-adapted aphid biotypes in pea. *Theor Appl Genet* 135(5):1511–1528.

Parihar AK, Dixit GP, Bohra A, Sen Gupta D, Singh AK et al. (2020) Genetic Advancement in dry pea (*Pisum sativum* L.): Retrospect and prospect. In: Gosal SS, Wani SH (eds) *Accelerated Plant Breeding*, vol. 3. Springer, Cham, pp. 283–341. https://doi.org/10.1007/978-3-030-47306-8_10

Parihar AK, Kumar J, Gupta DS, Lamichaney A, SJ SN et al. (2022a). Genomics enabled breeding strategies for major biotic stresses in pea (*Pisum sativum* L.). *Front Plant Sci* 13:861191.

Parihar AK, Yadav R, Lamichaney A, Mishra RK, Chandra A et al. (2022b) Field pea breeding. In: Yadava DK, Dikshit HK, Mishra GP, Tripathi S (eds) *Fundamentals of Field Crop Breeding*. Springer Nature, Singapore, pp. 1237–1321.

Parihar AK, Hazra KK, Lamichaney A, Dixit GP, Singh D et al. (2022c) Characterizing plant trait (s) for improved heat tolerance in field pea (*Pisum sativum* L.) under subtropical climate. *Int J Biometeorol* 66(6):1267–1281.

Parihar AK, Dixit GP, Lamichaney A, Das A, Tripathi K et al. (2023) Nutrient-dense pea (*Pisum sativum* L.): Genetics and genomics mediated developments. In: Kole C (eds) *Compendium of Crop Genome Designing for Nutraceuticals*. Springer Nature, Singapore. https://doi.org/10.1007/978-981-19-3627-2_25-1

Park PJ, Butte AJ, Kohane IS (2002) Comparing expression profiles of genes with similar promoter regions. *Bioinformatics* 18(12):1576–1584.

Pavan S, Schiavulli A, Appiano M, Marcotrigiano AR, Cillo F et al. (2011) Pea powdery mildew er1 resistance is associated to loss-of-function mutations at a *MLO* homologous locus. *Theor Appl Genet* 123:1425–1431.

Pavan S, Schiavulli A, Appiano M, Miacola C, Visser RG et al. (2013) Identification of a complete set of functional markers for the selection of *er1* powdery mildew resistance in *Pisum sativum* L. *Mol Breed* 31:247–253.

Pereira G, Marques C, Ribeiro R, Formiga S, Dâmaso M et al. (2010) Identification of DNA markers linked to an induced mutated gene conferring resistance to powdery mildew in pea (*Pisum sativum* L.). *Euphytica* 171:327–335.

Pilet-Nayel M, Muehlbauer F, McGee R, Kraft J, Baranger A et al. (2002) Quantitative trait loci for partial resistance to Aphanomyces root rot in pea. *Theor Appl Genet* 106:28–39.

Pilet-Nayel ML, Muehlbauer FJ, McGee RJ, Kraft JM, Baranger A et al. (2005) Consistent quantitative trait loci in pea for partial resistance to *Aphanomyces euteiches* isolates from the United States and France. *Phytopathology* 95(11):1287–1293.

Piya S, Pantalone V, Zadegan SB, Shipp S, Lakhssassi N et al. (2023) Soybean gene co-expression network analysis identifies two co-regulated gene modules associated with nodule formation and development. *Mol Plant Pathol* 24(6):628–636.

Praça-Fontes MM, Carvalho CR, Clarindo WR (2014) Karyotype revised of Pisum sativum using chromosomal DNA amount. *Pl Syst Evol* 300:1621–1626.

Premkumar A, Kumar M, Chauhan R (2020) Improvement of rice variety for biotic and abiotic stress resistance using the technique of allele mining. *J Pharmacog Phytochem* 9(5):1888–1893.

Prioul S, Frankewitz A, Deniot G, Morin G, Baranger A (2004) Mapping of quantitative trait loci for partial resistance to *Mycosphaerella pinodes* in pea (*Pisum sativum* L.), at the seedling and adult plant stages. *Theor Appl Genet* 108(7):1322–1334.

Pritchard JK, Stephens M, Donnelly P (2000) Inference of population structure using multilocus genotype data. *Genetics* 155:945–959.

Rai R, Singh AK, Singh BD, Joshi AK, Chand R et al. (2011) Molecular mapping for resistance to pea rust caused by *Uromyces fabae* (Pers.) de-Bary. *Theor Appl Genet* 123:803–813.

Raghavan C, Naredo MEB, Wang H, Atienza G, Liu B et al. (2007) Rapid method for detecting SNPs on agarose gels and its application in candidate gene mapping. *Mol Breed* 19:87–101.

Rameau C, Bodelin C, Cadier D, Grandjean O, Miard F et al. (1997) New ramosus mutants at loci Rms1, Rms3 and Rms4 resulting from the mutation breeding program at Versailles. *Pisum Genet* 29:7–12.

Rashid M, He G, Guanxiao Y, Khurram Z (2011) Relevance of tilling in plant genomics. *Aust J Crop Sci* 5(4):411–420.

Repinski SL, Kwak M, Gepts P (2012) The common bean growth habit gene *PvTFL1y* is a functional homolog of Arabidopsis *TFL1*. *Theor Appl Genet* 124:1539–1547.

Rozen S, Skaletsky H (2000) Primer3 on the WWW for General Users and for Biologist Programmers. In: Misener S, Krawetz SA (eds) *Bioinformatics Methods and Protocols. Methods in Molecular Biology*™, vol. 132. Humana Press, Totowa, NJ. https://doi.org/10.1385/1-59259-192-2:365

Sagan M, Duc G (1996) *Sym28* and *Sym29*, two new genes involved in regulation of nodulation in pea (*Pisum sativum* L.). *Symbiosis* 20:229–245.

Sainsbury F, Tattersall AD, Ambrose MJ, Turner L, Ellis THN et al. (2006) A crispa null mutant facilitates identification of a crispa-like pseudogene in pea. *Funct Plant Biol* 33:757–763. https://doi.org/10.1071/FP06090

Schmid CD, Praz V, Delorenzi M, Périer R, Bucher P (2004) The Eukaryotic promoter database EPD: The impact of in silico primer extension. *Nucleic Acids Res* 32(suppl-1):D82–D85.

Sehgal D, Mondal S, Crespo-Herrera L, Velu G, Juliana P et al. (2020) Haplotype-based, genome-wide association study reveals stable genomic regions for grain yield in CIMMYT spring bread wheat. *Front Genet* 11:589490.

Shahmuradov IA, Gammerman AJ, Hancock JM, Bramley PM, Solovyev VV (2003) PlantProm: A database of plant promoter sequences. *Nucleic Acids Res* 31(1):114–117.

Shannon CE, Weaver W (1963) *The Mathematical Theory of Communication*. The University of Illinois Press, Champaign, IL.

Shendure J, Ji H (2008) Next-generation DNA sequencing. *Nature Biotechnol* 26(10):1135–1145.

Shunmugam ASK, Liu X, Stonehouse R, Tar'an B, Bett KE et al. (2015) Mapping seed phytic acid concentration and iron bioavailability in a pea recombinant inbred line population. *Crop Sci* 55(2):828–836.

Sindhu A, Ramsay L, Sanderson LA, Stonehouse R, Li R et al. (2014) Gene-based SNP discovery and genetic mapping in pea. *Theor Appl Genet* 127:2225–2241.

Sinjushin A, Semenova E, Vishnyakova M (2022) Usage of morphological mutations for improvement of a garden pea (*Pisum sativum*): The experience of breeding in Russia. *Agronomy* 12(3):544.

Siol M, Jacquin F, Chabert-Martinello M, Smýkal P, Le Paslier MC et al. (2017) Patterns of genetic structure and linkage disequilibrium in a large collection of pea germplasm. *G3 Genes Genomes Genet* 7(8):2461–2471.

Smýkal P, Coyne CJ, Ford R, Redden R, Flavell A et al. (2008) Effort towards a world pea (*Pisum sativum* L.) germplasm core collection: The case for common markers and data compatibility. *Pisum Genet* 40:11–14.

Smýkal P, Coyne CJ, Redden R, Nigel M (2013) Peas. In: Singh M, Bisht IS (eds) *Genetic and Genomic Resources for Grain Legume Improvement*. Elsevier Insights, London, pp. 41–80.

Smýkal P, Trněný O, Brus J, Hanáček P, Rathore A et al. (2018) Genetic structure of wild pea (*Pisum sativum* subsp. elatius) populations in the Northern part of the Fertile Crescent reflects moderate cross-pollination and strong effect of geographic but not environmental distance. *PLoS One* 13(3):e0194056.

Sudheesh S, Lombardi M, Leonforte A, Cogan NO, Materne M et al. (2015a) Consensus genetic map construction for field pea (*Pisum sativum* L.), trait dissection of biotic and abiotic stress tolerance and development of a diagnostic marker for the er1 powdery mildew resistance gene. *Plant Mol Biol Rep* 33:1391–1403.

Sudheesh S, Rodda M, Kennedy P, Verma P, Leonforte A et al. (2015b) Construction of an integrated linkage map and trait dissection for bacterial blight resistance in field pea (*Pisum sativum* L.). *Mol Breed* 35:1–13.

Sullivan JA, Gray JC (2000) The pea light-independent photomorphogenesis1 mutant results from partial duplication of COP1 generating an internal promoter and producing two distinct transcripts. *Plant Cell* 12(10):1927–1937.

Sun H, Jiao WB, Krause K, Campoy JA, Goel M et al. (2022) Chromosome-scale and haplotype-resolved genome assembly of a tetraploid potato cultivar. *Nat Genet* 54(3):342–348.
Sun X, Yang T, Hao J, Zhang X, Ford R et al. (2014) SSR genetic linkage map construction of pea (Pisum sativum L.) based on Chinese native varieties. *The Crop J* 2(2–3):170–174.
Tafesse EG, Gali KK, Lachagari VR, Bueckert R, Warkentin TD (2020) Genome-wide association mapping for heat stress responsive traits in field pea. *Int J Mol Sci* 21(6):2043.
Tai PYP, Miller JD (2001) A core collection for *Saccharum spontaneum* L. from the world collection of sugarcane. *Crop Sci* 41(3):879–885.
Tamura K, Stecher G, Peterson D, Filipski A, Kumar S (2013) MEGA6: Molecular evolutionary genetics analysis version 6.0. *Mol Biol Evol* 30(12):2725–2729.
Tar'an B, Warkentin T, Somers DJ, Miranda D, Vandenberg A et al. (2004) Identification of quantitative trait loci for grain yield, seed protein concentration and maturity in field pea (*Pisum sativum* L.). *Euphytica* 136:297–306.
Tayeh N, Aluome C, Falque M, Jacquin F, Klein A et al. (2015) Development of two major resources for pea genomics: The GenoPea 13.2 K SNP Array and a high-density, high-resolution consensus genetic map. *Plant J* 84(6):1257–1273.
Tayeh N, Hofer J, Aubert G, Jacquin F, Turner L et al. (2023) Afila, the origin and nature of a major innovation in the history of pea breeding. *BioRxiv* 7.
Tello-Ruiz MK, Naithani S, Stein JC, Gupta P, Campbell M et al. (2018) Gramene 2018: Unifying comparative genomics and pathway resources for plant research. *Nucleic Acids Res* 46(D1):D1181–D1189.
Thachuk C, Crossa J, Franco J, Dreisigacker S, Warburton M et al. (2009) Core hunter: An algorithm for sampling genetic resources based on multiple genetic measures. *BMC Bioinformatics* 10:1–13.
Thornsberry JM, Goodman MM, Doebley J, Kresovich S, Nielsen D et al. (2001) Dwarf8 polymorphisms associate with variation in flowering time. *Nat Genet* 28(3):286–289.
Tian F, Yang DC, Meng YQ, Jin J, Gao G (2020) PlantRegMap: Charting functional regulatory maps in plants. *Nucleic Acids Res* 48(D1):D1104–D1113.
Till BJ, Zerr T, Comai L, Henikoff S (2006) A protocol for TILLING and ecotilling in plants and animals. *Nat Protocols* 1(5):2465–2477.
Timmerman-Vaughan GM, Frew TJ, Butler R, Murray S, Gilpin M et al. (2004) Validation of quantitative trait loci for Ascochyta blight resistance in pea (*Pisum sativum* L.), using populations from two crosses. *Theor Appl Genet* 109:1620–1631.
Timmerman-Vaughan GM, Frew TJ, Russell AC, Khan T, Butler R et al. (2002) QTL mapping of partial resistance to field epidemics of ascochyta blight of pea. *Crop Sci* 42(6):2100–2111.
Timmerman-Vaughan GM, McCallum JA, Frew TJ, Weeden NF, Russell AC (1996) Linkage mapping of quantitative trait loci controlling seed weight in pea (*Pisum sativum* L.). *Theor Appl Genet* 93:431–439.
Timmerman-Vaughan GM, Mills A, Whitfield C, Frew T, Butler R et al. (2005) Linkage mapping of QTL for seed yield, yield components, and developmental traits in pea. *Crop Sci* 45(4):1336–1344.
Timmerman-Vaughan GM, Moya L, Frew TJ, Murray SR, Crowhurst R (2016) Ascochyta blight disease of pea (*Pisum sativum* L.): Defence-related candidate genes associated with QTL regions and identification of epistatic QTL. *Theor Appl Genet* 129:879–896.
Tosh SM, Yada S (2010) Dietary fibres in pulse seeds and fractions: Characterization, functional attributes, and applications. *Food Res Int* 43(2):450–460.
Tran CT, Beissinger TM, Becker HC, Horneburg B (2023) Genetic diversity of pea (Pisum sativum L.) genotypes differing in leaf type using SNP markers. *Genet Resour Crop Evol* 70(4):1085–1095.
Triques K, Sturbois B, Gallais S, Dalmais M, Chauvin S et al. (2007) Characterization of Arabidopsis thaliana mismatch specific endonucleases: Application to mutation discovery by TILLING in pea. *Plant J* 51(6):1116–1125.
Tsubokura Y, Watanabe S, Xia Z, Kanamori H, Yamagata H et al. (2014) Natural variation in the genes responsible for maturity loci E1, E2, E3 and E4 in soybean. *Ann Bot* 113(3):429–441.
Tsyganov VE, Tsyganova AV (2020) Symbiotic regulatory genes controlling nodule development in *Pisum sativum* L. *Plants* 9(12):1741.
Upadhyaya HD, Pundir RPS, Gowda CLL, Reddy VG, Singh S (2009) Establishing a core collection of foxtail millet to enhance the utilization of germplasm of an underutilized crop. *Plant Genet Resour* 7(2):177–184.
Vanhala T, Normann KR, Lundström M, Weller JL, Leino MW, Hagenblad J (2016) Flowering time adaption in Swedish landrace pea (*Pisum sativum* L.). *BMC Genetics* 17(1):1–13.
Varshney RK, Graner A, Sorrells ME (2005) Genomics-assisted breeding for crop improvement. *Trends Plant Sci* 10:621–630.

Varshney RK, Nayak SN, May GD, Jackson SA (2009) Next-generation sequencing technologies and their implications for crop genetics and breeding. *Trends Biotechnol* 27:522–530.

Veerla S, Höglund M (2006) Analysis of promoter regions of co-expressed genes identified by microarray analysis. *BMC Bioinformatics* 7:1–15.

Vershinin AV, Allnutt TR, Knox MR, Ambrose MJ, Ellis TN (2003) Transposable elements reveal the impact of introgression, rather than transposition, in *Pisum* diversity, evolution, and domestication. *Mol Biol Evol* 20(12):2067–2075.

Wang O, Chin R, Cheng X, Wu MKY, Mao Q et al. (2019) Efficient and unique cobarcoding of second-generation sequencing reads from long DNA molecules enabling cost-effective and accurate sequencing, haplotyping, and *de novo* assembly. *Genome Res* 29(5):798–808.

Wang Z, Luo Y, Li X, Wang L, Xu S et al. (2008) Genetic control of floral zygomorphy in pea (*Pisum sativum* L.). *Proc Natl Acad Sci USA* 105(30):10414–10419.

Warkentin TD, Smýkal P, Coyne CJ, Weeden N, Domoney C et al. (2015) Pea. In: De Ron, Antonio M (ed) *Grain Legumes*. Springer, Cham, ISBN 978-1-4939-2797-5 (eBook), pp. 37–83.

Weeden NF (1998) A consensus linkage map for *Pisum sativum*. *Pisum Genet* 30:1–4.

Weeden NF, Porter L (2007) The genetic basis of Fusarium root rot tolerance in the 'Afghanistan' pea. *Pisum Genet* 39:35–36.

Weller JL, Liew LC, Hecht VF, Rajandran V, Laurie RE et al. (2012) A conserved molecular basis for photoperiod adaptation in two temperate legumes. *Proc Natl Acad Sci USA* 109(51):21158–21163.

Wu L, Fredua-Agyeman R, Hwang SF, Chang KF, Conner RL et al. (2021) Mapping QTL associated with partial resistance to Aphanomyces root rot in pea (*Pisum sativum* L.) using a 13.2 K SNP array and SSR markers. *Theor Appl Genet* 134(9):2965–2990.

Yang T, Fang L, Zhang X, Hu J, Bao S et al. (2015) High-throughput development of SSR markers from pea (*Pisum sativum* L.) based on next generation sequencing of a purified Chinese commercial variety. *PLoS One 1*0(10):e0139775.

Yang T, Liu R, Luo Y, Hu S, Wang D et al. (2022) Improved pea reference genome and pan-genome highlight genomic features and evolutionary characteristics. *Nat Genet* 54(10):1553–1563.

Ye J, Coulouris G, Zaretskaya I, Cutcutache I, Rozen S, Madden TL (2012) Primer-BLAST: A tool to design target-specific primers for polymerase chain reaction. *BMC Bioinformatics* 13:1–11.

Yu J, Pressoir G, Briggs WH, Vroh Bi I, Yamasaki M et al. (2006) A unified mixed-model method for association mapping that accounts for multiple levels of relatedness. *Nat Genet* 38(2):203–208.

Zhai H, Lü S, Wu H, Zhang Y, Zhang X et al. (2015) Diurnal expression pattern, allelic variation, and association analysis reveal functional features of the E1 gene in control of photoperiodic flowering in soybean. *PLoS One* 10(8):e0135909.

Zhou J, Gali KK, Jha AB, Tar'an B, Warkentin TD (2022) Identification of quantitative trait loci associated with seed protein concentration in a pea recombinant inbred line population. *Genes* 13(9):1531.

Zhou J, Wan Z, Gali KK, Jha AB, Nickerson MT et al. (2023) Quantitative trait loci associated with amino acid concentration and in vitro protein digestibility in pea (*Pisum sativum* L.). *Front Plant Sci* 14:1083086.

Zhu C, Gore M, Buckler ES, Yu J (2008) Status and prospects of association mapping in plants. *Plant Genome J* 1(1):5.

Zhu J, Zhang MQ (1999) SCPD: A promoter database of the yeast *Saccharomyces cerevisiae*. *Bioinformatics* 15(7):607–611.

Zhuang L, Ambrose M, Rameau C, Weng L, Yang J et al. (2012) LATHYROIDES, encoding a WUSCHEL-related homeobox1 transcriptional factor, controls organ lateral growth and regulates tendril and dorsal petal identities in garden pea (*Pisum sativum* L.). *Mol Plant* 5:1333–1345. https://doi.org/10.1093/mp/sss067

7 Allele Mining in Lentil

Jitendra Kumar, Debjyoti Sen Gupta, Arshad Husain, Ruchi Bansal, Harsh Kumar Dikshit, Muralidhar Aski, and Sachin Kumar

7.1 INTRODUCTION

Lentil (*Lens culinaris* Medik.) is a cool-season self-pollinated legume crop. It has 2n = 2x = 14 chromosomes and ~4 Gbp genome size (Ogutcen et al., 2018). Globally, lentil is grown in about 5.0 mha area, with an annual production of 5.86 mt and an average productivity of 1,305 kg/ha in the year 2020 over 52 countries (FAO, 2022). Dietary proteins, fiber, prebiotic carbohydrates, minerals, and vitamins are found in abundant amounts in its grain (Kumar et al., 2016). This crop has been identified as a solution to malnutrition problems and micronutrient deficiency for people of developing countries due to the presence of sufficient quantities of Fe, Zn, and folate in its grain (Kumar et al., 2016). Lentil consumption can bring health benefits to the people suffering with diabetes, obesity, and cardiovascular problems (Srivastava and Vasishtha, 2012). In this view, people are opting for lentil in their diets, leading to higher *per capita* consumption. Moreover, lentil-based cropping systems improve soil health similar to other legumes through symbiotic nitrogen fixation, and it is a useful commodity for improving the livelihood of smallholder farmers in dryland areas of developing countries (Kumar et al., 2013).

Traditional plant breeding methods used to develop improved cultivars have led to an increase in the current annual genetic gains of <0.7%, which cannot meet the growing demands of lentil. Therefore, it is needed to enhance the current productivity of lentils up to 2 t per ha among different biotic and abiotic constraints (Sharpe et al., 2013; Kumar et al., 2015). Further, high genotype and environmental interactions cause poor genetic gain in lentils through conventional breeding approaches (Kumar and Ali, 2006). Moreover, current climate-changing conditions are making this task more challenging. However, genomics-assisted breeding made it possible to increase the genetic gain in lentils through the precise and rapid accumulation of superior alleles in a breeding population (Kumar et al., 2015). Therefore, there is a need to relook at such superior alleles from existing plant genetic resources for gainful exploitation in the development of agronomically superior cultivars. These alleles could not be utilized in the breeding program as they were left behind during evolution and domestication.

In lentil, significant progress has been made to accumulate sequence and expression data, discovery and annotation of new genes, and draft genome sequences (Kumar et al., 2015; Lombardi et al., 2014; Wong et al., 2015; Khazaei et al., 2016; Dissanayake et al., 2020). These genomic resources provided opportunities to isolate useful alleles of genes conferring resistance to biotic and abiotic stresses, greater nutrient use efficiency, enhanced yield, and improved quality through allele-specific markers from a wide range of species. Thus, use of novel genomic tools for identifying allelic variation at candidate genes/loci responsible for a given trait among genotypes often refers to allele mining (dissection of naturally occurring variation). In lentil, during the past years, candidate genes have been identified for biotic and abiotic stress–related traits (Mustafa et al., 2009; Singh et al., 2019; Bhadauria et al., 2017a; Sudheesh et al., 2016; Sari et al., 2018; Cao et al., 2019; Dissanayake et al., 2020; Khorramdelazad et al., 2018; Kumar et al., 2021). These genes, along with available genetic resources, provide a way to use them for the identification of rare useful alleles for lentil improvement. In this chapter, we summarize the current status of allele mining in lentil.

DOI: 10.1201/9781003385059-7

7.2 AVAILABILITY OF GENETIC RESOURCES FOR MINING SUPERIOR ALLELES

Diverse genetic resources are one of the most important components for making genetic improvement in breeding programs. Existing genetic diversity conserved in gene bank can be used to mine desirable genes for utilization in lentil improvement. In lentil, 58,405 accessions are stored in different gene banks globally (Khazaei et al., 2016). The International Center for Agricultural Research in the Dry Areas (ICARDA) holds a collection of 13,907 and 603 accessions of cultivated and wild species of lentil, respectively (Singh and Chung, 2016). In India, the ICAR–National Bureau of Plant Genetic Resources (NBPGR) has 2,655 accessions (2,083 indigenous and 572 exotic) in the national gene bank (Singh and Chung, 2016). The second largest collection of lentil germplasm, comprising 5,254 accessions, is stored in the Australian Temperate Field Crop Collection. The Seed and Plant Improvement Institute of Iran and USDA separately contained 3,000 and 2,875 lentil accessions (Malhotra et al., 2019). In these collections, a sizable number of crop wild relatives (CWR) are also involved, which are useful sources for identifying the novel traits/alleles based on their historical record of adaptation to a diverse range of habitats.

Morphological, phenological, agronomic, and nutritional traits have been used to characterize the germplasm in the past years (Singh et al., 2014; Malhotra et al., 2019; Gorim and Vandenberg, 2017; Singh et al., 2020b). The core and FIGS (Focused Identification of Germplasm Strategy) sets have been developed for facilitating the accessibility and better use of germplasm in the breeding program. The United States Department of Agriculture (USDA) developed a core set of 287 accessions based on country of origin. This core set was evaluated for agro-morphological traits (Tullu et al., 2001). ICARDA used the FIGS strategy using robust geographical datasets and was found very successful in the identification of various adaptive traits, including heat and drought tolerance, and disease resistance in lentil (El Haddad et al., 2020).

Wild lentil species offer valuable resources, which provide opportunities to mine useful alleles which could not be used during domestication. Singh et al. (2014) evaluated 405 accessions of the wild *Lens* species (*L. culinaris* ssp. *culinaris*, *L. culinaris* ssp. *orientalis*, *L. culinaris* ssp. *odemensis*, *L. culinaris* ssp. *tomentosus*, *L. nigricans*, *L. ervoides*, and *L. lamottei*) for agronomic traits and disease reactions and identified useful sources for these traits. Further, a core set of 96 accessions was developed from these wild accessions and used to validate the nutritional and key biotic/abiotic traits (Kumar et al., 2018d; Singh et al., 2020b). Several other workers also assessed the genetic diversity available within wild and cultivated lentil germplasm for nutrition uptake and concentration (Se, Fe, and Zn), biotic and abiotic stresses, and agronomic traits (Thavarajah et al., 2017; Khazaei et al., 2017; Kumar et al., 2018a, 2018b, 2018c; Gorim and Vandenberg, 2017; Bhadauria et al., 2017a). These genetic resources have been used to identify QTLs/genes controlling important traits (Khazaei et al., 2017; Kumar et al., 2018a, 2018b; Polanco et al., 2019; Bhadauria et al., 2017a). Imaging-based advanced phenotyping techniques can accelerate the rapid and precise screening of lentil germplasm for the identification of useful donors for target traits (resistance/tolerance to drought, heat, wilt, rust, etc.). In lentil, advanced imaging-based phenotyping was used to identify resistant genotypes through quantitative evaluation of root rot (Marzouguiet al., 2019).

7.3 ALLELE MINING

It includes mining of sequence variations in coding regions/exons; non-coding (5′ UTR), including promoter, introns, and 3′ UTR; and regulatory regions of candidate genes that affect the transcript synthesis and accumulation, leading to changes in the phenotypic expression of the trait. This is also known as "true" allele mining (Kumar et al., 2010). Further, the identification of variations in regulatory regions and promoters of genes is termed "promoter mining." In lentil, different studies have been conducted to identify the variations within coding regions of genes through comparative mapping and other ways (Kumar et al., 2021). For example, gene-based markers developed from two major loci (*HR* and *ELF3*) controlling differences in photoperiod response between wild and domesticated peas led to the identification of an orthologous loci of *ELF3* in lentil (Weller et al., 2012). Another study has mined alleles among lentil species and 12 legume genera detected for expressed sequence tag (EST) sequences associated with different candidate genes involved in different metabolic activities (Singh et al., 2020a). Further, Kumar et al. (2021) reviewed the potential candidate genes for different traits in lentil that are useful resources for allele mining in lentil.

7.4 APPROACHES FOR MINING THE ALLELES

EcoTILLING and sequencing are two ways that are used to mine allele of the targeted genes. Naturally occurring variations for a gene that controls agronomically important trait can be identified through the EcoTILLING approach (Till et al., 2003). EcoTILLING utilizes advanced genotyping platforms, such as LICOR NEN Model 4300 DNA Analyzer, Transgenomic WAVE-HS denaturing high-performance liquid chromatography, ABI 377 sequencer, and eGene capillary electrophoresis systems, for detecting the variation within a candidate gene among diverse genotypes, leading to the identification of potential alleles associated with a trait. This has been applied in several crops (Perry et al., 2003; Caldwell et al., 2004; Comai et al., 2004; Henikoff et al., 2004; Yang et al., 2008; Suzuki et al., 2005). Among legume crops, this approach has been used in chickpeas and mined eight SNP alleles in the eight transcription factor genes regulating seed weight (Bajaj et al., 2016). Although the EcoTilling approach has not been used for allele mining in lentil, the availability of many candidate genes for different traits (Kumar et al., 2021) provides an opportunity to use this approach in allele mining.

Sequencing-based allele mining involves PCR-based amplification of target genes among diverse genotypes. These amplified DNA fragments of diverse genotypes are sequenced for the identification of allelic variation. However, advanced next-generation sequencing platforms have to speed up the generation of DNA sequences with increased throughput and accuracy (Kumar and Sen Gupta, 2020). These platforms helped mine alleles through re-sequencing, alignment of the sequence data, and their comparison with the reference genome (Margulies et al., 2005). Thus, sequencing-based allele mining helps analyze individuals for haplotype structure and diversity to infer the genetic association of allelic variation with traits. In lentil, *dtf6a* allele responsible for earliness in South Asian material was mined from a wide survey of >350 accessions with diverse origins using amplicon sequencing approach (Rajandran et al., 2022). In another study, next-generation sequencing–based targeted (tGBS) and transcriptome (GBS-t) genotyping-by-sequencing of 276 lentil accessions identified allelic variation for a range of candidate genes related to salt tolerance in lentil (Dissanayake et al., 2021).

7.5 USE OF ALLELE MINING

Superior alleles of candidate genes are discovered from diverse genetic resources, and studies of nucleotide changes defining an allele as superior are identified through allele mining, which provides insight into the molecular basis of novel trait variations. Besides the rate of evolution of alleles, allelic similarity/dissimilarity of a candidate gene and allelic synteny with other members of the family can also be studied. This helps discriminate related species at molecular level and identify allele-specific molecular markers which can be used for transferring the novel alleles through MAS or deployment through genetic engineering (GE).

7.5.1 Development of Allele-Specific Markers

Sequence variation from one allele to another allele of a candidate gene provides the opportunity for developing allele-specific markers. These markers help the precise transfer of the identified "superior and/or novel" alleles to the desirable background for developing improved cultivars. Such allele-specific markers have been developed for several genes, including the *Waxy* gene (five different alleles) (Mikami et al., 2008), starch synthase IIa (*SSIIa*) (Bao et al., 2006), *Pi-ta* (responsible for resistance to rice blast disease) (Huang et al., 2008; Wang et al., 2008b), *GS3* (grain size) (Takanokai et al., 2009), and *badh2* (rice fragrant) (Sakthivel et al., 2009a, 2009b) in rice. However, in lentil, allelic differences are needed to search among the genetic resources for targeted candidate genes identified through transcriptome analysis. The comparative mapping of *MLO* genes of other legume species with a lentil draft genome sequence has identified two likely candidates, *LcMLO1* and *LcMLO3*. These genes showed differences within and between the species only for SNPs and small indels in introns but produce identical amino acid sequences. However, two amino acid substitutions have been observed between *L. culinaris* and *L. orientalis* for the *MLO3* candidate gene, while amino acid substitutions and indels were observed in the carboxyl intracellular domain of this gene among three species (*L. odemensis*, *L. ervoides*, and *L. lamottei*). These genes belonging to the *MLO* gene family have been shown to be involved in controlling powdery mildew response in other species (Polanco et al., 2018) and, hence, can be useful for breeding powdery mildew resistance in lentil. Recently, kompetitive allele-specific PCR (KASP) markers were developed from 78 SNPs presented in unigenes (Wang et al., 2020). These allele-specific markers can be used to identify the candidate gene responsible for traits through association mapping, leading to the identification of allele-specific markers for mining the superior allele.

7.5.2 Unlocking Allelic Diversity of Candidate Gene(s) Available in the Germplasm Collection

Knowledge of allelic diversity at that candidate gene(s) that regulates the phenotype of interest paves the way for efficient utilization of available genetic resources. Therefore, the relationship among diverse accessions based on allelic diversity at a targeted candidate gene helps identify useful genetic resources for utilization in a breeding program. Thus, allele mining is an important way for unlocking the useful diversity in the collections of genetic resources in the world gene banks (Kaur et al., 2008). In lentil, sequences for candidate genes underlying the biotic and abiotic stress–related traits (Mustafa et al., 2009; Singh and Chung, 2016; Singh et al., 2019; Bhadauria et al., 2017b; Sudheesh et al., 2016; Sari et al., 2018; Cao et al., 2019; Dissanayake et al., 2020; Khorramdelazad et al., 2018; Kumar et al., 2021) are available that can facilitate a pragmatic gene-based use of genetic resources (Graner, 2006).

7.5.3 Allelic Similarity at Candidate Gene(s) and Establishing an Evolutionary Relationship

As allele mining provides the sequence information of candidate genes in a species/accession, this information can be used to identify similar loci/genes across the species/genera through syntenic relationships. These genes can be used to assess allelic diversity in the germplasm. For example, in lentil, two likely candidates, *LcMLO1* and *LcMLO3*, have been identified through the syntenic relationship of *MLO* gene sequences of other legume species. These genes have been used to identify allelic variation in other *Lens* species, and alleles had variations for SNPs and small indels in introns within and between the species. Alleles of *L. culinaris* and *L. orientalis* species differed at the *MLO3* candidate gene for two amino acid substitutions. However, allelic variation was observed for amino acid substitutions and indels in the carboxyl intracellular domain of this gene in *L. odemensis*, *L. ervoides*, and *L. lamottei* species. As the *MLO* gene family is involved in controlling powdery mildew response in other species, allelic variations for these genes can be useful for the development of powdery mildew–resistant cultivars in lentil (Polanco et al., 2018). This also helps establish the evolutionary lineage of a gene during the domestication of crop species.

7.6 CONSIDERATIONS FOR ALLELE MINING

Mining of a useful allele among the diverse germplasm accessions is required to consider some deliberations, which was discussed in detail earlier (Kumar et al., 2010) and summarized here in the context of the lentil crop. The selection of germplasm is important to mine the useful allele of a gene. Therefore, screening the entire available germplasm collection would certainly be helpful in finding rare alleles. However, it is difficult to screen huge germplasm collections efficiently. Therefore, representative core/mini-core collection of germplasm can be a way forward for discovering new alleles and allele combinations. Such starting genetic materials are already available for mining the novel alleles of candidate genes, which can be useful for making efforts toward allele mining in lentil (Singh et al., 2014; Tripathi et al., 2022). Further precision phenotyping and computational tools for reducing the problem of repetitiveness and genetic redundancy within the existing crop gene pools are needed for increasing the efficiency of allele mining. For example, a computational tool called "Power Core" has provided an opportunity to develop a set of germplasm having richness and reducing the redundancy of useful alleles (Park, 2007). This computational tool, along with the principal component score (PCS) strategy, has been used in lentil for developing a core set for efficient utilization in lentil improvement programs (Singh et al., 2014; Tripathi et al., 2022). Computational tools are also required for establishing the relationship between allelic nucleotide diversity and altered function of candidate genes and defining boundary of promoter/regions that vary from one gene to another (Brazma et al., 1998; Molina and Grotewold, 2005; Park et al., 2002; Veerla and Hoglund, 2006). Therefore, molecular genetics, statistics, and bioinformatics are needed to work together for allele mining through association genetics and comparative genomics. Further, this should be cost-effective and rapid for accelerating the allele mining.

7.7 CONCLUSION

In lentil, vast genetic resources are available for discovering useful alleles. However, these genetic resources are not used sincerely for mining the rare alleles of genes controlling agronomically important traits. However, limited efforts have been made to develop the core and mini-cores as starting materials for allele mining (Singh et al., 2014; Tripathi et al., 2022). The availability of candidate gene sequences is another essential requirement for the mining of alleles of those genes in gene pools. During the past years, significant progress has been made in the identification of candidate genes for several agronomically important traits in lentil (Mustafaet al., 2009; Singh and Chung, 2016; Singh et al., 2019; Bhadauria et al., 2017b; Sudheesh et al., 2016; Sari et al., 2018; Cao et al., 2019; Dissanayake et al., 2020; Khorramdelazad et al., 2018; Kumar et al., 2021). These genes provide an opportunity for identification of their useful alleles in germplasm and can be used further in lentil breeding programs through marker-assisted selection.

REFERENCES

Bajaj D, Srivastava R, Nath M, Tripathi S, Bharadwaj C et al. (2016) EcoTILLING-based association mapping efficiently delineates functionally relevant natural allelic variants of candidate genes governing agronomic traits in chickpea. *Front Plant Sci* 19:450.

Bao JS, Corke H, Sun M (2006) Nucleotide diversity in starch synthase IIa and validation of single nucleotide polymorphisms in relation to starch gelatinization temperature and other physicochemical properties in rice (*Oryza sativa* L.). *Theor Appl Genet* 113:1171–1183.

Bhadauria V, Ramsay L, Bett KE, Banniza S (2017a) QTL mapping reveals genetic determinants of fungal disease resistance in the wild lentil species *Lens ervoides*. *Sci Rep* 7(1):1–9.

Bhadauria V, Vijayan P, Wei Y, Banniza S (2017b) Transcriptome analysis reveals a complex interplay between resistance and effector genes during the compatible lentil-*Colletotrichum* lentis interaction. *Sci Rep* 7(1):1–13.

Brāzma A, Jonassen I, Vilo J, Ukkonen E (1998) Predicting gene regulatory elements *in silico* on a genomic scale. *Genomic Res* 8:1202–1215.

Caldwell DG, McCallum N, Shaw P, Muehlbauer GJ, Marshall DF et al. (2004) A structured mutant population for forward and reverse genetics in barley (*Hordeum vulgare* L.). *Plant J* 40:143–150.

Cao Z, Li L, Kapoor K, Banniza S (2019) Using a transcriptome sequencing approach to explore candidate resistance genes against Stemphylium blight in the wild lentil species *Lens ervoides*. *BMC Plant Biol* 19:1–16.

Comai L, Young K, Till BJ, Reynolds SH, Greene EA et al. (2004) Efficient discovery of DNA polymorphisms in natural populations by EcoTilling. *Plant J* 37:778–786.

Dissanayake R, Cogan NO, Smith KF, Kaur S (2021) Application of genomics to understand salt tolerance in lentil. *Genes* 12(3):332.

Dissanayake R, Braich S, Cogan NO, Smith K, Kaur S (2020) Characterization of genetic and allelic diversity amongst cultivated and wild lentil accessions for germplasm enhancement. *Front Genet* 11:546.

FAOSTAT (2022) *Statistical Databases*. Food and Agriculture Organization of the United Nations, Italy, July 30. http://www.fao.org/faostat/en/#data/QC

Gorim LY, Vandenberg A (2017) Evaluation of wild lentil species as genetic resources to improve drought tolerance in cultivated lentil. *Front Plant Sci* 8:1129.

Graner A (2006) Barley research at IPK. In: de Vicente MC, Glaszmann JC (eds) *Molecular Markers for Allele Mining*. IPGRI, Italy, p. 25.

Henikoff S, Till BJ, Comai L (2004) TILLING. Traditional mutagenesis meets functional genomics. *Plant Physiol* 135(2):630–636.

Huang CL, Hwang SY, Chiang YC, Lin TP (2008) Molecular evolution of the Pi-ta gene resistant to rice blast in wild rice (*Oryza rufipogon*). *Genetics* 179:1527–1538.

Kaur N, Street K, Mackay M, Yahiaoui N, Keller B (2008) Allele mining and sequence diversity at the wheat powdery mildew resistance locus *Pm3*. In: *Proceedings of the 11th International Wheat Genetics Symposium*, Sydney, pp. 1–3.

Khazaei H, Caron CT, Fedoruk M, Diapari M, Vandenberg A et al. (2016) Genetic diversity of cultivated lentil (*Lens culinaris* Medik.) and its relation to the world's agro-ecological zones. *Front Plant Sci* 7:1093.

Khazaei H, Podder R, Caron CT, Kundu SS, Diapari M et al. (2017) Marker–trait association analysis of iron and zinc concentration in lentil (*Lens culinaris* Medik.) seeds. *Plant Genome* 10(2):1–8. https://doi.org/10.3835/plantgenome2017.02.0007

Khorramdelazad M, Bar I, Whatmore P, Smetham G, Bhaaskaria V et al. (2018) Transcriptome profiling of lentil (*Lens culinaris*) through the first 24 hours of Ascochyta lentil infection reveals key defense response genes. *BMC Genomics* 19(1):108.

Kumar GR, Sakthivel K, Sundaram RM, Neeraja CN, Balachandran SM et al. (2010) Allele mining in crops: Prospects and potentials. *Biotechnol Adv* 28(4):451–461.

Kumar J, Basu PS, Gupta S, Dubey S, Gupta DS et al. (2018a) Physiological and molecular characterization for high temperature stress in *Lens culinaris*. *Funct Plant Biol* 45(4):474–487.

Kumar J, Gupta DS (2020) Prospects of next generation sequencing in lentil breeding. *Mol Biol Rep* 47(11):9043–9053.

Kumar J, Gupta DS, Baum M, Varshney RK, Kumar S (2021) Genomics-assisted lentil breeding: Current status and future strategies. *Legume Sci* 3(3):e71.

Kumar J, Gupta DS, Kumar S, Gupta S, Singh NP (2016) Current knowledge on genetic biofortification in lentil. *J Agric Food Chem* 64(33):6383–6396.

Kumar J, Gupta S, Biradar RS, Gupta P, Dubey S et al. (2018b) Association of functional markers with flowering time in lentil. *J Appl Genet* 59(1):9–21.

Kumar J, Gupta S, Gupta DS, Singh NP (2018c) Identification of QTLs for agronomic traits using association mapping in lentil. *Euphytica* 214(4):75.

Kumar S, Ali M (2006) GE interaction and its breeding implications in pulses. *Botanica* 56:31–36.

Kumar S, Barpete S, Kumar J, Gupta P, Sarker A (2013) Global lentil production: Constraints and strategies. *SATSA Mukhapatra-Annual Technical* 17:1–3.

Kumar S, Choudhary AK, Rana KS, Sarker A, Singh M (2018d) Bio-fortification potential of global wild annual lentil core collection. *PLoS One* 13(1):e0191122.

Kumar S, Rajendran K, Kumar J, Hamwieh A, Baum M (2015) Current knowledge in lentil genomics and its application for crop improvement. *Front Plant Sci* 6:78.

Lombardi M, Materne M, Cogan NO, Rodda M, Daetwyler HD et al. (2014) Assessment of genetic variation within a global collection of lentil (*Lens culinaris* Medik.) cultivars and landraces using SNP markers. *BMC Genet* 15(1):150.

Malhotra N, Panatu S, Singh B et al. (2019) Genetic resources: Collection, conservation, characterization and maintenance. In: Singh M (ed) *Lentils*. Academic Press, Cambridge, MA, pp. 21–41.

Margulies M, Egholm M, Altman WE, Attiya S, Bader JS et al. (2005) Genome sequencing in microfabricated high-density picolitre reactors. *Nature* 437:376–380.

Marzougui A, Ma Y, Zhang C, McGee RJ, Coyne CJ et al. (2019) Advanced imaging for quantitative evaluation of Aphanomyces root rot resistance in lentil. *Front Plant Sci* 10:383.

Mikami I, Uwatoko N, Ikeda Y, Yamaguchi J, Hirano HY et al. (2008) Allelic diversification at the WX locus in landraces of Asian rice. *Theor Appl Genet* 116:979–989.

Molina C, Grotewold E (2005) Genome wide analysis of Arabidopsis core promoters. *BMC Genomics* 6:25.

Mustafa BM, Coram TE, Pang EC, Taylor PW, Ford R (2009) A cDNA microarray approach to decipher lentil (*Lens culinaris*) responses to *Ascochyta lentis*. *Aust Plant Pathol* 38(6):617–631.

El Haddad N, Rajendran K, Smouni A, Es-Safi NE, Benbrahim N et al. (2020) Screening the FIGS set of lentil (*Lens culinaris* Medikus) germplasm for tolerance to terminal heat and combined drought-heat stress. *Agronomy* 10:1036.

Ogutcen E, Ramsay L, von Wettberg EB, Bett KE (2018) Capturing variation in *Lens* (Fabaceae): Development and utility of an exome capture array for lentil. *Appl Plant Sci* 6(7):e01165.

Park PJ, Butte AJ, Kohane IS (2002) Comparing expression profiles of genes with similar promoter regions. *Bioinformatics* 12:1576–1584.

Park YJ (2007) *Allele Mining Experience on Rice Germplasm Through Advanced M Strategy Using Modified Heuristic Search.* Plant and Animal Genomes XV Conference, January 13–17, San Diego, CA, W9.

Perry JA, Wang TL, Welham TJ, Gardner S, Pike JM et al. (2003) A TILLING reverse genetics tool and a web-accessible collection of mutants of the legume *Lotus japonicus*. *Plant Physiol* 131(3):866–871.

Polanco C, Sáenz de Miera LE, González AI, García P, Fratini R et al. (2019) Construction of a high-density interspecific (*Lens culinaris* × *L. odemensis*) genetic map based on functional markers for mapping morphological and agronomical traits, and QTLs affecting resistance to *Ascochyta* in lentil. *PLoS One* 14(3):e0214409.

Polanco C, Saenz de Miera LE, Bett K, Perez de la Vega M (2018) A genome-wide identification and comparative analysis of the lentil *MLO* genes. *PLoS One* 13(3):e0194945.

Rajandran V, Ortega R, Vander Schoor JK, Butler JB, Freeman JS et al. (2022) Genetic analysis of early phenology in lentil identifies distinct loci controlling component traits. *J Exp Bot* 73(12):3963–3977.

Sakthivel K, Shobha Rani N, Pandey MK, Sivaranjani AK, Neeraja CN et al. (2009a) Development of a simple functional marker for fragrance in rice and its validation in Indian Basmati and non-Basmati fragrant rice varieties. *Mol Breed* 24:185–190.

Sakthivel K, Sundaram RM, Rani NS, Balachandran SM, Neeraja CNN (2009b) Genetic and molecular basis of fragrance in rice. *Biotechnol Adv* 27:468–473.

Sari E, Bhadauria V, Ramsay L, Borhan MH, Lichtenzveig J et al. (2018) Defense responses of lentil (*Lens culinaris*) genotypes carrying non-allelic ascochyta blight resistance genes to *Ascochyta lentis* infection. *PLoS One* 13(9):e0204124.

Sharpe AG, Ramsay L, Sanderson LA, Fedoruk MJ, Clarke WE et al. (2013) Ancient orphan crop joins modern era: Gene-based SNP discovery and mapping in lentil. *BMC Genomics* 14(1):192.

Singh D, Singh CK, Taunk J, Jadon V, Pal M et al. (2019) Genome wide transcriptome analysis reveals vital role of heat responsive genes in regulatory mechanisms of lentil (*Lens culinaris* Medikus). *Sci Rep* 9(1):1–19.

Singh D, Singh CK, Tribuvan KU, Tyagi P, Taunk J et al. (2020a) Development, characterization, and cross species/genera transferability of novel EST-SSR markers in lentil, with their molecular applications. *Plant Mol Biol Rep* 38(1):114–129.

Singh M, Bisht IS, Kumar S, Dutta M, Bansal KC et al. (2014) Global wild annual *Lens* collection: A potential resource for lentil genetic base broadening and yield enhancement. *PLoS One* 9(9):e107781.

Singh M, Kumar S, Basandrai AK, Basandrai D, Malhotra N et al. (2020b) Evaluation and identification of wild lentil accessions for enhancing genetic gains of cultivated varieties. *PLoS One* 15(3):e0229554.

Singh RJ, Chung GH (2016) Landmark research for pulses improvement. *Indian J Genet Plant Breed* 76(4):399–409.

Srivastava RP, Vasishtha H (2012) Saponins and lectins of Indian chickpeas (*Cicer arietinum*) and lentils (*Lens culinaris*). *Indian J Agric Biochem* 25(1):44–47.

Sudheesh S, Verma P, Forster JW, Cogan NO, Kaur S (2016) Generation and characterization of a reference transcriptome for lentil (*Lens culinaris* Medik.). *Int J Mol Sci* 17(11):1887.

Suzuki T, Eiguchi M, Satoh H, Kumamaru T, Kurata N et al. (2005) A modified TILLING system for rice mutant screening. *Rice Genet Newsl* 22:89.

Takano-Kai N, Jiang H, Kubo T, Sweeney M, Matsumoto T et al. (2009) Evolutionary history of GS3, a gene conferring grain length in rice. *Genetics* 182:1323–1334.

Thavarajah D, Abare A, Mapa I, Coyne CJ, Thavarajah P et al. (2017) Selecting lentil accessions for global selenium biofortification. *Plants* 6(3):34.

Till BJ, Reynolds SH, Greene EA, Codomo CA, Enns LC et al. (2003) Large-scale discovery of induced point mutations with high-throughput TILLING. *Genome Res* 13:524–530.

Tripathi K, Kumari J, Gore PG, Mishra DC, Singh AK et al. (2022) Agro-Morphological characterization of lentil germplasm of Indian national genebank and development of a core set for efficient utilization in lentil improvement programs. *Front Plant Sci* 12:751429.

Tullu AB, Kusmenoglu IS, McPhee KE, Muehlbauer FJ (2001) Characterization of core collection of lentil germplasm for phenology, morphology, seed and straw yields. *Genet Resour Crop Evol* 48(2):143–152.

Veerla S, Hoglund M (2006) Analysis of promoter regions of co-expressed genes identified by microarray analysis. *BMC Bioinform* 7:384.

Wang XU, Jia Y, Shu QY, Wu DI (2008b) Haplotype diversity at the Pi-ta locus in cultivated rice and its wild relatives. *Phytopathololy* 98:1305–1311.

Wang D, Yang T, Liu R, Li N, Wang X et al. (2020) RNA-Seq analysis and development of SSR and KASP markers in lentil (*Lens culinaris* Medikus subsp. *culinaris*). *Crop J* 8(6):953–965.

Weller JL, Liew LC, Hecht VF, Rajandran V, Laurie RE et al. (2012) A conserved molecular basis for photoperiod adaptation in two temperate legumes. *Proc Natl Acad Sci* 109(51):21158–21163.

Wong MM, Gujaria-Verma N, Ramsay L, Yuan HY, Caron C et al. (2015) Classification and characterization of species within the genus *Lens* using genotyping-by-sequencing (GBS). *PLoS One* 10(3):e0122025.

Yang S, Gu T, Pan C, Feng Z, Ding J et al. (2008) Genetic variation of NBS-LRR class resistance genes in rice lines. *Theor Appl Genet* 116:165–177.

8 Allele Mining in Mung Bean (*Vigna radiata* [L.] Walczak)

Kuldeep Tripathi, K. R. Ramya, P. S. Basavaraj, B. Manu, Kuldeep Kumar, Sudhir Kumar Jha, Gayacharan, Harsh Kumar Dikshit, Aditya Pratap, and Gyan Prakash Mishra

8.1 INTRODUCTION

Genetic diversity or variation in crop plants is the essential feature required for plant breeding activities to achieve continuous crop improvement. The variability of the plant genetic resources or crop germplasm assures the researchers to engage in exploiting the available genetic variations within their gene pool for the development of novel cultivars/varieties with desirable characteristics, with good economic returns and improved nutritional profile. Currently, utilizing nucleotide sequence variation to identify desirable traits contributed by those variations will be the major step up in crop improvement programs. Allele mining is one of the effective approaches to find out the variation or polymorphism in the crop plants occurring due to point mutations or induced mutations. The word *allele* comes from the Greek word *allelos*, meaning "each other," that is used to describe the variant forms of gene which represents the different phenotypes. Identification of allelic variation that results in the desirable trait or phenotype is the utmost challenge to be faced by plant breeders for utilizing large-scale collection of plant genetic resources of different crop plants at the various gene banks or the naturally existing crop wild relatives. The variation may arise naturally or may be induced through the use of mutagens. Mutation is typically a rapid, heritable alteration in a gene that results in the allelic diversity which is already present in organisms or any crop species. By inducing specific gene mutations that may subsequently be examined for phenotypes, allele mining (reverse genetics) is an effective technique for revealing gene function and creating novel variations (Till et al., 2007). By changing the structure and function of the encoded protein, mutations in the coding region can have a profound effect on the phenotype. The mung bean, or *Vigna radiata*, commonly called as green gram or moong, is a significant pulse crop that is consumed worldwide but is particularly well-liked in Asian nations. Mung bean is known to originate in India, and our country is the largest producer of mung bean in the world. It is an important food and cash crop as it supplies an essential protein supplement and income to farming. In addition, it has been the favorite pulse of many farmers because of the short duration, low input requirement, and it can also be used as fodder for cattle. Further, it is well suited to crop rotation and is one of the choices of crop in cereal pulse crop rotation in rice fallow regions of India. It has historically been utilized as a pulse, sprout, immature pod, and fodder in traditional medicine (the entire plant after harvest). It is well-known as a functional food for promoting human health because of its well-known excellent protein, dietary fiber, minerals, vitamins, and substantial concentrations of bioactive substances. It has been shown to have hepatoprotective and immune-modulating properties, as well as the ability to reduce hyperglycemia, hyperlipemia, and hypertension. (Hou et al., 2019). Research programs concerning the mung bean have to be accelerated since it is considered as an under-resourced crop in many countries. An international mung bean improvement program was established in 2016 globally, and in India, an all India coordinated research project (AICRP) on MULLaRP (mung bean, urdbean, lentil, lathyrus, rajma, and peas) for pulses, including mung bean, was initiated at various pulse

DOI: 10.1201/9781003385059-8

research institutes and state agricultural universities handling legume crops to improve the breeding activities of different pulse crops. Breeding programs require an understanding of the genetic diversity within the mung bean and with other species of *Vigna*. According to study results in a numerical taxonomic analysis, this genus exhibits significant genetic variations, with *Vigna* species from Africa having a substantially greater variance than those from Asia (Fatokun et al., 1993). Therefore, it is hoped that allele mining will significantly speed up finding mutations by allowing fast detection and characterization of allelic variants (Latha et al., 2004) in mung bean.

8.2 ALLELE MINING AND STRATEGIES

Allele mining is the best approach to analyze the existing allelic variants of candidate genes with favorable agro-morphological traits along with enhanced resistance-specific alleles. It helps in the identification of new haplotypes and the development of allele-specific markers (Ashkani et al., 2015). The factors which are considered in allele mining include knowledge about the genetic basis and genome sequences of a particular crop plant and the availability of phenotyping and sequencing platforms, genomic resources, and efficient bioinformatic tools.

Various works on genome investigations of different crops, such as linkage gene mapping, the discovery of quantitative trait loci, map-based cloning, the creation of physical maps, high-density genetic linkage map incorporating pertinent markers, and association studies which encompasses the whole genome or a candidate gene, are crucial, and various molecular linkage maps are also published (Lambrides et al., 2000; Humphry et al., 2002). Based on the genetic distance and recombination frequency in the linkage map, variation on the chromosomes can be identified, and in functional linkage maps, genes of specific interest can be identified. Various strategies like MutMap, comparative mapping, and fine mapping are used to identify the mutant allele showing single-nucleotide polymorphism (SNP variation). The QTL can be identified for the target traits with their responsible gene. The frequency of rare alleles in the population will be assured by using biparental populations for the QTL analysis. In the association mapping, there are two approaches, *viz.*, genome-wide and candidate gene approaches, and these will identify the polymorphism or variation within a gene that is responsible for the phenotype in the concerned trait spanning the genetic variants of the entire genome by genome-wide association studies (GWAS) or a relatively few genes by candidate gene approach. *Gene annotation* is the process of identifying genes and regulatory sequences, along with their functions; simply, it is the description of the gene and its products, such as ribonucleic acid (RNA) or protein. *In silico* gene prediction/gene hunting/gene finding refers to the identification of genes from genome sequence analysis. These many forward genetics tools are there in the way to allele mining. Compared to forward genetics (trait to genotype), the reverse genetics (from genotype to phenotype) method often requires less time. Due to the ubiquitous accessibility of sequencing data, scientists may quickly develop reverse genetic approaches to identify gene function. The sequence variations in the plant populations can be identified by (1) targeting induced local lesions in genomes (TILLING); (2) EcoTILLING method, which is a modified version of TILLING; and (3) allele mining based on sequencing and the polymerase chain reaction (PCR)–based allele mining. Mung bean is a significant grain legume crop in Asian agriculture, especially in India, Southeast Asia, and East Asia, where it is used as a pulse to supplement cereal-based diets with a significant amount of digestible protein. Researchers studying mung bean have access to techniques to boost production and improve quality thanks to the advent of molecular plant improvement technology.

8.2.1 TILLING

TILLING is the most effective method for detecting mutations in plant genomes, which capitalizes on the ability of chemical mutagenesis to produce induced mutations in a population. The discovery of many mutations within a specified region of the genome distinguishes the TILLING strategy from transgenic methods. These mutations make up allelic series that enable structure and functional investigations as well as the possibility of a variety of symptoms, from mild to strong. TILLING is a non-transgenic technique that introduces and identifies unique genetic variations in genes that influence important properties in domesticated crops. It has been created to make use of the most recent discoveries regarding the deoxyribose nucleic acid (DNA) sequence and to research the functions of particular genes. It offers a high frequency of point mutations that are randomly distributed throughout the genome, and high throughput screening of nucleotide polymorphisms in a specific sequence.

The heteroduplex analysis could help in the identification of chromosomal rearrangements or chromosomal breakdown and the heteroduplexes in the genome of a plant due to any point mutations induced by this technique (Till et al., 2003). The schematic representation of allele mining is given in Figure 8.1.

8.2.2 EcoTILLING

An approach to measuring the degree of natural variation in certain crop where polymorphisms present naturally is EcoTILLING. The process is very similar to TILLING, with the exception that the mutations are found from naturally occurring alleles rather than being induced on purpose (Comai and Henikoff, 2006). Furthermore, this technique used in candidate gene mapping and the quick, inexpensive screening of novel allelic variations of existing candidates have both been thoroughly investigated, indicating that it could be a useful technique in crop improvement programs (Raghavan et al., 2007). This technique involves the use of PCR to amplify alleles from various genotypes, followed by DNA sequencing to detect nucleotide variation. Even if it is asserted that EcoTILLING can be carried out for a small portion of the cost of SNP/haplotyping techniques, the procedure's wide applicability is in doubt due to the complicated equipment, specialized knowledge, and requirement for confirmation sequencing. Sequencing-based allele mining requires less-expensive equipment and fewer arduous steps, but it has higher sequencing expenses. However, each technology has pros and cons that must be carefully assessed against the requirements of the intended use.

8.2.3 NGS Technologies in Allele Mining

The emergence of "highly scalable" methods in recent years has also sparked the creation of next-generation sequencing (NGS) machines with higher precision and throughput, where re-sequencing, sequence alignment, and comparison of their results with reference genomes are carried out. 454 Life Sciences commercialized the first of its kind, which relied on pyrosequencing and did not require cloning (Margulies et al., 2005). Sequencing-based allele mining may eventually lead to rapid allelic data at a lower cost and longer read length. The mining of the target genes in the selected accessions by sequencing would significantly speed up precision breeding tactics for agricultural crop plants. Faster sequencing platforms have made the sequencing procedure easier and less expensive.

8.3 BIOINFORMATIC TOOLS

Generation of sequence data at the unprecedented rate through high-throughput technologies accelerated the initiation of genome sequencing in many crops. This leads to the development of computer programs and software such that it can handle large sequence data, thereby analyzing and storing sufficient information and being user-friendly. Detection of polymorphism in the genotypes by comparative analysis of sequences marks the beginning of the allele mining process. There are many software programs that can handle complex nucleotide data, predict putative structural or functional elements of complex macromolecules, forecast changes in amino acid composition

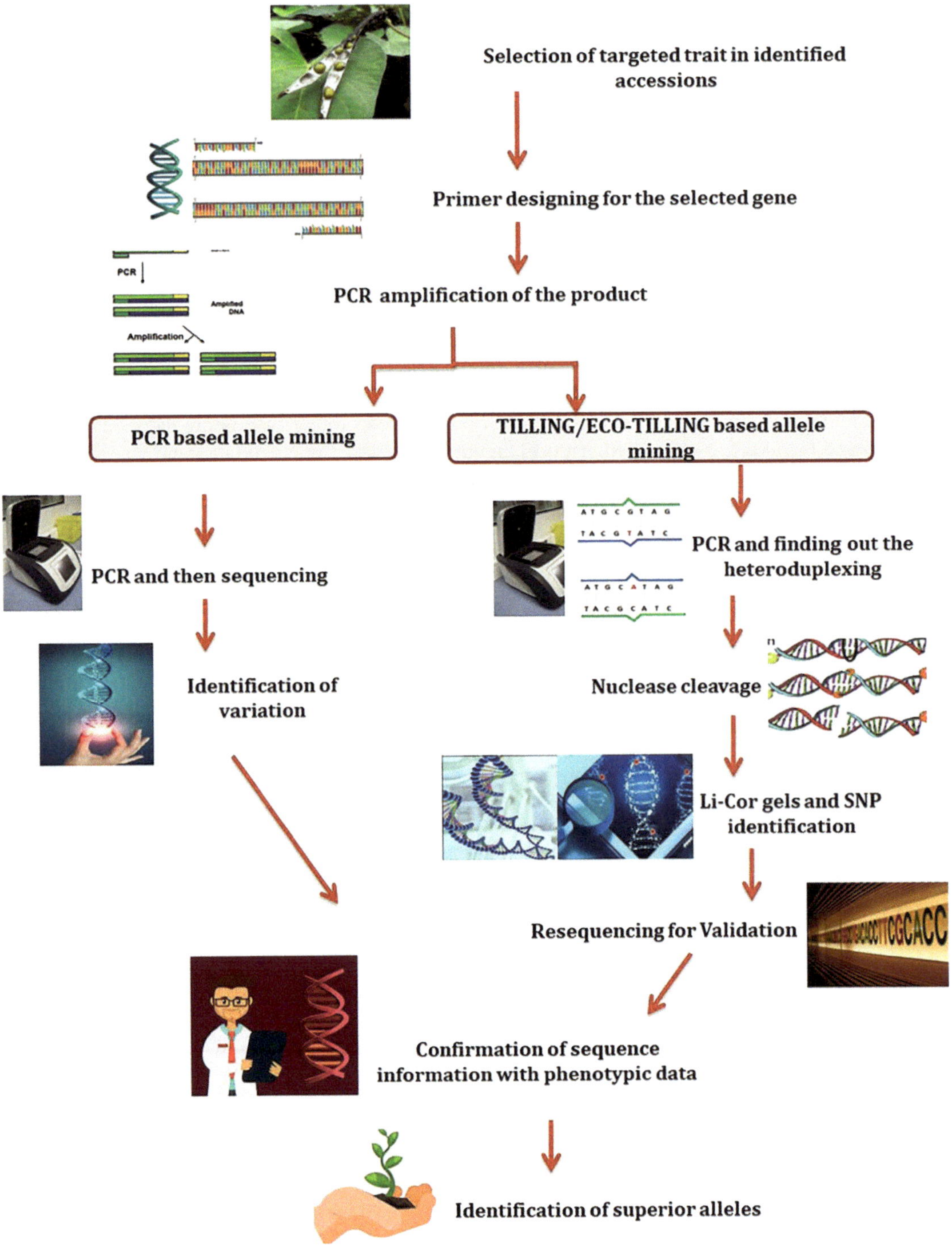

FIGURE 8.1 Schematic representation of process of allele mining.

that will affect the structure or functionality of encoded proteins, locate sequence polymorphisms, and predict transcription factor binding sites. The different online platforms which are used in allele mining are given in Table 8.1. Using Torrent Suite 1.5 or another bioinformatic tool after library sequencing, sequence data can be filtered, trimmed, and aligned using the relevant reference sequences, and defined coordinates retrieved from a gene bank. After the sequences have been aligned, SNP can be filtered from sequenced contigs using online processing technologies like Variant Caller pipeline. The PARSESNP tool (http://www.proweb.org/parsesnp/) / / BALL-SNP/BAM matcher was used to evaluate the identified sequence variations. This program offers information on the location as well as the specifics of amino acid alterations and the severity of mutations. These pipelines also offer details on the increase or decrease in restriction sites brought on by induced polymorphisms. Each amino acid's addition, deletion, and substitution could have

TABLE 8.1
Online Tools in Allele Mining (TILLING)

Sl. No.	Online Tools	Link
1	QualitySNP	https://www.bioinformatics.nl/tools/snpweb/
2	PupaSNP	https://bioinformaticshome.com/tools/descriptions/PupaSNP_Finder.html
3	SniPlay	https://sniplay.southgreen.fr/cgi-bin/home.cgi
4	SNPrank	https://bio.tools/snprank
5	SNPServer	https://bioinformaticshome.com/tools/descriptions/SNPServer.html
6	CERENKOV2	https://bio.tools/CERENKOV2
7	plantCARE	https://bioinformatics.psb.ugent.be/webtools/plantcare/html/
8	TRANSFAC	https://bio.tools/transfac
9	MEME	https://meme-suite.org/meme/
10	JASPAR	https://jaspar.genereg.net/
11	SNP2CAPS	http://pgrc.ipk-gatersleben.de/snp2caps/
Tools Used in Allele Mining (Used in Linkage Analysis, QTL Mapping, Association Mapping, and Gene Prediction)		
1	G-Mendel	https://gaow.github.io/genetic-analysis-software/g/g-mendel/
2	MultiMap	https://bio.tools/multimap
3	AntMap	http://lbm.ab.a.u-tokyo.ac.jp/~iwata/antmap/
4	MultiPool	https://bio.tools/snprank
5	MapPop	https://gaow.github.io/genetic-analysis-software/m/mappop/
6	MapMarker/QTL	http://www2.hawaii.edu/~durrell/Software/B14F5F09–6238–43DA-B95F-958D4FB709AD.html
7	PLABQTL	ftp://ftp.uni-hohenheim.de/pub/plabqtl
8	QTL Cartographer	https://brcwebportal.cos.ncsu.edu/qtlcart/
9	GeneNetwork	http://www.genenetwork.org/
10	The QTL cCafe	http://www.biosciences-labs.bham.ac.uk/Kearsey/applet.html
11	TASSEL	https://www.maizegenetics.net/tassel
12	GAPIT	https://zzlab.net/GAPIT/
13	EasyMap	https://bio.tools/easymap
14	EMMAX	https://bioinformaticshome.com/tools/gwas/descriptions/EMMAX.html#gsc.tab=0
15	GenABEL	https://bio.tools/GENABEL
16	GENESCAN	https://bio.tools/genscan
17	GENEPARSER	https://bio.tools/geneparser
18	NNPP	https://www.fruitfly.org/seq_tools/promoter.html

an impact on how proteins operate (Sim et al., 2012). SIFT, another web-based tool, was utilized to identify functional mutations since it makes predictions about how amino acid changes will affect protein structure and function based on sequence homology and amino acid physical properties. The legume bioinformatics database, which is the repository of stored information regarding the protein or nucleotide sequence, is the legume information system (LIS) (https://www.hsls.pitt.edu/obrc/index.php?page=URL1106589199). It is important to note that computation-based predictions are dependent on premade algorithms, many of which were developed using previously reported sequences that are readily available in public databases. Therefore, wet lab experimental techniques should always be used to check such predictions.

8.4 GENE POOLS IN MUNG BEAN

Fabaceae (Leguminosae) is the third largest family of the angiosperms and is represented by about 600 genera and 12,000 species. The genus *Vigna* comprises 82 described species with six subgenera. Mostly, the subgenera *Ceratotopis* has domesticated in Asia and consists of 21 species, of which seven are domesticated, including mung bean, moth bean, adzuki bean, black gram, creole bean, minni payaru, and rice bean (Takahashi et al., 2016). Among these, there are four pulses which are important and cultivated as an economic crop in many parts of the world, namely, adzuki bean, black gram, moth bean, and mung bean (Tomooka et al., 2014). The morphological diversity of different species of *Vigna* is given in Figure 8.2. The different species of *Vigna* with its potential uses and its distribution according to ICAR publication ("Wild Relatives of Cultivated Plants in India" Pradheep et al., 2014) are given in Table 8.2. Mung bean originated in India, and it is an important source of protein, phosphorus, and several other minerals in the human diet. This crop can intercross with other *Vigna* species, such as *V. mungo*, *V. radiata var. sublobata*, *V. radiata var. setulosa*, *V. trilobata*, *V. trinervia*,

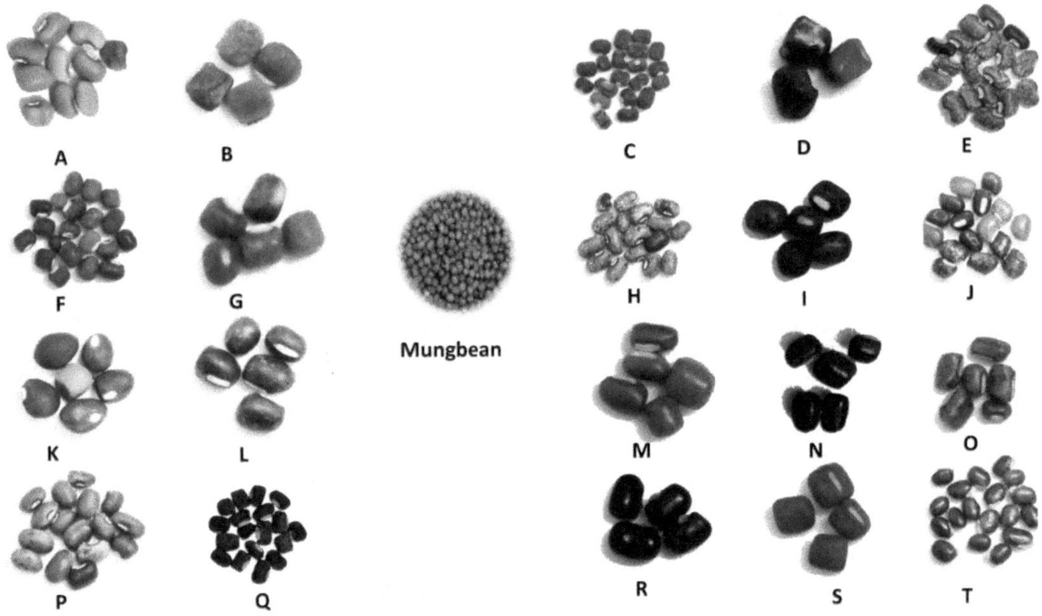

FIGURE 8.2 Morpho-diversity of different *Vigna* species.

Note: (a) *Vigna unguiculata*; (b) *Vigna sublobata*; (c) *Vigna vexillata*; (d) *Vigna hainiana*; (e) *Vigna unguiculata* sp. *stenophylla*; (f) *Vigna stipulaceae*; (g) *Vigna trilobata*; (h) *Vigna unguiculata* sp. *dekindtiana*; (i) *Vigna aconitifolia*; (j) *Vigna luteola*; (k) *Vigna subterranea*; (l) *Vigna nipponensis*; (m) *Vigna dalzelliana*; (n) *Vigna trinervia*; (o) *Vigna minima*; (p) *Vigna vexillata*; (q) *Vigna mungo*; (r) *Vigna pilosa*; (s) *Vigna bourneae*; (t) *Vigna oblongifolia*; centre: mung bean (*Vigna radiata*).

TABLE 8.2
Different Species of *Vigna* in India with Its Uses

Species	Characteristics	Distribution
V. radiata	Cultivated pulse crop and important source of protein	Originated in India; usually occurs in Northern western ghats and eastern ghats which has profuse pod bearing tendency (Dikshit et al. 2007); Grown in all parts of the country
V. silvestris	Tolerant to MYMV and it is resistant to bruchids.	Distributed in the Maharashtra, Goa, Karnataka, Madhya Pradesh and Rajasthan
V. gracilis, V. ambacensis, V. oblongifolia, V. heterophylla, V. parkeri, V. laurentii, V. racemosa, V. frutescens, V. multinervis and V. membranaceae	Belongs to the *section Vigna* and close to cowpea	All these are Introduced for academic purposes (Vijaykumar et al. 2010)
V. sahyadriana	Morphologically close to *V. silvestris*	Wild occurs in Northern western ghats (Aitawade et al. 2012)
V. trinervia	Var. *trinervia* is bruchid resistant	Var. *trinervia* occurs in southern Indian parts (Tamooka et al. 2000); Var *bournea* in Karnataka, Tamil Nadu, Kerala (Seehalak et al. 2006);
V. radiata var *setulosa*	Belongs to primary genepool. It has a profuse pod bearing tendency	North western ghats and eastern ghats (Dikshit et al. 2007)
V. sublobata	Belongs to primary genepool. Resistant to YMV and is tolerant to drought, salinity (Ignacimuthu S and Babu CR, 1987) and alkalinity, has high photosynthetic efficiency and high methionine content (Babu et al. 1988), resistant to bruchids (Bisht et al. 2005).	Distributed in all parts of mungbean growing areas as wild
V. subramaniana	Belongs to secondary genepool. Resistant to bruchids (Tamooka et al. 2000).	Wild occurs in Madhya Pradesh, Tamil Nadu, Punjab, Uttar Pradesh
V. unguiculata	Wide known pulse, fodder and vegetable crop. Crossable with *V. radiata* and produce fertile hybrids	Widely distributed cultigen
V. mungo	Well known and grown pulse crop next to mungbean. Crossable with *V.radiata* and produce fertile hybrids. Notable for arabinogalacton content which gives its mucilaginous property.	Widely distributed cultigen
V. luteola	Has Forage value. Karyologically similar to *V.unguiculata* and resistant to cowpea pod sucking bugs (Laghetti et al. 1998) and storage weevil and is close to *V. marina* (Maxted et al. 2004)	Widely present as in the sandy and rocky coastal areas of West Bengal
V. marina	Tolerant to drought and salinity (Padulosi S and Ng N Q, 1990)	In coastal regions of Andaman and Nicobor of Islands, Kerala, Odisha, Tamil Nadu.
V. acontifolia	Both fodder and pulse crop. It is the most drought tolerant species (Hasanuzzaman et al. 2003).	Originated from India (throughout the country up to 1200 m; in dry areas.

(Continued)

TABLE 8.2 *(Continued)*
Different Species of *Vigna* in India with Its Uses

Species	Characteristics	Distribution
V. indica	Close to *V. aridicola*. Belongs to secondary genepool.	Present as wild
V. trilobata	Belongs to secondary genepool; semi domesticated cover crop and fodder crop (Singh et al. 2005)	Distributed throughout India
V. minima	Belongs to Primary genepool. Fodder crop	Distributed in hilly regions as wild
V. umbellata var *umbellata*	Belongs to secondary genepool. Crossable with *V.radiata* and produce fertile hybrids. It is used as pulse in Northeast India [states of Assam, Meghalaya, Manipur, Mizoram, Nagaland, and Tripura] and fodder in Western Himalaya, East India [parts of Bihar, Jharkhand, Odisha, West Bengal], East India [parts of Bihar, Jharkhand, Odisha, West Bengal]. It is resistant to mungbean yellow mosaic virus, bruchids and bacterial leaf spot (Arora et al. 1980). It has the highest yield potential among *ceratotropis* species.	Widely grown nowadays for a research purpose. Domesticated probably in Thailand and Myanmar (Seehalak et al. 2006); Present throughout India, especially hills.
V. umbellata var *gracilis*	Belongs to secondary genepool and progenitor of *V. umbellata*	Occurs in Himalayas and North eastern regions
V. dalzelliana	Semi domesticated as a cover crop and fodder crop in India.	Distributed in peninsular India [latitude below Vindhya Range], central India [parts of Madhya Pradesh and adjoining areas], Rajasthan, East India [parts of Bihar, Jharkhand, Odisha, West Bengal], Sikkim, Andaman and Nicobar Islands, Uttar Pradesh
V. clarkei	Wild species; fodder crop	Occurs mostly in Eastern Himalayas and Northeast India [states of Assam, Meghalaya, Manipur, Mizoram, Nagaland, and Tripura]
V. grahamiana	Wild species; fodder crop	Present in Kerala, Karnataka, Tamil Nadu and Odisha
V. prainiana	Wild species; fodder crop	Distributed in eastern Himalayas and Uttarakhand
V. vexillata	Belongs to tertiary genepool. It has high protein and resistant to cowpea pod sucking bugs and bruchids (Laghetti et al. 1998); as a tuber crop (Singh et al. 2013).	Semi domesticated and occurs throughout India in North Eastern India, Eastern parts of India, Karnataka, Kerala, Tamil Nadu, peninsular India [latitude below Vindhya Range], central India [parts of Madhya Pradesh and adjoining areas]
V. adenantha, V. hosei, V. kirkii, V. longifolia, V. grandiflora and V. reticulata	Introduced for academic purpose	*V. adenantha* is naturalized throughout the India and *V. hosei* has been introduced in Tamil Nadu, Kerala, Odisha
V. subterranea	Pods are subterraneal. Low-cost source of complex carbohydrates, plant-based protein; crossable with cowpea (Bagemann S, 1997)	Origin is Africa and introduced in India for research purposes.

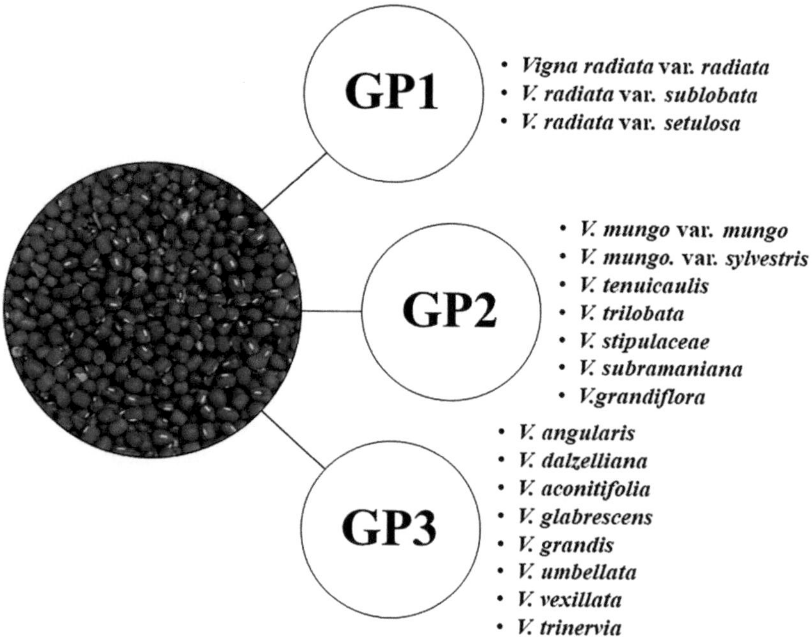

FIGURE 8.3 Gene pool of mung bean.

Source: Pradheep et al., 2014; Pratap et al. (2014); Archak et al. (2020).

V. hainiana, and *V. dalzelliana*, and there have also been instances of crossability barrier issues, particularly because of incompatibility in chromosomal pairing (Smýkal et al., 2015). The gene pool concept was given by Harlan and de Wet in 1971. According to this, there are three different types of gene pool, namely, primary (GP1), secondary (GP2), and tertiary (GP3) gene pool. *Primary gene pool* includes the cultivated species and its close relatives. The species in this gene pool are easily crossable and produce fertile offspring; in the secondary gene pool, the species can be crossed with GP1 with little effort and produce partial fertile offspring on the crossing. The tertiary gene pool consists of species that are not crossable with GP1/GP2 but where crossing can done with much difficulty by using species techniques like wide crossing, bridge crosses, etc., and it produces nonviable offspring, and the embryo can be saved with the embryo rescue technique. The quaternary gene pool is another classification which includes the clonal and recombination products of the plant or organisms. The different species of mung bean which are grouped among different gene pools (Pradheep et al., 2014; Pratap et al., 2014; Tamooka et al., 2014; Archak et al., 2020) are given in Figure 8.3.

8.5 ALLELE MINING TOOLS AND TECHNIQUES

Relationships among the members of the Fabaceae have been greatly aided by molecular marker technologies, such as restricted fragment length polymorphism (RFLPs), random amplified polymorphic DNA (RAPDs), amplified fragment length polymorphism (AFLPs), and microsatellites markers or simple sequence repeats (SSRs). This is due to their reproducibility, multiallelic nature, and codominant nature. This developed and identified genomic resources which significantly impacted the comprehension of structure and evolution of the mung bean genome. The estimated genome size of mung bean is nearly comparable to the genome sizes of other different *Vigna* species (Somta and Srinives, 2007). This results in greater understanding, given the boost and flourishment to the candidate gene discovery. The *de novo* synthesis sequence data from 18 *Vigna* species was a major accomplishment in the *Vigna* genomic research in 2014, and subsequently, whole-genome sequence of mung bean was published in 2014. The study of genomes was advanced (Chen et al., 2015). Apart from this, genome-wide SNP markers identified by Illumina HiSeq sequencing enriched the molecular resources (Van et al., 2013), by evaluating the diversity of mung bean available in the United States of America to perform a genotyping-by-sequencing (GBS), and GWAS performed for various nutrients based on a seven-mineral analysis identified a total of 6,486 SNPs and found that there is feasibility of 43 marker and trait associations with nutrients like calcium, iron, etc. in mung bean grain produced (Wu et al., 2020); GWAS for PUtE traits in mung bean was studied and found candidate genes controlling PUtE by regulation of nutrient uptake and the root architectural development pathways (Reddy et al., 2021). Lee et al. (2021) identified genetic factors responsible for the compound raceme of mung bean, while Liu et al. (2022a) identified genic male sterility (GMS)–related gene *VrCYCA1* in mung bean. *VrFRO8*, a related gene for salt tolerance in mung bean, was identified by Liu et al. (2022b). The reference and pangenome assembly of mung bean collection in China by sequencing was done by Liu et al. (2022c). The entire genome of numerous crop species has now been sequenced as a result of developments in sequencing technology. With the ultimate goal of crop development, this has made it possible to perform significant research on structural and functional features. The genomic research in many major legume species has accelerated the research on under-resourced crops which lack the sequence data, by the use of invention of widely used legume markers (Gepts et al., 2005). Plant breeding is becoming simpler and more successful because of the exchange of genetic and genomic data between species made feasible by translational genomics techniques (Stacey and Vanden Bosch, 2005). A key candidate gene, *Vradi05g03810*, for bruchid resistance was identified, and it is due to the deletion in the coding region in bruchid-resistant lines (Chen et al., 2022). Liu et al. (2022c) carried out the exemplary work that the reference genome of mung bean using Vrad_JL7 was constructed, and it has the size of about 479.35 Mb. The first pangenome of mung bean was constructed, which regulates the photoperiodic flowering pathway. GWAS was carried for mung bean mini-core by the World Vegetable Center for 100-seed weight using 6,506 SNP markers, and all these were located in the chromosome 8 (Akhtar et al., 2021). Mutation in *VrTAF5 gene* (which encodes for TAF5) induces the *Cercospora* resistance observed in V4718 (Yundaeng et al., 2020). Genome sequence of mung bean by re-sequencing was constructed and identified candidate genes for synchronous pod maturity by Ha et al. (2021). P uptake efficiency (PUpE) and P utilization efficiency (PUtE) traits in mung bean were studied with GWAS approach and identified 61 protein-coding genes, and of these, six genes had a high expression (*VRADI08G00070*, *VRADI05G20860*, *VRADI06G12490*, *VRADI01G04370*, *VRADI09G09030*, and *VRADI08G20910*) in root, shoot, apical meristem, and leaf (Reddy et al., 2021). With genotyping-by-sequencing (GBS) platform, the possible candidate gene, QTL *qMYMV4–1*, was identified for MYMV resistance (Mathivathana et al., 2019). Identification of candidate genes in mung bean for improving low-phosphorus tolerance, which includes *VRADI11G08340*, *VRADI01G05520*, and *VRADI04G10750*, with missense SNPs in coding sequence region (Reddy et al., 2020). Diversity panel constituted by genotyping of 16 wild accessions and identification of over 22,000 SNPs accelerate mung bean research (Noble et al., 2018). GWAS was undertaken to mine the genes involved in salt stress tolerance by the World Vegetable Center with the developed mung bean mini-core collection and identified genes such as *Vradi07g01630*, *Vradi09g09510*, and *Vradi09g09600* that have salinity tolerance (Breria et al., 2020).

8.6 APPLICATION OF ALLELE MINING IN MUNG BEAN

Researchers put their efforts into recombining the beneficial alleles into a common background in order to develop an elite cultivar with greater endurance. The large germplasm collections possess huge genetic diversity for any particular loci, but not all allelic variations have significant phenotypic effect. The characterization of suitable allelic variants with larger phenotypic effects is the utmost requirement for it. As the term is self-explanatory, allele mining aims for the allelic variation at a particular locus. Natural selection is the major driving force for allelic variations. Usually, non-synonymous variation that has a high effect on the final products is the main driver for evolution, divergence, and speciation. Generally, for particular loci, in natural conditions, only those alleles having a desirable impact on the plant's survival is selected, for example, antinutritional factor genes are naturally selected as they impart the plants resistance to several predators. On the other hand, allelic selection by humans is completely different, as they majorly look for traits with high economic importance, like yield (Kumar et al., 2010). Both types of selection are major players in genome evolution, sometimes also leading to complete loss/gain of genes, giving rise to the concept of pangenome.

Since many diseases are governed by recessive alleles, and because the word *dominance* has a positive connotation, the dominant phenotype is also called "superior" with respect to fitness. It is important to distinguish between the terms "superior allele" and "dominant allele," as the former refers to the allele that best fits human expectations in terms of desirable phenotypes. Once the loci are sequenced from diverse germplasm collections, several advanced bioinformatics tools related to variant calling and alignment are deployed to know the allelic status, followed by association studies to get true marker–trait association, as not all allelic variants are casual. Finally, *allele mining* is the collective term which includes the identification of variations in the plant populations that can be effectively used in the crop improvement program. Some of the applications of allele mining are given in Figure 8.4.

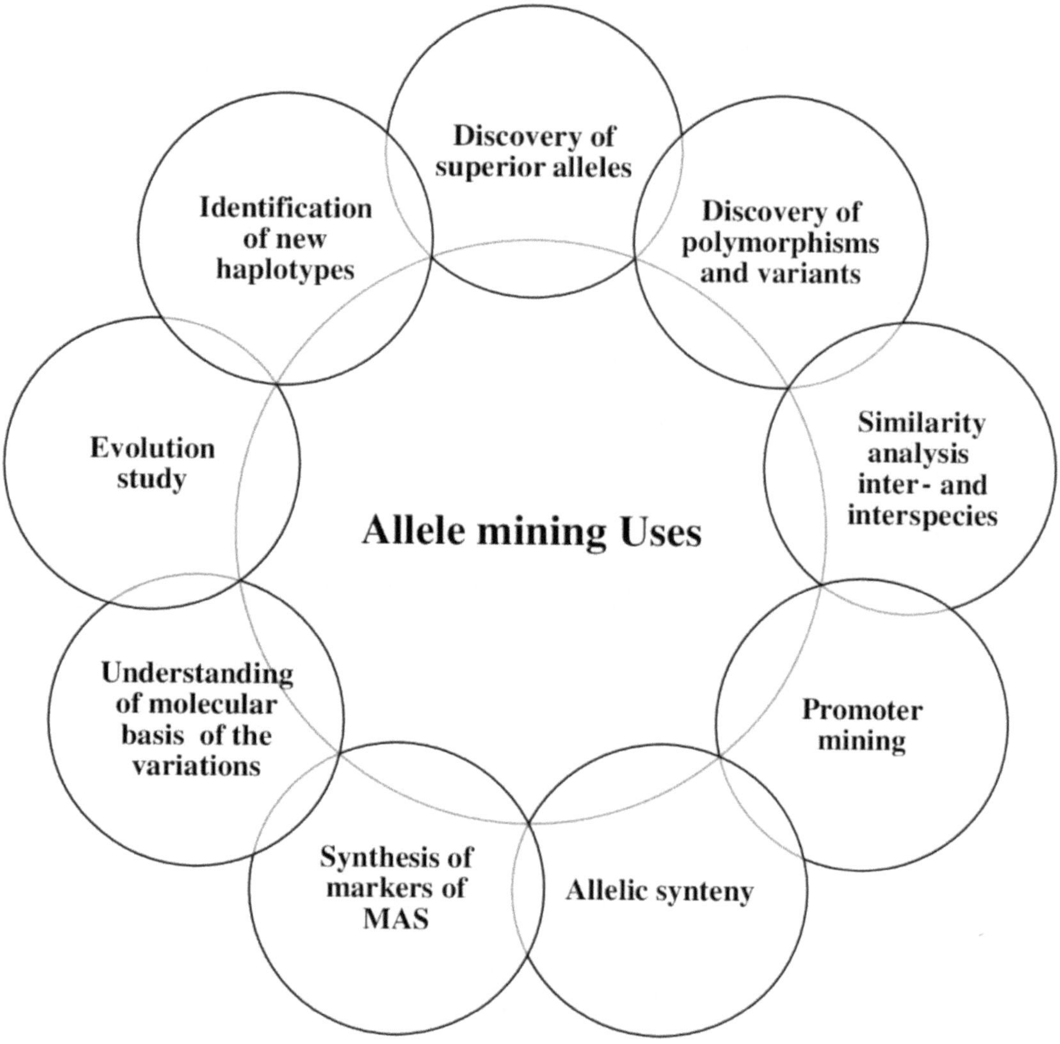

FIGURE 8.4 Applications of allele mining in crop improvement.

A lot of efforts have been made in searching for superior alleles for several biotic and abiotic stresses, as well as traits of agronomic importance in mung bean. Many works by various researchers of the world have enabled and improved mung bean breeding activities globally. Reddy et al. (2020) have identified nine SNPs for traits like root dry weight and total leaf surface area, QTLs for seed hardiness and seed weight (Humphry et al., 2005), mungbean yellow mosaic India virus (MYMIV) (Kitsanachandee et al., 2013), *Cercospora* leaf spot (Chankaew et al., 2013), and powdery mildew (Kasettranan et al., 2010). Several other QTL mapping and association mapping studies conducted on *Vigna* and allied species have been described by Manu et al. (2022).

8.6.1 Allele Mining in Discerning New Haplotypes

Crop genome sequencing has resulted in enormous amounts of sequence data being made accessible to the general population (Singh et al., 2021). Natural allelic variation at candidate genes affecting important agricultural properties may be dissected using allele mining, a promising method that might be used in crop enhancement projects. Allele evolution can be tracked, and new haplotypes and trait-specific markers can be discovered. Many worldwide agricultural research organizations have begun allele mining studies after realizing the enormous potential of this technique.

Genome-wide association studies can better capture allelic variation in the case of quantitative traits, like flowering, than biparental mapping populations, where only a few recombination events are intended (Kumar et al., 2022). However, the effects of haplotypes from multiple founders on phenotypes are not considered (Ogawa et al., 2018). Also, in a polygenic trait, an allele never functions alone; it has a cumulative effect on the phenotype, along with the other counterpart gene, giving rise to the concept of haplotype-based breeding. Haplotypes are the genes, or markers, that segregate together due to linkage disequilibrium. Therefore, with the advent of the genomics era, association mapping has advanced to the search of superior haplotypes after the suitable marker–trait associations are known. Gene, allele variability, and diversity at candidate genes of interest in particular must be understood in order to fully use gene banks.

Handling the complete germplasm, whether for traditional plant breeding or allele mining, is a daunting task that needs sampling procedures to reduce it to a manageable scale while keeping variety. The Asian Vegetable Research and Development Center (AVRDC) has created a core collection of mung bean lines consisting of 1,481 entries (Schafleitner et al., 2015). Furthermore, the core collection was characterized for 20 SSR markers for developing a mini-core set of 289 accessions. Previously, this was elucidated by Upadhyaya et al. (2006). The goal now is to combine data from whole-genome sequencing, SNP assays, QTL mapping, and trait variation to use developed genomic resources more effectively in mung bean breeding.

The knowledge gap between genes and gene function must also be filled in order to develop new productive plant ideotypes. Trait-enhancing alleles have been found in genes, loci, and germplasm sources; thus, it is important to document exceptional and superior alleles, as well as their behavior in appropriate genetic settings and scenarios, as efficient data mining bioinformatic tools are available now and are anticipated to overtake other crop improvement techniques for producing or identifying novel alleles and taking center stage in all crop improvement activities.

8.6.2 Allele Mining for Biotic Stress Tolerance

Mung bean can be incorporated in crop rotation because of its short duration and as a major pulse since it has high protein and can be grown in every season and on any type of soil. However, biotic and abiotic stresses constantly pose a threat to their productivity. Furthermore, issues related to climate change may exacerbate the situation and reduce mung bean productivity. However, because of its advantages for human nutrition, adaptability to various cropping systems, and biological ability to fix nitrogen, cultivation area and output have been able to grow. Traditional breeding techniques have historically made a significant contribution to the creation of mung bean varieties that are resilient to both biological and non-biological stress, but there are still many challenges to overcome, such as identification of donor sources for various stress and economically important diseases (Nair et al., 2019). With the quick development of sequencing technologies, genomic resources and tools have become an essential aspect of agricultural improvement for numerous crops, including mung bean. Marker-based selection has significantly replaced traditional phenotype-based selection over the past ten years. Modern molecular methods and technologies, like marker-assisted selection (MAS), marker-assisted backcross breeding (MABB), GWAS, and genomic selection (GS), aid in the quick and precise identification and transfer of genes, alleles, QTLs linked to desired traits. The era of NGS technologies has made SNP and allele mining simple and straightforward. Additionally, it aids in the conservation, characterization, and evaluation of the genetic resources for mung bean. Breeders have the chance to use these genetic resources to mine novel alleles for creating mung bean cultivars that are climate-resilient.

Many crops, including mung bean, are affected by stored pests of bruchids, or seed weevils (Srinives et al., 2007). According to Somta et al. (2008), antibiosis in the cotyledon is the primary cause of bruchid resistance, and the *locus Br* that controls resistance has some modifications (Somta et al., 2007). Combining transcriptomic, proteomic, and gene expression analysis allowed the identification of the genes *g34458* and *g39185* as potential bruchid resistance genes (Lin et al., 2016). The Br locus VrPGIP2 showed strong bruchid resistance than VrPGIP1 by *SNP* mutation at the coding region was discovered owing to sequence-based allele mining (Chotechung et al., 2016). Chen et al. (2022) subsequently mapped and found *Vr04g00919*, a potential gene for bruchid resistance. Powdery mildew caused by *Erysiphe vignae* sp. *nov.* causes economic losses in mung bean production at various stages of the crop (Kelly et al., 2021), and *MLO* (*mildew locus O*) genes are associated with resistance/susceptibility to powdery mildew disease in mung bean and other crops. Resistance to powdery mildew (PM) is associated with loss of *MLO* genes (Kusch and Panstruga, 2017). To date, more than 18 loci conferring resistance to PM have been identified from different sources in mung bean (Rispail and Rubiales, 2016). The candidate gene *VrMLO12* associated with PM resistance was identified (Yundaeng et al., 2020). Therefore, there is ample scope to identify additional candidate genes for tolerance to PM. In addition to powdery mildew, *Cercospora* leaf spot caused by *Cercospora canescens* is also an important foliar disease in mung bean causing significant yield losses (Somta et al., 2022). Though this disease is of great importance in terms of economic loss, there are only few studies at genetics and genomics level against CLS in mung bean. A gene *VrTAF5* mined from resistant germplasm, *V4718*, on chromosome 6 of mung bean confers tolerance to CLS (Yundaeng et al., 2021). Singh et al. (2020) identified 1,097 alleles against MYMIV in diverse mung bean genotypes through GWAS, and still it is the biggest menace to the mung bean cultivation by affecting the yield, and this can be overlooked by new, emerging gene editing techniques, Mishra et al. (2020). Further, Rohilla et al. (2022) identified candidate genes for yellow mosaic disease (YMD) in a 80 diverse mung bean genotypes. Sudha et al. (2022) performed transcriptome analysis to identify differentially expressed genes and candidate genes(s) for YMD and found *Vradi04g06840* as a candidate gene, which leads to further study.

8.6.3 Abiotic Stress Tolerance

Mung bean production is also affected by a plethora of abiotic stress factors, like drought, salinity, heat, water logging, and nutrient deficiency. In the recent past, genetic resources for tolerance against these abiotic stress factors have been identified. The different candidate genes/alleles responsible for various biotic and abiotic stress tolerances in mung bean are given in Table 8.3.

Mung bean is one of several short-duration grain legumes that can be grown well across seasons and soils. Hence, they have more chances to be prone to different kinds of biotic and abiotic stresses. Classical breeding approaches have led to the development of high-yielding mung bean

TABLE 8.3
Candidate Genes/Alleles for Various Stress Tolerances in Mung Bean

Candidate Genes	Crop Species	Trait	Reference
Biotic Stress			
g34458 and *g39185*	*Vigna radiata*	Bruchid resistance	Lin et al., 2016
Vr04g00919	*Vigna radiata*	Bruchid resistance	Chen et al., 2022
VrMLO12	*Vigna radiata*	Powdery mildew resistance	Yundaeng et al., 2020
Diverse alleles	*Vigna radiata*	Mung bean yellow mosaic India virus	Singh et al., 2020
VrTAF5	*Vigna radiata*	*Cercospora* leaf spot	Yundaeng et al., 2021
Vradi04g06840	*Vigna radiata*	YMD resistance	Sudha et al., 2022
Abiotic Stress			
Vradi09g09510, Vradi09g09600	*Vigna radiata*	Salt tolerance at germination stage	Breria et al., 2020
VRADI11G08340, VRADI01G05520, VRADI04G10750	*Vigna radiata*	PUtE (Phosphorus Utilization Efficiency)	Reddy et al., 2020
VabZIP6, VabZIP34, VabZIP50, VrbZIP	*Vigna angularis*	Deficit moisture stress tolerance	Wang et al., 2018
VrHsfA6a, VrHsfA6b	*Vigna radiata*	Drought stress	Li et al., 2019
VrSKP1	*Vigna radiata*	Drought tolerance	Bharadwaj et al., 2018
VaWRKY61	*Vigna angularis*	Osmotic adjustment	Srivastava et al., 2018
Coda	*Vigna radiata*	Abiotic stress tolerance	Baloda et al. (2017)

cultivars. Population is growing in an alarming manner, with increased production constraints; thus, to safeguard food security, we need to develop climate-smart and high-yielding mung bean cultivars. Nevertheless, classical breeding tools are time-consuming and are associated with linkage drag, and many of the traits are under the control of polygenes; manipulation of polygene through conventional breeding is a very challenging task. Thus, we have to integrate genomics tools and resources to fast-track product development in mung bean breeding programs.

8.6.4 Allele Mining for Agronomically Desirable Traits

Through conventional methods, breeding for complex traits of agronomic importance is difficult, since they are governed by many genes and their expression is influenced by the environment; hence, it is very difficult to breed for them. The variation seen for agronomic traits is very limited, and it is a product of genotype and the environment. This is further complicated by various biotic and abiotic stresses affecting the yield potential. Allele mining offers a huge scope and opportunity to understand/discover novel genetic variations for complex traits that will accelerate the breeding crops. Noble et al. (2018) characterized mung bean diversity panel to dissect genetic basis of important agronomic traits. GWAS for seed mineral concentration in mung bean was carried out by Wu et al. (2020) and to understand the mechanism behind the PUpE and PUtE by Reddy et al. (2021). Various agro-morphological studies were undertaken to find out the best accessions by phenotypic diversity assessment (Gayacharan et al., 2020) and the selection of good mutant lines for yield and duration induced by gamma rays (Sarkar and Kundagrami, 2018).

8.6.5 Allele Mining for Climate Change

Traditional plant breeding involves selection of superior genotypes based on their phenotypic expressions. Selection is dependent on genetic recombination (genotype), environment, and their interactions. Selection based on phenotype during optimum climatic condition has helped in developing many high-yielding cultivars. However, in a climate change scenario, various abiotic stresses like heat, drought, waterlogging, and salinity have affected the global food security. An estimated crop loss of about 50% due to abiotic stresses globally (Rasool et al., 2013; Rodziewicz et al., 2014). Godfray et al. (2010) predicted that from the presently available limited genetic and environmental resources, there will be double the rate of higher demand for food production by 2030. Challenges faced by scientists include increasing the output or yield productivity of the superior varieties and, at the same time, making the cultivar climate-resilient (Frona et al., 2019). The introduction of high-yielding varieties (HYV) has narrowed the genetic base of a crop and has attained yield plateau. It is important to widen the genetic base of the HYVs to break the yield plateau by exploring the genomic resources. Exploring the superior genes/alleles will help in overcoming challenges posed by climate change by incorporating the superior genes/alleles in the HYVs. New breeding techniques and approaches that are ecologically sustainable and help in attaining humanitarian goals are the need of the hour. With the NGS technologies, enormous information on the genome of different crops has been generated. The sequence information generated will be of great help in dissecting the naturally occurring allelic variations or helps in identifying the superior alleles for candidate genes governing the major trait.

Regulatory regions of a gene play a vital role in the expression of a gene; hence, allele mining also includes studying of non-coding and the regulatory regions of a candidate gene and also covers the functional variations of relevance in the gene. Promoters' being one of the key regions where transcription factors bind has been regarded as the major target for allele mining, which is also called *promoter mining*. NGS-based techniques paved the way for millions of nucleotide polymorphisms to be accessible (Ganal et al., 2012; Bassi et al., 2016). NGS platforms have provided the advantage of high marker density with genome coverage at reduced cost (Przewieslik et al., 2019), with high resolution and prediction in GS.

Using SNP genotyping techniques, rare/novel alleles have been discovered in pulse crops. Raizada et al. (2021) have studied diversity analysis through high-resolution melting (HRM) analysis and discovered 13 genic SNPs responsible for different stress and defense responses.

Majority of the important traits are complex in nature, owing to their polygenic control. These complex traits are influenced by the environment. SNP-based GWAS and SNP-based genomic selections have some limitations, such as SNPs being biallelic, high chances of false positives and negatives, chance of missing desirable rare alleles, epistatic interactions being undetected, reduced power in detection of marker–trait association, and low accuracy in genomic estimated breeding values (GEBV) prediction. To overcome the drawbacks of SNP-based genomic selections, haplotype-based GS can be used. Haplotype-based approaches are multiallelic, with high polymorphism content, better control of false positives and negatives, identification of rare alleles, capture of epistatic interactions, high power in detection of marker–trait association, and improved accuracy in the GEBV prediction for genotype selection.

The idea of haplotype-based breeding described by Varshney et al. (2020) has given the importance of hastening the production of designer cultivars. In contrast to marker-assisted backcross (MABC), which reduces genetic diversity and generates bottleneck effects while taking years and generations to transfer superior genes, this strategy has many advantages, as it targets to transfer superior haplotypes underpinning genetic variants that are in strong linkage disequilibrium (LD) with the putative genomic areas linked to the phenotypes of interest (Varshney et al., 2021). Creating superior cultivars for future climates will require assembling gene(s) dispersed throughout the genome, and effective methods for doing so will depend on strategies that take advantage of haplotypes. Bhat et al. (2021) addressed how to mine haplotypes and haplotype-assisted genomic selection, as well as the characteristics and its uses in crop breeding. To ensure the safety of the world's food supply, improvement tactics that effectively use superior haplotypes to speed the breeding progress will be essential.

8.7 CONCLUSION

The era of genomics with the development in technologies and software makes the allele mining process easier, with greater precision. This can be utilized in the mung bean breeding program for the effective use of this nutrient-rich crop.

REFERENCES

Aitawade MM, Sutar SP, Rao SR, Malik SK, Yadav SR et al. (2012) Section *Ceratotropis* of subgenus *Ceratotropis* of *Vigna* (Leguminosae-Papilionoideae) in India with a new species from Northern Western Ghats. *Rheedea* 22(1):20–27.

Akhtar MA, Aslam M, Schafleitner R, Ahsan M, Murtaza G (2021) Population structure and association mapping for hundred seed weight in mungbean minicore. *bioRxiv* 5.

Archak S, Gupta K, Gupta V, Tyagi V, Singh K (2020) Mungbean genetic resources and utilization. In: Nair RM, Schafleitner R, Lee S-H (eds) *The Mungbean Genome*. Springer, Cham, Switzerland, pp. 9–25.

Arora RK, Chandel KPS, Joshi BS, Pant KC (1980) Rice bean: Tribal pulse of Eastern India. *Econ Bot* 34:260–263.

Ashkani S, Yusop MR, Shabanimofrad M, Azadi A, Ghasemzadeh A et al. (2015) Allele mining strategies: Principles and utilisation for blast resistance genes in rice (*Oryza sativa* L.). *Curr Iss Mol Biol* 17(1):57–74.

Babu CR, Sharma SK, Chatterjee SR, Abrol YP (1988) Seed protein and amino acid composition of Wild *Vigna radiata* var. *sublobata* (Fabaceae) and Two Cultigens, *V. mungo* and *V. radiata*. *Econ Bot* 42:54–61.

Baloda A, Madanpotra S (2017) Transformation of mungbean plants for salt and drought tolerance by introducing a gene for an osmoprotectant glycine betaine. *J Plant Stress Physiol* 3:5–11.

Bassi FM, Bentley AR, Charmet G, Ortiz R, Crossa J (2016) Breeding schemes for the implementation of genomic selection in wheat (*Triticum* spp.). *Plant Sci* 242:23–36.

Begemann F (1997) An experiment to cross Bombara groundnut and cowpea. In: Heller J, Begemann F, Mushonga J (eds) *Promoting the Conservation and Use of Underutilized and Neglected Crops 9*. Institute of Plant Genetics and Crop Plant Research, Gatersleben/IPGRI, Rome, Italy, pp. 135–137.

Bharadwaj N (2018) Morpho-physiological responses in different mungbean genotypes under drought stress. *Res J Recent Sci* 7(7):1–5.

Bhat JA, Yu D, Bohra A, Ganie SA, Varshney RK (2021) Features and applications of haplotypes in crop breeding. *Commun Biol* 4(1):Article 1266.

Bisht IS, Bhat KV, Latha M, Abraham Z, Dikshit N et al. (2005) Distribution, diversity and species relationships of wild *Vigna* species in mungo-radiata complex. *Indian J Plant Genet Resour* 18(2):169–179.

Breria CM, Hsieh CH, Yen TB, Yen JY, Noble TJ et al. (2020) A SNP based genome-wide association study to mine genetic loci associated to salinity tolerance in mungbean (*Vigna radiata* L.). *Genes* 11(7):759.

Chankaew S, Somta P, Isemura T, Tomooka N, Kaga A et al. (2013) Quantitative trait locus mapping reveals conservation of major and minor loci for powdery mildew resistance in four sources of resistance in mungbean [*Vigna radiata* (L.) Wilczek]. *Mol Breed* 32(1):121–130.

Chen H, Wang L, Wang S, Liu C, Blair MW et al. (2015) Transcriptome sequencing of mungbean (*Vigna radiata* L.) genes and the identification of EST-SSR markers. *PLoS One* 10(4):e0120273.

Chen T, Hu L, Wang S, Wang L, Cheng X et al. (2022) Construction of high-density genetic map and identification of a bruchid resistance locus in mungbean (*Vigna radiata* L.). *Front Genet* 13.

Chotechung S, Somta P, Chen J, Yimram T, Chen X et al. (2016) A gene encoding a polygalacturonase-inhibiting protein (PGIP) is a candidate gene for bruchid (*Coleoptera: bruchidae*) resistance in mungbean (*Vigna radiata*). *Theor Appl Genet* 129(9):1673–1683.

Comai L, Henikoff S (2006) TILLING: Practical single nucleotide mutation discovery. *Plant J* 45(4):684–694.

Dikshit HK, Jhang T, Singh NK, Koundal KR, Bansal KC et al. (2007) Genetic differentiation of *Vigna* species by RAPD, URP and SSR markers. *Biol Planta* 51(3):451–457.

Fatokun CA, Danesh, Young ND, Stewart EL (1993) Molecular taxonomic relationships in the genus *Vigna* based on RFLP analysis. *Theor Appl Genet* 86(1):97–104.

Fróna D, Szenderák J, Harangi-Rákos M (2019) The challenge of feeding the world. *Sustainability* 11(20):5816.

Ganal MW, Polley A, Graner EM, Plieske J, Wieseke R et al. (2012) Large SNP arrays for genotyping in crop plants. *J Biosci* 37(5):821–828.

Gayacharan, Tripathi K, Meena SK, Panwar BS, Lal H et al. (2020) Understanding genetic variability in the mungbean (*Vigna radiata* L.) genepool. *Ann Appl Biol* 177(3):346–357.

Gepts P, Beavis WD, Brummer EC, Shoemaker RC, Stalker HT et al. (2005) Legumes as a model plant family. Genomics for food and feed report of the cross-legume advances through genomics Conference. *Plant Physiol* 137(4):1228–1235.

Godfray HCJ, Beddington JR, Crute IR, Haddad L, Lawrence D et al. (2010) Food security: The challenge of feeding 9 billion people. *Science* 327(5967):812–818.

Ha J, Satyawan D, Jeong H, Lee E, Cho KH et al. (2021) A near-complete genome sequence of mungbean (*Vigna radiata* L.) provides key insights into the modern breeding program. *Plant Genome* 14(3):e20121.

Harlan JR, de Wet JM (1971) Toward a rational classification of cultivated plants. *Taxon* 20(4):509–517.

Hasanuzzaman SM, Dhillon BS, Saxena S, Upadhyaya MP, Joshi BK et al. (2003) *Plant Genetic Resources in SAARC Countries: Their Conservation and Management*. SAARC Agricultural Information Centre (SAIC), Dhaka, Bangladesh.

Hou D, Yousaf L, Xue Y, Hu J, Wu et al. (2019) Mungbean (*Vigna radiata* L.): Bioactive polyphenols, polysaccharides, peptides, and health benefits. *Nutrients* 11(6):1238.

Humphry M, Konduri V, Lambrides C, Magner T, McIntyre C et al. (2002) Development of a mungbean (*Vigna radiata*) RFLP linkage map and its comparison with lablab (*Lablab purpureus*) reveals a high level of colinearity between the two genomes. *Theor Appl Genet* 105(1):160–166.

Humphry ME, Lambrides CJ, Chapman SC, Aitken EAB, Imrie BC et al. (2005) Relationships between hard seededness and seed weight in mungbean (*Vigna radiata*) assessed by QTL analysis. *Plant Breed* 124(3):292–298.

Ignacimuthu S, Babu CR (1987) *Vigna radiata* var. *sublobata* (Fabaceae): Economically useful wild relative of urd and mungbeans. *Econ Bot* 41(3):418–422.

Kasettranan W, Somta P, Srinives P (2010) Mapping of quantitative trait loci controlling powdery mildew resistance in mungbean (*Vigna radiata* (L.) Wilczek). *J Crop Sci Biotechnol* 13(3):155–161.

Kelly LA, Vaghef N, Bransgrove K, Fechner NA, Stuart K et al. (2021) One crop disease, how many pathogens? *Podosphaeraxanthii* and *Erysiphevignae* sp. *nov*. identified as the two species that cause powdery mildew of mungbean (*Vigna radiata*) and black gram (*V. mungo*) in Australia. *Phytopathology* 111(7):1193–1206.

Kitsanachandee R, Somta P, Chatchawankanphanich O, Akhtar KP, Shah TM et al. (2013) Detection of quantitative trait loci for mungbean yellow mosaic India virus (MYMIV) resistance in mungbean (*Vigna radiata* (L.) Wilczek) in India and Pakistan. *Breed Sci* 63(4):367–373.

Kumar GR, Sakthivel K, Sundaram RM, Neeraja CN, Balachandran et al. (2010) Allele mining in crops: Prospects and potentials. *Biotechnol Adv* 28(4):451–461.

Kumar K, Anjoy P, Sahu S, Durgesh K, Das A et al. (2022) Single trait versus principal component-based association analysis for flowering related traits in pigeon pea. *Sci Rep* 12(1):1–15.

Kusch S, Panstruga R (2017) *MLO*-based resistance: An apparently universal "weapon" to defeat powdery mildew disease. *Mol Plant-Micr Interact* 30(3):179–189.

Laghetti G, Pienaar BL, Padulosi S, Perrino P (1998) *Ecogeographical Distribution of Vigna savi in Southern Africa and Some Areas of the Mediterranean Basin*. Plant Genetic Resources Newsletter (IPGRI/FAO), Bulletin des Ressources Phytogenetiques (IPGRI/FAO). Noticiario de Recursos Fitogeneticos (IPGRI/FAO), Rome, Italy.

Lambrides CJ, Lawn RJ, Godwin ID, Manners J, Imrie BC (2000) Two genetic linkage maps of mungbean using RFLP and RAPD markers. *Aust J Agric Res* 51(4):415–425.

Latha R, Rubia L, Bennett J, Swaminathan MS (2004) Allele mining for stress tolerance genes in *Oryza* species and related germplasm. *Mol Biotechnol* 27(2):101–108.

Lee E, Yang X, Ha J, Kim MY, Park KY et al. (2021) Identification of a locus controlling compound raceme inflorescence in mungbean [*Vigna radiata* (L.) R. Wilczek]. *Front Genet* 12:642518.

Li S, Wang R, Jin H, Ding Y, Cai C (2019) Molecular characterization and expression profile analysis of heat shock transcription factors in mungbean. *Front Genet* 9:736.

Lin WJ, Ko CY, Liu MS, Kuo CY, Wu DC et al. (2016) Transcriptomic and proteomic research to explore bruchid-resistant genes in mungbean isogenic lines. *J Agric Food Chem* 64(34):6648–6658.

Liu J, Lin Y, Chen J, Xue C, Wu R et al. (2022a) Identification and clarification of *VrCYCA1*: A key genic male sterility-related gene in mungbean by multi-omics analysis. *Agriculture* 12(5):686.

Liu J, Xue C, Lin Y, Yan Q, Chen J et al. (2022b) Genetic analysis and identification of *VrFRO8*, a salt tolerance-related gene in mungbean. *Gene* 836:146658.

Liu C, Wang Y, Peng J, Fan B, Xu D et al. (2022c) High-quality genome assembly and pan-genome studies facilitate genetic discovery in mungbean and its improvement. *Plant Commun* 100352.

Manu B, Biradar R, Sabale PR, Kumar K, Aski MS et al. (2022) Genomic designing for abiotic stress tolerance in mungbean and urdbean. In: Kole C (ed) *Genomic Designing for Abiotic Stress Resistant Pulse Crops*. Springer, Cham, Switzerland, pp. 271–343.

Margulies M, Egholm M, Altman WE, Attiya S, Bader JS et al. (2005) Genome sequencing in microfabricated high-density picolitre reactors. *Nature* 437(7057):376–380.

Mathivathana MK, Murukarthick J, Karthikeyan A, Jang W, Dhasarathan M et al. (2019) Detection of QTLs associated with mungbean yellow mosaic virus (MYMV) resistance using the interspecific cross of *Vigna radiata*× *Vigna umbellata*. *J Appl Genet* 60(3):255–268.

Maxted N, Mabuza-Diamini P, Moss H, Padulosi S, Jarvis A et al. (2004) *An Ecogeographic Study African Vigna*. IPGRI, Italy, p. 454.

Mishra GP, Dikshit HK, Ramesh SV, Tripathi K, Kumar RR et al. (2020) Yellow mosaic disease (YMD) of mungbean (*Vigna radiata* (L.) Wilczek): Current status and management opportunities. *Front Plant Sci* 11:918.

Nair RM, Pandey AK, War AR, Hanumantharao B, Shwe T et al. (2019) Biotic and abiotic constraints in mungbean production – progress in genetic improvement. *Front Plant Sci* 10:1340.

Noble TJ, Tao Y, Mace ES, Williams B, Jordan DR et al. (2018) Characterization of linkage disequilibrium and population structure in a mungbean diversity panel. *Front Plant Sci* 8:2102.

Ogawa D, Yamamoto E, Ohtani T, Kanno N, Tsunematsu H et al. (2018) Haplotype-based allele mining in the Japan-MAGIC rice population. *Sci Rep* 8(1):1–11.

Padulosi S, Ng NQ (1990) *Wild Vigna Species in Africa: Their Collection and Potential Utilization. Cowpea Genetic Resources*. Seminar held on February 6–7 (1989) International Institute of Tropical Agriculture, Ibadan, Nigeria, pp. 58–77.

Pradheep K, Bansal KC, Bhandari DC (2014) *Wild Relatives of Cultivated Plants in India*. Directorate of Knowledge Management in Agriculture, Indian Council of Agricultural Research, New Delhi, India.

Pratap A, Kumar J (eds) (2014) *Alien Gene Transfer in Crop Plants, Volume 2: Achievements and Impacts*. Springer, Singapore.

Przewieslik-Allen AM, Burridge AJ, Wilkinson PA, Winfield MO, Shaw DS et al. (2019) Developing a high-throughput SNP-based marker system to facilitate the introgression of traits from Aegilops species into bread wheat (*Triticum aestivum*). *Front Plant Sci* 9:1993.

Raghavan C, Naredo M, Elizabeth B, Wang H, Atienza G et al. (2007) Rapid method for detecting SNPs on agarose gels and its application in candidate gene mapping. *Mol Breed* 19(2):87–101.

Raizada A, Souframanien J (2021) SNP genotyping and diversity analysis based on genic-SNPs through high resolution melting (HRM) analysis in blackgram [*Vigna mungo* (L.) Hepper]. *Genet Resour Crop Evol* 68(4):1331–1343.

Rasool S, Ahmad A, Siddiqi TO, Ahmad P (2013) Changes in growth, lipid peroxidation and some key antioxidant enzymes in chickpea genotypes under salt stress. *Acta Physiol Planta* 35(4):1039–1050.

Reddy VRP, Das S, Dikshit HK, Mishra GP, Aski M et al. (2020) Genome-wide association analysis for phosphorus use efficiency traits in mungbean (*Vigna radiata* L. Wilczek) using genotyping by sequencing approach. *Front Plant Sci* 11:537766.

Reddy VRP, Das S, Dikshit HK, Mishra GP, Aski MS et al. (2021) Genetic dissection of phosphorous uptake and utilization efficiency traits using GWAS in mungbean. *Agronomy* 11(7):1401.

Rispail N, Rubiales D (2016) Genome-wide identification and comparison of legume *MLO* gene family. *Sci Rep* 6(1):1–12.

Rodziewicz P, Swarcewicz B, Chmielewska K, Wojakowska A, Stobiecki M (2014) Influence of abiotic stresses on plant proteome and metabolome changes. *Acta Physiol Planta* 36(1):1–19.

Rohilla V, Yadav RK, Poonia A, Sheoran R, Kumari G et al. (2022) Association mapping for yield attributing traits and yellow mosaic disease resistance in mung bean [*Vigna radiata* (L.) Wilczek]. *Front Plant Sci* 12:749439.

Sarkar M, Kundagrami S (2018) Selection of high yielding, extra short duration lines of mungbean derived through gamma radiation. *Indian J Genet* 78(2):233–241.

Schafleitner R, Nair RM, Rathor A, Wang YW, Lin CY et al. (2015) The AVRDC–the world vegetable center mungbean (*Vigna radiata*) core and mini core collections. *BMC Genomics* 16(1):1–11.

Seehalak W, Tomooka N, Waranyuwat A, Thipyapong P, Laosuwan P et al. (2006) Genetic diversity of the *Vigna* germplasm from Thailand and neighboring regions revealed by AFLP analysis. *Genetic Resour Crop Evol* 53(5):1043–1059.

Sim NL, Kumar P, Hu J, Henikoff S, Schneider G et al. (2012) SIFT web server: Predicting effects of amino acid substitutions on proteins. *Nucleic Acids Res* 40(W1):W452–W457.

Singh M, Bisht IS, Sardana S, Gautam NK, Hussain Z et al. (2005) Asiatic *Vigna*. In: Dhillon BS, Saxena S, Agarwal A, Tyagi RK (eds) *Plant Genetic Resources: Food Crops*. Narosa Publishing House, New Delhi, India, pp. 257–301.

Singh AK, Rana RS, Mal B, Singh B, Agrawal RC (2013) *Cultivated Plants and Their Wild Relatives in India: An Inventory*. Protection of Plant Varieties & Farmers Rights Authority, New Delhi, India, p. 215.

Singh CM, Pratap A, Gupta S, Biradar RS, Singh NP (2020) Association mapping for mungbean yellow mosaic India virus resistance in mungbean (*Vigna radiata* L. Wilczek). *3 Biotech* 10(2):1–12.

Singh AK, Kumar K, Singh LB, Ahuja A, Choudhury S et al. (2021) Omics technology: Role and future in providing biotic and abiotic stress tolerance to plants. In: Sharma A (ed) *Microbes and Signaling Biomolecules Against Plant Stress*. Springer, Singapore, pp. 151–168.

Smýkal P, Coyne CJ, Ambrose MJ, Maxted N, Schaefer H et al. (2015) Legume crops phylogeny and genetic diversity for science and breeding. *Crit Rev Plant Sci* 34(1–3):43–104.

Somta P, Ammaranan C, Ooi PC, Srinives P (2007) Inheritance of seed resistance to bruchids in cultivated mungbean (*Vigna radiate* L. Wilczek). *Euphytica* 155(1):47–55.

Somta C, Somta P, Tomooka N, Ooi PC, Vaughan DA et al. (2008) Characterization of new sources of mungbean (*Vigna radiata* L. Wilczek) resistance to bruchids, *Callosobruchus*spp. (Coleoptera: Bruchidae). *J Stored Prod Res* 44(4):316–321.

Somta P, Laosatit K, Yuan X, Chen X (2022) Thirty years of mungbean genome research: Where do we stand and what have we learned? *Front Plant Sci* 13.

Somta P, Srinives P (2007) Genome research in mungbean [*Vigna radiata* (L.) Wilczek] and blackgram [*V. mungo* (L.) Hepper]. *Sci Asia* 33(1):69–74.

Srinives P, Somta P, Somta C (2007) Genetics and breeding of resistance to bruchids (*Callosobruchus* spp.) in *Vigna* crops: A review. *NU Int J Sci* 4(1):1–17.

Srivastava R, Kumar S, Kobayashi Y, Kusunoki K, Tripathi P et al. (2018) Comparative genome-wide analysis of WRKY transcription factors in two Asian legume crops: Adzuki bean and Mungbean. *Sci Rep* 8(1):1–19.

Stacey G, Vanden Bosch K (2005) "Translational" legume biology: Models to crops. *Plant Physiol* 137(4):1173–117.

Sudha M, Karthikeyan A, Madhumitha B, Veera Ranjani R, Kanimoli Mathivathana M et al. (2022) Dynamic transcriptome profiling of mungbean genotypes unveil the genes respond to the infection of mungbean yellow mosaic virus. *Pathogens* 11(2):190.

Takahashi Y, Somta P, Muto C, Iseki K, Naito K et al. (2016) Novel genetic resources in the genus Vigna unveiled from gene bank accessions. *PLoS One*, 11(1), e0147568.

Till BJ, Cooper J, Tai TH, Colowit P, Greene EA et al. (2007) Discovery of chemically induced mutations in rice by TILLING. *BMC Plant Biol* 7(1):1–12.

Till BJ, Reynolds SH, Greene EA, Codomo CA, Enns LC et al. (2003) Large-scale discovery of induced point mutations with high-throughput TILLING. *Genome Res* 13(3):524–530.

Tomooka N, Kashiwaba K, Vaughan DA, Ishimoto M, Egawa Y (2000) The effectiveness of evaluating wild species: Searching for sources of resistance to bruchid beetles in the genus *Vigna* subgenus *Ceratotropis*. *Euphytica* 115(1):27–41.

Tomooka N, Naito K, Kaga A, Sakai H, Isemura T et al. (2014) Evolution, domestication and neo-domestication of the genus *Vigna*. *Plant Genet Resour* 12(S1):S168–S171.

Upadhyaya HD, Gowda CLL, Buhariwalla HK, Crouch JH (2006) Efficient use of crop germplasm resources: Identifying useful germplasm for crop improvement through core and mini-core collections and molecular marker approaches. *Plant Genet Resour* 4(1):25–35.

Van K, Kang YJ, Han KS, Lee YH, Gwag JG et al. (2013) Genome-wide SNP discovery in mungbean by Illumina HiSeq. *Theor Appl Genet* 126(8):2017–2027.

Varshney RK, Sinha P, Singh VK, Kumar A, Zhang Q et al. (2020) 5Gs for crop genetic improvement. *Curr Opin Plant Biol* 56:190–196.

Varshney RK, Bohra A, Yu J, Graner A, Zhang Q et al. (2021) Designing future crops: Genomics-assisted breeding comes of age. *Trends Plant Sci* 26(6):631–649.

Vijaykumar A, Saini A, Jawali N (2010) Phylogenetic analysis of subgenus *Vigna* species using nuclear ribosomal RNA ITS: Evidence of hybridization among *Vigna unguiculata* subspecies. *J Hered* 101(2):177–188.

Wang L, Zhu J, Li X, Wang S, Wu J (2018) Salt and drought stress and ABA responses related to *bZIP* genes from *V. radiata* and *V. angularis*. *Gene* 651:152–160.

Wu X, Islam AF, Limpot N, Mackasmiel L, Mierzwa J et al. (2020) Genome-wide SNP identification and association mapping for seed mineral concentration in mungbean (*Vigna radiata* L.). *Front Genet* 11:656.
Yundaeng C, Somta P, Chen J, Yuan X, Chankaew S et al. (2021) Fine mapping of *QTL* conferring Cercospora leaf spot disease resistance in mungbean revealed *TAF5* as candidate gene for the resistance. *Theor Appl Genet* 134(2):701–714.
Yundaeng C, Somta P, Chen J, Yuan X, Chankaew S et al. (2020) Candidate gene mapping reveals *VrMLO12 (MLO Clade II)* is associated with powdery mildew resistance in mungbean (*Vigna radiata* [L.] Wilczek). *Plant Sci* 298:110594.

9 Allele Mining for Seed Nutrient Content in Green Gram

Sutanu Sarkar, Aditya Pratap Singh, and Somnath Bhattacharyya

9.1 INTRODUCTION

Green gram (*Vigna radiata* [L.] Wilczek) is an important pulse crop for most Asian countries, such as India, Bangladesh, China, Indonesia, Pakistan, Thailand, and Vietnam, with increasing importance in the Western world as well (Nair et al., 2013). It is a diploid species having chromosome number 2n = 2x = 22 (Karpechenko, 1925), with a small genome size of 579 mb. Green gram is native to the Indo-Burma region of the Hindustan center (Vavilov, 1926). It covers an area of over 6 million hectares worldwide (8.5% of the global pulse area), with a productivity of over 3 million tons (5% of global pulse production) (Schafleitner et al., 2016). India produced 2.34 million tons of green gram from about 9.40 lakh ha during 2020–2021 (Anonymous, 2020). Green gram is a highly nutritious, easily cooked, palatable legume (Rehman et al., 2019). Green gram proteins are easily digestible, and the seeds contain a wide array of micronutrients (Kollárová et al., 2010). The seeds are consumed in cooked form and as vegetable sprouts (Kim et al., 2015). Decorticated seed flour is used in commercial products like flatbreads, noodles, and sweets (Dahiya et al., 2015). Besides proteins, fatty acids, dietary fibers, and minerals, green gram seeds also have antinutritional factors, like phenolics, saponins, trypsin inhibitors, phytic acid, tannin, oligosaccharides, and hemagglutinins (**Table 9.1**). Nevertheless, green gram can play a significant role in micronutrient deficiency prevention (Dwivedi et al., 2012).

Nutritional deficiencies in humans can cause stunted growth and development, decreased resistance to diseases, and higher mortality rates in children (Caulfield et al., 2006). More than 75% of the global population is not getting an adequate intake of one or more mineral nutrients, with many having deficiency of iron and zinc in particular (White and Broadley, 2009). Thus, mineral deficiencies are of prime concern for global health (Soetan et al., 2010), especially in developing countries, where grain crops are the primary source of essential nutrients (Dwivedi et al., 2012). Although mineral nutrient concentrations are higher in legumes than in cereals or root crops, their bioavailability and total concentration still need improvement (Blair, 2013). This can be achieved through genetic improvement for higher mineral acquisition, the minimal concentration of antinutritional factors, and a balanced nutrient profile (Pfeiffer and McClafferty, 2007; Dwivedi et al., 2012). Designing climate-resilient stable varieties will address the problems of food and nutritional security. In view of developing nutrient-enriched green gram varieties, untapped genetic variation in wild relatives and landraces could be exploited; one of the ways is "allele mining."

Alleles are different forms of DNA sequence (a single base or a segment of bases) at a given genomic location that affect a particular phenotype. Thus, different phenotypes arise depending on allelic combinations. Researchers try to incorporate new alleles into crop varieties to improve yield and quality traits. Mining a crop's gene pool for superior novel alleles is becoming feasible in the current "omics" era. There are huge amounts of distinct alleles in gene banks worldwide, which may be useful for crop breeding. With allele mining, it is possible to explore this hidden diversity (Varshney et al., 2005; Kumar et al., 2010), which can be caused by naturally occurring sequence variation in coding or regulatory regions of genes. Numerous genetic variation types may alter the

DOI: 10.1201/9781003385059-9

TABLE 9.1
Nutrient Composition of Green Gram

Particulars	Value	References
Total carbohydrates (g/100 g dry matter)	46.13 ± 0.064	Longvah et al. (2017)
Crude protein (g/100 g dry weight)	24.55	Nagrale et al. (2018)
Amino acids (g/100 g protein)		
Phenylalanine	4.62	Ullah et al. (2014)
Valine	4.16	
Threonine	2.78	
Tryptophan	0.09	
Methionine	1.22	
Leucine	6.66	
Isoleucine	4.04	
Lysine	3.49	
Histidine	1.79	
Arginine	5.29	
Tyrosine	2.57	
Cysteine	0.1	
Glycine	3.56	
Proline	1.86	
Serine	4.25	
Alanine	3.65	
Aspartate	12.46	
Glutamate	21.33	
Fatty acids (mg/100 g of dry seeds)		
Oleic acid (C18:1)	2.52 ± 0.31	Pal et al. (2010)
Linoleic acid (C18:2)	18.18 ± 0.35	
Linolenic acid (C18:3)	9.5 ± 0.12	
Monounsaturated fatty acid	11.37	
Polyunsaturated fatty acid	43.86	
Saturated fatty acid	44.78	
Dietary fibers (g/100 g dry matter)		
Insoluble dietary fiber	14.59 ± 0.42	Longvah et al. (2017)
Soluble dietary fiber	2.44 ± 0.15	
Total dietary fiber	17.04 ± 0.38	
Vitamins (mg/100 g of dry matter)		
Thiamine	0.45 ± 0.27	Longvah et al. (2017)
Riboflavin	0.27 ± 0.011	
Niacin	2.6 ± 0.13	
Pantothenic acid	2.02 ± 0.24	
Pyridoxine	0.35 ± 0.034	
Biotin	1.35 ± 0.16	
Folic acid	145 ± 5.4	
Ascorbic acid	3.98 ± 0.00	Oburuoga and Anyika (2012)
Vitamin A	17.22 ± 0.02	

TABLE 9.1 ***(Continued)***
Nutrient Composition of Green Gram

Particulars	Value	References
Minerals (mg/100 g dry matter)		
Sodium	12.48 ± 0.07	Longvah et al. (2017)
Potassium	1,177 ± 74.0.3	
Calcium	92.43 ± 10.68	
Phosphorus	353 ± 33.6	
Magnesium	198 ± 39.2	
Iron	4.89 ± 0.46	
Manganese	1.05 ± 0.08	
Zinc	2.67 ± 0.13	
Copper	1.00 ± 0.14	
Antinutritional factors (dry weight basis)		
Total phenolics (%)	0.30 ± 0.04	Tresina et al. (2014)
Trypsin inhibitor activity (TIU/mg protein)	36.20	
Tannins (%)	0.50 ± 0.03	
Phytic acid (%)	0.37 ± 0.00	
Saponins (%)	1.51 ± 0.16	Siddhuraju et al. (2002)
Phytohemagglutinating activity (HU/g sample)	2,670	Mubarak (2005)

phenotypic changes, including single-nucleotide polymorphisms (SNPs) and insertions and deletions (InDels). These changes may affect protein content, structure, or function (Kumar et al., 2010). The rapid advancements in genomics in recent years have led to the accumulation of vast amounts of sequence information as well as the rapid evolution of bioinformatics tools that make it possible to identify, characterize, isolate, and deploy genetic variation sources previously unknown or underutilized. The identification of genes responsible for nutritional traits can be accomplished through the use of quantitative trait loci (QTLs) analyses or genome-wide association studies (GWAS) using molecular markers and high-throughput sequencing techniques (Blair, 2013).

There have been previous quantitative trait locus (QTL) studies in common bean (Blair, 2013), lentil (Ates et al., 2018), *Medicago truncatula* (Sankaran et al., 2009), and pea (Ma et al., 2017) that used mapping populations for mineral nutrients. In green gram, GWAS has not been used for micronutrient concentration (Kwon et al., 2012). Although green gram has a high economic and nutritional value, its expansion is limited by a lack of genetic resources that facilitate crop adaptation (Das et al., 2018). As a priority for plant breeding, the selection of new varieties of green gram needs to be based on modern marker tools (Burlyaeva et al., 2019). In modern plant breeding of legumes, genetic markers are an essential component. There have been a few genetic studies on green gram, but a whole-genome sequence has allowed researchers to investigate the genes of this important legume (Kang et al., 2014).

9.2 GENETIC RESOURCES FOR ALLELE MINING

In addition to having a narrow genetic base, improved green gram cultivars lack adaptability to changing growth conditions in different agroecological environments. It is possible to develop improved cultivars in crop plants using germplasm resources which contain valuable genes. It is, therefore, imperative to better utilize the genetic potential of germplasm accessions in gene banks. As part of its genetic enhancement/pre-breeding studies, the National Bureau of Plant Genetic Resources (NBPGR) in India uses the Bureau's core collection as a starting point to study diverse parents from the cultivated gene pool for green gram. By transferring desired genes from germplasm into elite genotypes, germplasm enhancement attempts to broaden the genetic base of breeding materials. A selection of mild and decentralized material is maintained at target sites throughout the country. A set of 102 progenies was advanced to the F_5 generation for further selection and use. Bisht et al. (2005) reported improved yield traits and plant types in a few selected enhanced progenies. It is clear from the study that germplasm accessions conserved in gene banks have the potential to effectively be used as part of large-scale base-broadening programs for green gram.

The Asian Vegetable Research and Development Center (AVRDC), presently known as World Vegetable Center, maintains 5,616 accessions of green gram (Bisht and Singh, 2013). There are characterized accessions that can be exchanged and utilized. Depending on the geographical region, the accessions have morphologically diverse traits. There are multiple germplasm collections for green gram maintained worldwide, including those of the Chinese Academy of Agricultural Sciences and of the University of Georgia (Ebert, 2013). Additionally, the University of the Philippines and the Rural Development Administration (RDA), Korea, hold duplicates of the World Vegetables Center's green gram germplasm collection. These germplasm accessions are used for introducing desirable traits in breeding programs.

Genomic data for model and crop legumes are now being generated with the assistance of next-generation sequencing (NGS) platforms and high-throughput genotyping technologies. Green gram genome sequence assemblies were mainly produced on these platforms (Kang et al., 2014). It is also becoming more accessible to mine superior alleles using whole-genome sequencing data. There are several genome sequences of green gram available, including the whole genome/transcriptome database (Kang et al., 2014), the chloroplast genome database (Tangphatsornruang et al., 2010), and the mitochondrial genome database (Alverson et al., 2011). These resources may be tremendously useful in designing nutrient-rich green gram varieties. In addition, several QTLs in green gram have been identified, which can be utilized to improve yields and related traits (Sompong et al., 2012; Singh et al., 2017).

9.3 DEVELOPMENTS IN GREEN GRAM NUTRIGENOMICS

Human civilization is adjusting to too many threats, such as unstable climate, pandemics, terrorism, mechanization, poverty, etc., which affect agricultural production in terms of yield and food and nutritional security. We have to intake food for survival, but at the same time, nutritious food is essential to lead a healthy life. Since their incorporation into agriculture, grain legumes have been crucial in supplying proteins at least to the poor man's table. In terms of worldwide consumption, especially in Asia, green gram sprouts are widely preferred due to their richness in vitamins and minerals (Ranadhir et al., 2004). They are a rich source of proteins and easily digestible carbohydrates as well. In the following, we will discuss the nutrigenomics of green gram seeds.

9.3.1 Carbohydrate

On the basis of the digestibility of legume, there are three types of starches: readily digestible starch (RDS), slowly digestible starch (SDS), and resistant starch (RS); the content of these starches are 4.2 to 10.9%, 16.9 to 40%, and 50.3 to 78.9%, respectively. The digestibility of starch is negatively correlated to starch granule diameter and molecular weight of amylopectin and amylose. RDS is associated with a more rapid elevation of postprandial plasma glucose as it is quickly and completely digested in the small intestine. SDS is the most desirable form of dietary starch, which is found among legumes following this order: green gram > chickpea > field pea > lentil > black gram > pigeon pea. On the other hand, RS content follows this order: pigeon pea > lentil > black gram > field pea > chickpea > green gram. The total digestibility of starches (RDS + SDS) is highest in green gram and lowest in pigeon peas. Legume starches have a lower digestibility than cereal starches due to lower hydrolysis indices (HI) and glycemic index (GI). Green gram starch is most suitable for malnourished patients among other legumes due to its highest digestibility and GI. For diabetic patients, pigeon pea starch is suitable as it contains the lowest digestibility and GI (Sandhu and Lim, 2008).

Starch and noodle industries always prefer green gram seeds with high seed starch content. Masari et al. (2017) reported genetic control of green gram seed starch content for the first time. They developed 123 F_2 individuals by crossing V6087AG (high seed starch) with V5020BY (low seed starch) genotypes of green gram; female parent V6087AG and male parent V5020BY possessed 46.0% and 37.8% seed starch content, respectively. No transgressive segregation for the trait was identified. A positive and moderate correlation was observed between starch content and 100-seed weight. They found broad-sense heritability (h^2) of green gram seed starch content of more than 80% by analyzing F_3 and F_4 seeds. The h^2 estimates for 100-seed weight were 83.0% in the F_2 population (by analyzing F_3 seeds) and 67.3% in the F_3 population (by analyzing F_4 seeds). So selection of green gram with bolder seeds or higher seed weight could accompany high seed starch content.

Among 1,417 simple sequence repeat (SSR) markers screened for polymorphism between V6087AG and V5020BY, 106 were found to be polymorphic. Bulk segregant analysis (BSA) revealed that only SSR marker CEDG092 on green gram chromosome 8 was associated with the starch content. With the help of 23 SSR markers, including CEDG092, and also implementing software QTL IciMapping 4.1, a major QTL for seed starch content was identified both in the F_2 and F_3 populations; here, they followed inclusive composite interval mapping (ICIM) described by Li et al. (2007). SSR markers Vr08-SSR113 and Vr08-SSR114 flank the QTL, which was named *qSSC8.1*; it explained about 12.34 to 13.84% of the total variation of the seed starch content in the two populations. Alleles from the parent V6087AG increased seed starch content at this QTL. Therefore, vr08-SSR113 and Vr08-SSR114 should be potential markers in green gram breeding for improving seed starch content. It should also be noted that *qSSC8.1* does not co-localize with the seed weight QTL in green gram.

Based on location on the green gram reference genome (Kang et al., 2014), approximately 385 Kbp regions, comprising 21 annotated genes, were marked in between Vr08-SSR113 and Vr08-SSR114, although these 21 genes do not encode enzymes involved in starch biosynthesis in the plant. Nonetheless, this genome region can be used as the target for fine mapping to identify genes controlling seed starch content in green gram.

9.3.2 Protein

The storage globulins are reported to be found as major seed proteins in which ~3% are basic 7S type, ~90% are 8S (i.e., vicilin type), and ~8% are 11S (i.e., legumin type) (Mendoza et al., 2001; Itoh et al., 2006). 8S globulin has great potential to be used as an ingredient in food and non-food systems through protein engineering (Garcia et al., 2006). However, Dahiya et al. (2013) reported that green gram genotypes with higher crude protein content in the seeds are nutritionally poor due to the unavailability of good-quality protein.

For the first time, Yu et al. (2020) identified 1,662 and 1,795 peptides in green gram seeds and sprouts, respectively, in performing a peptidomics approach. The increment in peptide number during sprouting is due to the hydrolysis of storage proteins into peptides by the activated green gram proteases. Yu et al. (2020) also identified 1,918 and 1,955 proteins in green gram seeds and sprouts, respectively, through differential protein analysis and comparative proteomic profiles and specified 240 and 277 proteins only; these proteins are primarily utilized in various physiological processes during germination. Finally, they selected 260 differential proteins based on relatively higher abundance, among which 149 and 111 proteins are for seed sprouts, respectively.

Late embryogenesis abundant (LEA) proteins in green gram seeds are well-known as they are associated with desiccation tolerance during storage and stress conditions. Nandhini et al. (2019)

identified and sequenced a *VrLEA2* gene in green gram. Later, Likhith et al. (2021) performed genome-wide identification of the LEA gene families in green gram.

9.3.3 Minerals

According to Dahiya et al. (2013), large amounts of any particular mineral can influence the absorption of other minerals. Nair et al. (2015) screened two green gram genotypes (CN 9–5 and Harsha), which could double the seed iron (Fe) content when grown on soil with increased Fe availability. The mechanism of up-taking the extra Fe into the seeds from the root needs to be investigated; genes controlling Fe reductase activity might work in this case, as Fe is up-taken through an iron reduction mechanism, namely, ferric iron to ferrous iron. Ascorbic acid prevents the formation of less-soluble ferric compounds by reducing and promoting Fe absorption. Seed germination in green gram significantly increased ascorbic acid content. Hence, germinated green gram seeds could be a better diet for higher Fe absorption. Cooking of, either germinated or fermented, green gram seeds enhances Fe bioavailability (Barkoti and Bains, 2007).

Aneja et al. (2012), for the first time, used sequence-related amplified polymorphism (SRAP) markers to assess molecular diversity among 21 green gram genotypes varying in Fe and zinc (Zn) content in seeds. Working with 92 green gram genotypes, Singh et al. (2018) identified a few promising cultivars having high Fe, Zn, protein, and higher yield. Those cultivars could be used in marker-assisted breeding programs for Fe and Zn biofortification. They found a positive correlation between seed Fe and Zn concentration; this indicated that a high Fe content is linked to high Zn content. Although both Fe and Zn exhibited little positive correlation with seed protein content, as revealed by their study, the authors also found negative associations between seed yield with protein and zinc content in green gram seeds.

Singh et al. (2017) randomly selected 120 F_6 recombinant inbred lines (RILs) to identify and map QTLs associated with iron and zinc content; the RILs were developed from ML776 × Sattya green gram genotypes of Indian origin. Previously, Taunk et al. (2012) identified those two parents differing in contrasting Fe and Zn content in seed and recommended for utilizing in breeding purposes. ML776 has higher Fe and Zn content, whereas parent Sattya was selected for lower Fe and Zn content. The genotyping of those 120 RILs (Singh et al., 2017) was done by using 26 polymorphic SSR markers, such as MBSSRG5, MBSSRG7, Bng95, Bng91, BMd-16, BMd22, CEDG118, PVBR229, PVBR215, PVBR113, BMd-27, BMd33, BM185, BMd-12, BMd41, BM210, PVBR269, CEDG279, PVBR82, CEDG248, BM202, BM138, BM184, CEDG008, CEDG111, and BM141, and linkage maps were constructed. By composite interval mapping analysis using the software Win QTL Cartographer 2.5, the authors (Singh et al., 2017) identified 2 and 15 QTLs for iron and zinc content in green gram seeds, respectively; these QTLs were distributed in chromosomes 4, 6, 7, and 11. This was the first report on mapping QTLs for Fe and Zn in green gram. **Table 9.2** depicts the chromosomal location of the QTLs for Fe and Zn content in green gram seeds, along with the flanking SSR markers.

Among two QTLs for green gram seed iron content, one came from parent Sattya, and one came from parent ML776 in the aforesaid green gram cross; these two QTLs were mapped on chromosome numbers 4 and 11 and accounted for 21% and 13% phenotypic variation, respectively. Among the other 15 QTLs associated with green gram seed zinc content, three QTLs in chromosome 4, five QTLs on chromosome 11, and one QTL on chromosome 6 accounted for phenotypic variation range 11% to 22%, 4% to 24%, and 20%, respectively; these QTLs were contributed by the parent ML776. Similarly, two QTLs on chromosome 7 and four QTLs on chromosome 6 accounted for 14% to 38% and 13% to 34% phenotypic variation, respectively. The parent Sattya was the contributor to these nine QTLs. In chromosomes 4 and 11, the overlapping QTLs are found for Fe and Zn concentration; *qZn-4–3* and *qFe-4–1* in between two SSR markers, PVBR82 and BM210, where qZn-11–2 and qFe-11–1 between BM141 and BM184. Those might be tightly linked to each other, or these might be QTLs with pleiotropic effects. Singh et al. (2017) found SSR marker PVBR82 to

TABLE 9.2
Genetic Location of QTLs Associated with Iron and Zinc Concentration

Seed Mineral Trait	Chromosome Number	QTLs	Flanking SSR Marker Pairs
Fe	4	*qFe-4–1*	PVBR82, BM210
	11	*qFe-11–1*	BM141, BM184
Zn	4	*qZn-4–1*	PVBR269, BM185
		qZn-4–2	BM185, PVBR82
		qZn-4–3	PVBR82, BM210
	6	*qZn-6–1*	Bng95, MBSSRG7
		qZn-6–2	MBSSRG5, PVBR113
		qZn-6–3	BMd33, CEDG248
		qZn-6–4	CEDG248, BMd12
		qZn-6–5	BMd12, PVBR215
	7	*qZn-7–1*	PVBR82, BM138
		qZn-7–2	BM138, PVBR229
	11	*qZn-11–1*	BMd22, BM141
		qZn-11–2	BM141, BM184
		qZn-11–3	BM184, BM202
		qZn-11–4	BM184, BM202
		qZn-11–5	BMd27, Bng91

Source: Reconstructed from Singh et al. (2017).

be the most consistent in QTLs for both Fe and Zn and recommended the SSR marker to be utilized in marker-assisted breeding for biofortification in green gram. The same authors recommended five promising QTLs, namely, *qFe-11–1*, *qZn-11–4*, *qZn-11–5*, *qZn-4–1*, and *qZn-4–2*.

Wu et al. (2020) performed GWAS in 95 cultivated green gram genotypes collected from 13 countries in view of discovering marker–trait associations (MTAs). They found 6,484 SNPs based on genotyping-by-sequencing (GBS). They followed the digestion/ligation technique suggested by Rohland and Reich (2012). The single-end sequencing was then performed to construct a GBS database to identify SNPs. The SNP loci were confirmed through *Vigna radiata genome* v1.0 from the Legume Information System (LIS), implementing Burrows–Wheeler aligner (BWA) algorithm as described by Li and Durbin (2009). A physical map of the newly identified SNPs was constructed using R v.3.3.1 software, and the SNPs were scored using vcf tools. Finally, 43 SNP markers strongly associated with a concentration of seven minerals, such as phosphorous (P), potassium (K), sulfur (S), calcium (Ca), Fe, Zn, and manganese (Mn), in green gram seeds were identified, explaining approximately 22% of the overall variation. **Table 9.3** provides the relevant information as found by Wu et al. (2020). They also calculated pairwise correlation coefficients for the seed minerals; positive correlations could be speculated among the seven minerals of green gram seeds from their study.

For macronutrients like K, P, and S, 26 SNP markers are associated and distributed on green gram chromosomes 1, 5, 7, 8, and 9. There are 4 common SNP markers associated with both P and K, while 1 marker is common for both P and S. Remaining minerals like Fe, Zn, and Mn are associated with 17 different SNP markers and distributed on chromosomes 1, 6, 7, 8, and 9. Three SNP loci, namely, SVR06_21630747, SVR06_22114556, and SVR06_21903341, identified on the 4 Mb region of chromosome 6 were related to Fe, accumulation, transport at ion, and metabolism and, therefore, could be utilized for selection purpose in seed Fe breeding. Interestingly, the SNP locus SVR06_21903341 is a homolog of the yellow stripe1-like (*YSL*) gene. The *YSL* gene family is directly involved in Fe uptake and transportation from the rhizosphere into the seeds, and this gene

TABLE 9.3
SNP Markers Associated with Seed Mineral Concentrations in Green Gram

Nutrient	Green Gram Chromosome Number	Marker Name
Phosphorus	1	*SVR01_1387204, SVR01_1392172, SVR01_1401760
	5	SVR05_25211011, SVR07_49754942
	7	SVR07_47951933, SVR07_49780103, SVR07_49780128
Potassium	1	SVR01_1387204, SVR01_1392172, SVR01_1401760
	5	SVR05_25211011
	7	SVR07_33824920, SVR07_33825176
	8	SVR08_45006252, SVR08_45006217, SVR08_38988189
Iron	6	SVR06_21630747, SVR06_21903341, SVR06_22114556, SVR06_22271143, SVR06_22461411
Zinc	1	SVR01_8497478
	6	SVR06_2352372, SVR06_2334764
	7	SVR07_13191880, SVR07_13740249
	8	SVR08_12313446
	9	SVR09_21001271
Sulfur	7	SVR07_49754942, SVR07_44411453, SVR07_41344562,
	8	SVR08_36596365, SVR08_36596397, SVR08_36072603, SVR08_36006946, SVR08_43903167
	9	SVR09_7847978
Manganese	1	SVR01_22929122
	8	SVR08_18615379, SVR08_18582837
	9	SVR09_20969884, SVR09_20995726

* SVR: SNP for *V. radiata*.
Source: Reconstructed from Wu et al. (2020).

family is also induced in response to Fe deficiency (Kobayashi and Nishizawa, 2012). Islam and Blair (2018) tagged *YSL* homologous gene by an SNP marker MBkSNP_38 in a set of 94 green gram genotypes using kompetitive allele specific PCR (KASP)–based SNP markers. Wu et al. (2020) also found a gene, SVR07_13191880, on chromosome 7 associated with Zn accumulation; that gene is found to be similar to *YSL7*. They also tagged the purple acid phosphatase gene augmenting symbiotic nitrogen fixation by SNP marker MBkSNP_39.

Reddy et al. (2021) employed a GBS-based GWAS approach to dissect the complexity of PUpE (P uptake efficiency) and PUtE (P utilization efficiency) traits at the genetic level in 120 diverse genotypes of green gram. They identified 116 SNPs in 61 protein-coding genes; 16 genes have been found to enhance the traits mentioned earlier. In addition, they also identified 6 genes, namely, VRADI01G04370, VRADI05G20860, VRADI06G12490, VRADI08G20910, VRADI08G00070, and VRADI09G09030, having high expression in root, shoot apical meristem, and leaf; these 6 genes regulate the P uptake and P utilization efficiency in green gram.

9.3.4 Antinutritional Factors

Antinutrients in green gram seeds are phytic acid or phytate, tannins, hemagglutinins, and polyphenols. The genotypes with higher mineral content tend to possess lower concentrations of antinutritional factors.

9.3.4.1 Phytic Acid

Phytic acid (PA) is my-inositol-1, 2, 3, 4, 5, 6-hexaphosphate, and it presents as a mineral complex. Most of the organic P in seeds is stored in the form of PA. PA inhibits the absorption of protein and minerals such as Fe, Zn, Ca, Mn, and Mg. It harms Fe and Zn bioavailability in plant-based diets. It is insoluble in the human intestine due to a shortage of phytase enzyme and is therefore regarded as antinutritional factor, so it is one of the main reasons for human mineral deficiencies in the population following legume and cereal-based staples. Monogastric animals consuming PA-enriched food and feed excrete a significant fraction of phytate salts, a water pollutant. PA in animal waste can contribute to water pollution. However, PA reduces free radical production and acts as an antioxidant. It also has anti-inflammatory and hypocholesterolemic properties (Sompong et al., 2010).

Sompong et al. (2010) investigated the inheritance of seed P compound and phytate contents by analyzing F_2 population derived from a cross between V1658BBR and V1141BG 2; these two green gram genotypes possessed low and high PA contents, respectively. High h^2 for seed P content was observed, which was estimated as 80.8, 78.6, and 80.7%, respectively, for total P (TP), inorganic P (IP), and phytate P (PAP). The F_2 population (analysis of F_3 seeds) segregated 9:7 ratio high to low TP and PAP in seeds; two major independent genes, designated as *mlpa1* and *mlpa2*, with duplicated recessive epistasis, are there to control the content of P compounds. Transgressive segregants for TP and PAP were also detected, indicating the presence of many modifying genes for the traits. This finding could help in green gram breeding for desirable seed PA.

For the first time, Sompong et al. (2012) constructed the QTL map for PA and P contents in seeds and 3-day-old seedlings of green gram. A high-PA-containing green gram cultivar from Iran, V1725BG, was crossed to develop a low-PA-containing AusTRCF321925; the latter was a wild green gram species of Australia. F_4 seeds were analyzed for assaying PAP, IP, and TP from seed and seedling, and also, transgressive segregants were identified for those traits. PAP and IP from both seed and seedling were correlated positively to TP. The positive correlations among seed and seedling P fractions, namely, PAP, IP, and TP, were observed. Only seedling PAP was found to be positively correlated to seed weight, but the rest all showed a negative correlation to seed weight. As the authors also observed a positive association of P fractions with both days to flowering and days to maturity, plant phenology could also be considered to study genetics regarding seed mineral content in green gram.

Out of 991 SSR markers used, only 101 were polymorphic between V1725BG and AusTRCF321925; among these 101 polymorphic SSRs, 44 were green gram SSR markers, and the remaining 31, 17, and 9 were the SSR markers of azuki bean, cowpea, and common bean, respectively. Thirteen QTLs were associated with seed and seedling PAP, IP, and TP. In **Table 9.4**, the chromosomal location of these QTLs has been presented.

Seven (2 + 4 + 1) QTLs for seed PAP, IP, and TP, distributed in five chromosomes, explain 3.43% to 11.24% phenotypic variation. Six (3 + 1 + 2) QTLs for seedling PAP, IP, and TP, distributed in three green gram chromosomes, explain 1.38% to 13.82% phenotypic variation. QTLs such as SDIP1.1, SDIP8.1, SDIP11.1, SLPAP7.1, SLPAP8.1, and SLTP4.1 are contributed by the parent AusTRCF321925; hence, alleles for these QTLs are responsible for reduction of the respective P compounds, hence could be utilized in low-phytate green gram breeding. It is observed from **Table 9.4** that two QTLs on chromosome 4, SLPAP4.1 and SLTP4.1, are flaked by SSR markers CEDG088 and CEDG091. Another two QTLs on chromosome 7, SLPAP7.2 and SLTP7.1, are flaked by SSR IAC98 and Bmd26. Two QTLs, SDPAP4.1 and SDTP4.1, co-localize on chromosome 4, and MB-SSR179 shows linkage to both of them and could be a potential marker. Two QTLs for P compounds in seed and seedling, namely, SLPAP8.1 and SDIP8.1, co-localize chromosome 8, and SSR marker IAC100 is found to be linked to them; therefore, it could also be utilized as a potential marker. However, the authors suggested selecting different QTLs for low PAP in seed and seedling; they also recommended that these results be carefully implemented in a breeding program due to overlapping QTLs for seed and seedling P with QTLs for seed size, flowering, and maturity.

TABLE 9.4
Genetic Location of QTLs Associated with P Fraction Traits in Green Gram

Trait Regarding P Fractions	Chromosome Number	QTLs	Flanking SSR Marker Pairs
Seed PAP	4	*SDPAP4.1	CEDG139, MB-SSR179
	11	SDPAP11.1	BM141, VR222
Seed IP	1	SDIP1.1	Satt713, VR102
	8	SDIP8.1	CEDG257, IAC100
	10	SDIP10.1	VR226, CEDG097
	11	sSDIP11.1	VR216, CEDG168
Seed TP	4	SDTP4.1	Bmd25, MB-SSR179
Seedling PAP	7	*SLPAP7.1	MB-SSR217, GBssr-MB7
	7	SLPAP7.2	IAC98, Bmd26
	8	SLPAP8.1	MB58, IAC100
Seedling IP	4	SLPAP4.1	CEDG088, CEDG091
Seedling TP	4	SLTP4.1	CEDG088, CEDG091
	7	SLTP7.1	IAC98, Bmd26

* *SD*: seed; *SL*: seedling.
Source: Adopted from Sompong et al. (2012).

9.3.4.2 Phenolic Compounds

Phenolics are natural phytochemicals that possess antioxidant activity. Predominantly, *phenolics* mean phenolic acids and flavonoids. Phenolic acids include hydroxybenzoic and hydroxycinnamic acids; flavonoids include flavone, flavanones, flavonols, flavanol, and isoflavones (Kim et al., 2013). Quercetin and ferulic acid increase the amount of bioavailable Fe in humans. But some other phenolics, like gallic acid and catechin, decrease Fe bioavailability, that is, they are the inhibitors. A dark-colored green gram seed coat is associated with higher polyphenol content and carotenoids (Dahiya et al., 2013, 2015).

9.3.4.3 Other Factors

Seed hardness in green gram was associated with seed fiber content. Seed moisture content is also related to seeds' physical properties, like seed dimensions, sphericity, porosity, bulk, and true density (Dahiya et al., 2013, 2015). One should also know that seeds of some wild green gram genotypes show bruchid resistance (Sarkar et al., 2011) due to the presence of insecticidal cyclopeptide alkaloids, vignatic acids (Sugawara et al., 1996). Kaga and Ishimoto (1998) mapped the vignatic acid gene, *Va*, to a single locus and 0.2 cM away from the Bruchid resistance gene *Br*.

9.3.5 Human Consumption

Home traditional processes such as dehulling, soaking, germination, boiling, autoclaving and microwave cooking, etc. cause a reduction in the concentration of lysine, tryptophan, threonine, and sulfur-containing amino acids of green gram seed while enhancing the concentration of leucine, isoleucine, valine, and aromatic amino acids. Germination causes phytic acid reduction, which will help in the retention of all minerals. Germination and cooking processes cause significant reduction in fat, carbohydrate fractions, antinutritional factors, and total ash contents (Mubarak, 2005). Barakoti and Bains (2007) also experimented with a few household processes in green gram, such as dehulling, pressure cooking, germination, and fermentation, and found reduced ash, phytin phosphorus, and polyphenols in all the cases. After cooking and dehulling, there was a significant reduction in neutral detergent fiber content, whereas fiber enhancement was observed after germination and fermentation. Keeping the fact in mind, a green gram–based diet should be developed to improve the mineral status of the human population. Moreover, breeding for low-PA crops, including green gram, must be a potential way to enhance the nutritional value of crop-based products.

9.4 CHALLENGES AND FUTURE PROSPECTS

The term *nutrigenomics* is self-explanatory; clear-cut demarcation is there if you consider human and plant systems. We depend on plant produce for our daily cuisines and nourishment; hence, there is a relationship between our genes and our diet. Nutrigenomics and also nutrigenetics – these two fields are unraveling this complex relationship separately; however, both are trying to optimize human health and physiology via nutritional intervention. On the other hand, plant nutrigenomics study elucidates characterization and utilization of plant genes and their alleles responsible for nutrient upliftment or lowering of antinutrient content in the plant part we consume; in a nutshell, it deals mainly with enhancing plant nutrient bioavailability to us in combatting malnutrition worldwide. Supply of nutrient-rich diets to each one of the global population is a challenge in the present day. Plant breeding plays a major role in biofortification and lowering the antinutrients in the consumable part of the crops without compromising yield.

We all know pulses are the premiere source of nutrients for most of us, especially vegans. The green gram has worldwide acceptability and preference over other pulses. It is a farmer-friendly crop, as its cultivation is not season-bound and has short seed-to-seed tenure. To date, we noticed scanty research findings regarding nutrient improvement of green gram through allele mining. Exploration of the wild green gram germplasms with better seed nutrient traits and simultaneous inclusion into the breeding program is necessary to combat the research gap. The immense potentiality of allele mining and marker-assisted breeding is well established in crop improvement programs and is completely accepted by the scientific community due to the absence of biosafety regulations, unlike transgenic crop breeding. Green gram breeders must intensify their research efforts to develop dense seed nutrient and/or reduced seed antinutrient genotypes; marker-assisted breeding will expedite the process. Many of the researchers, viz., Tangphatsornruang et al. (2009); Moe et al. (2011); Van et al. (2013); Chen et al. (2015); and Sokolkova et al. (2020), used high-throughput methodologies such as the prediction of potential microRNA, RNA sequence technology, and Illumina-based transcriptome evaluation for investigating a huge number of SSR and SNP markers in green gram; these markers must be utilized for speeding up the green gram breeding program, especially enhancement of seed nutrient content.

9.5 CONCLUSION

Green gram is a widely consumed food in Asia and other areas of the world, and its consumption is only likely to rise. Due to the continually expanding demand–supply imbalance for legumes, there will be a severe shortage of supply for the world's expanding population in the upcoming years. The only opportunity is to put all the efforts into creating enhanced, high-yielding cultivars that are resistant to or tolerant of the main stressors, particularly in light of climate change. Additionally, the nutritional enrichment of legume crops is crucial. By identifying the genes responsible for various phenotypes and enabling the selection of plants based on genotyping data, cost-effective sequencing technologies have ushered in a new age in genomics and breeding. Breeders must incorporate the knowledge produced by all "omics" investigations into their breeding processes in order to transition away from "chance breeding" and toward "knowledge-based breeding." A number of novel genomic technologies have recently been created to support trait mapping, gene identification, and trait enhancement projects in green gram. Additionally, critical information from transcriptomics, proteomics, metabolomics, and epigenomics has dramatically helped the scientific community in developing a better understanding of the traits and crops, leading to the development of an effective strategy for achieving higher genetic gains in a shorter amount of time. The existing and emerging technology will certainly support legume improvement efforts in a more efficient, convenient, and time-saving way.

REFERENCES

Alverson AJ, Zhuo S, Rice DW, Sloan DB, Palmer JD (2011) The mitochondrial genome of the legume *Vigna radiata* and the analysis of recombination across short mitochondrial repeats. *PLoS One* 6(1):e16404.

Aneja B, Yadav NR, Chawla V, Yadav RC (2012) Sequence-related amplified polymorphism (SRAP) molecular marker system and its applications in crop improvement. *Mol Breed* 30(4):1635–1648.

Anonymous. Green Gram Outlook (2020) Agricultural Market Intelligence Centre, PJTSAU, Government of India, Ministry of Agriculture and Farmers Welfare, Department of Agriculture, Cooperation and Farmers Welfare, Directorate of Economics and Statistics, 3rd Advance Estimates, 2019–20:1–3, July.

Ates D, Aldemir S, Yagmur B, Kahraman A, Ozkan H et al. (2018) QTL mapping of genome regions controlling manganese uptake in lentil seed. *G3: Genes Genomes Genet* 8(5):1409–1416.

Barakoti L, Bains K (2007) Effect of household processing on the in vitro bioavailability of iron in mungbean (*Vigna radiata*). *Food Nutr Bull* 28(1):18–22.

Bisht IS, Bhat KV, Lakhanpaul S, Latha M, Jayan PK et al. (2005) Diversity and genetic resources of wild *Vigna* species in India. *Genet Resour Crop Evol* 52(1):53–68.

Bisht IS, Singh M (2013) 10. Asian Vigna. In: Singh M, Upadhyaya HD, Bisht IS (eds) *Genetic and Genomic Resources of Grain Legume Improvement*. Elsevier, London & Waltham, MA, pp. 237–267.

Blair MW (2013) Mineral biofortification strategies for food staples: The example of common bean. *J Agric Food Chem* 61(35):8287–8294.

Burlyaeva M, Vishnyakova M, Gurkina M, Kozlov K, Lee CR et al. (2019) Collections of mungbean [*Vigna radiata*) (L.) R. Wilczek] and urdbean [*V. mungo* (L.) Hepper] in Vavilov Institute (VIR): Traits diversity and trends in the breeding process over the last 100 years. *Genet Resour Crop Evol* 66(4):767–781.

Caulfield LE, Richard SA, Rivera JA, Musgrove P, Black RE (2006) Stunting, wasting, and micronutrient deficiency disorders. In: Jamison DT, Breman JG, Measham AR et al. (eds) *Disease Control Priorities in Developing Countries*, 2nd edn. The International Bank for Reconstruction and Development/The World Bank, Washington, DC, pp. 551–567.

Chen H, Wang L, Wang S, Liu C, Blair MW et al. (2015) Transcriptome sequencing of mung bean (*Vigna radiata* L.) genes and the identification of EST-SSR markers. *PLoS One* 10(4):e0120273.

Dahiya PK, Linnemann AR, Nout MJR, Van Boekel MAJS, Grewal RB (2013) Nutrient composition of selected newly bred and established mung bean varieties. *LWT-Food Sci Technol* 54(1):249–256.

Dahiya PK, Linnemann AR, Van Boekel MAJS, Khetarpaul N, Grewal RB et al. (2015) Mung bean: Technological and nutritional potential. *Crit Rev Food Sci Nutr* 55(5):670–688.

Das A, Singh P, Kushwah NS, Thakur S, Rathore M et al. (2018) Genetic and genomic approaches for improvement in mungbean (*Vigna radiata* L.). In: Wani S, Jain M (eds) *Pulse Improvement*. Springer, Cham, pp. 175–189.

Dwivedi SL, Sahrawat KL, Rai KN, Blair MW, Andersson MS et al. (2012) Nutritionally enhanced staple food crops. In: Janick J (ed) *Plant Breeding Reviews*. Wiley Sons, Inc, Hoboken, NJ, pp. 169–292.

Ebert AW (2013) Ex situ conservation of plant genetic resources of major vegetables. In: Normah M, Chin H, Reed B (eds) *Conservation of Tropical Plant Species*. Springer, New York, NY, pp. 373–417.

Garcia RN, Adachi M, Mendoza EMT, Bernardo AEN, Utsumi S (2006) Physicochemical properties of native and recombinant mungbean (*Vigna radiata* L. Wilczek) 8S globulins and the effects of the N-linked glycans. *J Agric Food Chem* 54(16):6005–6010.

Islam AF, Blair MW (2018) Molecular characterization of mung bean germplasm from the USDA core collection using newly developed KASP-based SNP markers. *Crop Sci* 58(4):1659–1670.

Itoh T, Garcia RN, Adachi M, Maruyama Y, Mendoza EMT et al. (2006) Structure of 8Sα globulin, the major seed storage protein of mung bean. *Acta Crystallograph Sect D Biol Crystallograph* 62(7):824–832.

Kaga A, Ishimoto M (1998) Genetic localization of a bruchid resistance gene and its relationship to insecticidal cyclopeptide alkaloids, the vignatic acids, in mungbean (*Vigna radiata* L. Wilczek). *Mol Gen Genet* 258(4):378–384.

Kang YJ, Kim SK, Kim MY, Lestari P, Kim KH et al. (2014) Genome sequence of mungbean and insights into evolution within *Vigna* species. *Nat Commun* 5(1):1–9.

Karpechenko GD (1925) Chromosomes of Phaseolinae. *Bull Appl Bot* 14:143–148.

Kim JK, Kim EH, Lee OK, Park SY, Lee B et al. (2013) Variation and correlation analysis of phenolic compounds in mungbean (*Vigna radiata* L.) varieties. *Food Chem* 141(3):2988–2997.

Kim SK, Nair RM, Lee J, Lee SH (2015) Genomic resources in mungbean for future breeding programs. *Front Plant Sci* 6:626.

Kobayashi T, Nishizawa NK (2012) Iron uptake, translocation, and regulation in higher plants. *Annu Rev Plant Biol* 63(1):131–152.

Kollárová K, Vatehová Z, Slováková LU, Lišková D (2010) Interaction of galactoglucomannan oligosaccharides with auxin in mung bean primary root. *Plant Physiol Biochem* 48(6):401–406.

Kumar GR, Sakthivel K, Sundaram RM, Neeraja CN, Balachandran SM et al. (2010) Allele mining in crops: Prospects and potentials. *Biotechnol Adv* 28(4):451–461.

Kwon SJ, Brown AF, Hu J, McGee R, Watt C et al. (2012) Genetic diversity, population structure and genome-wide marker-trait association analysis emphasizing seed nutrients of the USDA pea (*Pisum sativum* L.) core collection. *Genes Genom* 34(3):305–320.

Li H, Durbin R (2009) Fast and accurate short read alignment with Burrows–Wheeler transform. *Bioinformatics* 25(14):1754–1760.

Li H, Ye G, Wang J (2007) A modified algorithm for the improvement of composite interval mapping. *Genetics* 175(1):361–374.

Likhith RK, Alagarasan G, Muthurajan R, Parasuraman B, Subramanian R (2021) Genome wide identification of mungbean (*Vigna radiata* [L.] R. Wilczek) late embryogenesis abundant (LEA) protein gene family. *ISR J Plant Sci* 69(1–2):79–86.

Longvah T, Anantan I, Bhaskarachary K, Venkaiah K, Longvah T (2017) *Indian food composition tables.* National Institute of Nutrition, Indian Council of Medical Research, Hyderabad, pp. 2–58.

Ma Y, Coyne CJ, Grusak MA, Mazourek M, Cheng P et al. (2017) Genome-wide SNP identification, linkage map construction and QTL mapping for seed mineral concentrations and contents in pea (*Pisum sativum* L.). *BMC Plant Biol* 17(1):1–17.

Masari A, Kaewwongwal A, Somta P, Srinives P (2017) Inheritance and a major quantitative trait locus of seed starch content in mungbean (*Vigna radiata* (L.) Wilczek. *Euphytica* 213(8):1–8.

Mendoza EMT, Adachi M, Bernardo AEN, Utsumi S (2001) Mungbean [*Vigna radiata* (L.) Wilczek] globulins: Purification and characterization. *J Agricultural Food Chem* 49(3):1552–1558.

Moe KT, Chung JW, Cho YI, Moon JK, Ku JH et al. (2011) Sequence information on simple sequence repeats and single nucleotide polymorphisms through transcriptome analysis of mungbean. *J Integr Plant Biol* 53(1):63–73.

Mubarak AE (2005) Nutritional composition and antinutritional factors of mung bean seeds (*Phaseolus aureus*) as affected by some home traditional processes. *Food Chem* 89(4):489–495.

Nagrale SC, Patil AN, Tayade N, Jadhav PV, Wakode YS (2018) Proximate composition and estimation of mineral content from different mungbean (*Vigna radiata* (L). Wliczek) genotypes. *J Pharmacog Phytochem* 7(4):3434–3436.

Nair RM, Thavarajah P, Giri RR, Ledesma D, Yang RY et al. (2015) Mineral and phenolic concentrations of mungbean [*Vigna radiata* (L.) R. Wilczek var. *radiata*] grown in semi-arid tropical India. *J Food Compos Analys* 39:23–32.

Nair RM, Yang RY, Easdown WJ, Thavarajah D, Thavarajah P et al. (2013) Biofortification of mungbean (*Vigna radiata*) as a whole food to enhance human health. *J Sci Food Agric* 93(8):1805–1813.

Nandhini U, Raveendran M, Rajesh S (2019) Isolation and Characterization of a Late Embryogenesis Abundant (LEA) Protein Coding Gene (VrLEA2) from Mung Bean (Vigna radiata (L) Wilczek). *Int J Curr Microbiol Appl Sci* 8(6):2289–2296.

Oburuoga AC, Anyika JU (2012) Nutrient and antinutrient composition of mungbean (*Vigna radiata*), acha (*Digitaria exilis*) and crayfish (*Astacus fluviatilis*) flours. *Pak J Nutr* 11(9):841–844.

Pal M, Brahmachary RL, Ghosh M (2010) Comparative studies on physicochemical and biochemical characteristics of scented and non-scented strains of mung beans (*Vigna radiata*) of Indian origin. *Legume Res* 33:1–9.

Pfeiffer WH, McClafferty B (2007) HarvestPlus: Breeding crops for better nutrition. *Crop Sci* 47:S88.

Randhir R, Lin YT, Shetty K (2004) Stimulation of phenolics, antioxidant and antimicrobial activities in dark germinated mung bean sprouts in response to peptide and phytochemical elicitors. *Process Biochem* 39(5):637–646.

Reddy VRP, Das S, Dikshit HK, Mishra GP, Aski MS et al. (2021) Genetic dissection of phosphorous uptake and utilization efficiency traits using GWAS in mungbean. *Agronomy* 11(7):1401.

Rehman HM, Cooper JW, Lam HM, Yang SH (2019) Legume biofortification is an underexploited strategy for combatting hidden hunger. *Plant Cell Environ* 42(1):52–70.

Rohland N, Reich D (2012) Cost-effective, high-throughput DNA sequencing libraries for multiplexed target capture. *Genome Res* 22(5):939–946.
Sandhu KS, Lim ST (2008) Digestibility of legume starches as influenced by their physical and structural properties. *Carbohyd Polymers* 71(2):245–252.
Sankaran RP, Huguet T, Grusak MA (2009) Identification of QTL affecting seed mineral concentrations and content in the model legume *Medicago truncatula*. *Theor Appl Genet* 119(2):241–253.
Sarkar S, Ghosh S, Chatterjee M, Das P, Lahari T et al. (2011) Molecular markers linked with bruchid resistance in *Vigna radiata* var. Sublobata and their validation. *J Plant Biochem Biotechnol* 20:155–160.
Schafleitner R, Huang SM, Chu SH, Yen JY, Lin CY et al. (2016) Identification of single nucleotide polymorphism markers associated with resistance to bruchids (*Callosobruchus* spp.) in wild mungbean (*Vigna radiata* var. sublobata) and cultivated *V. radiata* through genotyping by sequencing and quantitative trait locus analysis. *BMC Plant Biol* 16(1):1–15.
Siddhuraju P, Osoniyi O, Makkar HPS, Becker K (2002) Effect of soaking and ionising radiation on various antinutritional factors of seeds from different species of an unconventional legume, Sesbania and a common legume, green gram (*Vigna radiata*). *Food Chem* 79(3):273–281.
Singh R, Van HAW, Kumar R, Visser RGF (2018) Genetic variation and correlation studies between micronutrient (Fe & Zn), protein content and yield attributing traits in mungbean (*Vigna. radiata* L.). *Legume Res* 41(2):167–174.
Singh V, Yadav RK, Yadav NR, Yadav R, Malik RS et al. (2017) Identification of genomic Regions/genes for high iron and zinc content and cross transferability of SSR markers in mungbean (*Vigna radiata* L.). *Legume Res* 40(6):1004–1011.
Soetan K, Olaiya C, Oyewole O (2010) The importance of mineral elements for humans, domestic animals and plants-a review. *Afr J Food Sci* 4(5):200–222.
Sokolkova A, Burlyaeva M, Valiannikova T, Vishnyakova M, Schafleitner R et al. (2020) Genome-wide association study in accessions of the mini-core collection of mungbean (*Vigna radiata*) from the world vegetable gene bank (Taiwan). *BMC Plant Biol* 20(1):1–9.
Sompong U, Kaewprasit C, Nakasathien S, Srinives P (2010) Inheritance of seed phytate in mungbean (*Vigna radiata* (L.) Wilczek). *Euphytica* 171(3):389–396.
Sompong U, Somta P, Raboy V, Srinives P (2012) Mapping of quantitative trait loci for phytic acid and phosphorus contents in seed and seedling of mungbean (*Vigna radiata* (L.) Wilczek). *Breed Sci* 62(1):87–92.
Sugawara F, Ishimoto M, Le-Van N, Koshino H, Uzawa J et al. (1996) Insecticidal peptide from mungbean: A resistant factor against infestation with azuki bean weevil. *J Agri Food Chem* 44(10):3360–3364.
Tangphatsornruang S, Sangsrakru D, Chanprasert J, Uthaipaisanwong P, Yoocha T et al. (2010) The chloroplast genome sequence of mungbean (*Vigna radiata*) determined by high-throughput pyrosequencing: Structural organization and phylogenetic relationships. *DNA Res* 17(1):11–22.
Tangphatsornruang S, Somta P, Uthaipaisanwong P, Chanprasert J, Sangsrakru D et al. (2009) Characterization of microsatellites and gene contents from genome shotgun sequences of mungbean (*Vigna radiata* (L.) Wilczek). *BMC Plant Biol* 9(1):1–12.
Taunk J, Yadav NR, Yadav RC, Kumar R (2012) Genetic diversity among green gram [*Vigna radiata* (L.) Wilczek] genotypes varying in micronutrient (Fe and Zn) content using RAPD markers. *Indian J Biotechnol* 11:48–53.
Tresina PS, Daffodil ED, Lincy P, Mohan VR (2014) Assessment of biochemical composition and nutritional potential of three varieties of *Vigna radiata* (L.) Wilezek. *Biolife* 2(2):655–667.
Ullah R, Ullah Z, Al-Deyab SS, Adnan M, Tariq A (2014) Nutritional assessment and antioxidant activities of different varieties of *Vigna radiata*. *Sci World J* 2014:871753.
Van K, Kang YJ, Han KS, Lee YH, Gwag JG et al. (2013) Genome-wide SNP discovery in mungbean by Illumina HiSeq. *Theor Appl Genet* 126(8):2017–2027.
Varshney RK, Graner A, Sorrells ME (2005) Genomics-assisted breeding for crop improvement. *Trends Plant Sci* 10(12):621–630.
Vavilov NI (1926) Centers of origin of cultivated plants. NI Vavilov origin and geography of cultivated plants (Translated by Doris Löve) February 1994. *Arch Nat Hist* 21(1):142.
White PJ, Broadley MR (2009) Biofortification of crops with seven mineral elements often lacking in human diets–iron, zinc, copper, calcium, magnesium, selenium and iodine. *New Phytol* 182(1):49–84.
Wu X, Islam AF, Limpot N, Mackasmiel L, Mierzwa J et al. (2020) Genome-wide SNP identification and association mapping for seed mineral concentration in mung bean (*Vigna radiata* L.). *Front Genet* 11:656.
Yu W, Zhang G, Wang W, Jiang C, Cao L (2020) Identification and comparison of proteomic and peptide profiles of mung bean seeds and sprouts. *BMC Chem* 14(1):1–12.

10 Genetic and Genomic Resources for Allele Mining and Genetic Improvement of Grass Pea

Neetu Singh Kushwah, Ashok Kumar Parihar, Md. Mosiur Rahman, Sarita Pandey, Abhimanyu Sarkar, and Chittaranjan Kole

10.1 INTRODUCTION

The grass pea (*Lathyrus sativus* L.), a cool-season grain legume widely produced in the Mediterranean region, sub-Saharan Africa (SSA), and South Asia (SA), is a stress-resilient crop with excellent nutritional content, and it is considered as a potential source of traits for breeding for climate change adaptation/mitigation. It is also referred to as a crop that is appropriate for more sustainable production strategies, such as intercropping (Gonçalves et al., 2022). One to two weeks before harvest, it is often broadcast-seeded into standing rice crops. This enables grass pea to successfully utilize the remaining moisture left behind after rice harvest (Joshi, 1998). The grass pea originated in Southwest and Central Asia before spreading to the Eastern Mediterranean (Kislev, 1989). Grass pea is a major crop in several countries, like Bangladesh, Pakistan, India, Nepal, and Ethiopia, and it is also grown to a lesser extent in many European countries, the Middle East, Northern Africa, China, Chile, and Brazil (Campbell, 1997; Mikic´ et al., 2011). Production of grass peas has dropped in the Mediterranean (Polignano et al., 2009; Rubiales et al., 2020) but has improved in Bangladesh and Ethiopia. Indeed, grass pea was recently ranked 19th among the 21 most important crop species in Ethiopia, and 22nd among the top 24 in South Asia (Kumar et al., 2013). It is used for different purposes, including food, feed, and fodder, due to its nutritional characteristics. Grass pea is widely cultivated in South Asian countries for both grain and feed purposes (Kaul et al., 1986), though the crop has grown in popularity for use as animal feed rather than human consumption. Lathyrus animal feed is typically consisting of ground or split grain or flour and is mostly used to feed lactation cattle or other draft livestock (Campbell, 1997). Nutrient-dense food crops with low water requirements, such as Lathyrus, are projected to play an important role in reducing world hunger. However, the presence of a neurotoxic is the primary explanation for this lack of research effort. Development of cultivars with superior balances in key amino acids and diet may be important to increase nutritional value without affecting this promising crop's multiple stress tolerance.

In this work, we discuss a more integrative approach to allele mining which covers the use of allele mining for variety improvement–based breeding objectives. We discuss various genomic approaches for grass pea as a source of resilient characteristics capable of surviving environmental challenges as a sustainable agriculture model crop. Finally, we emphasize breeding strategies and genetic studies such as GWAS (genome-wide association) to investigate potential candidate genes and mechanisms underpinning agronomically significant features.

DOI: 10.1201/9781003385059-10

10.2 ORIGIN AND DISTRIBUTION

Grass pea, belonging to the genus *Lathyrus*, is part of a group encompassing a total of 187 species (Sarkar et al., 2019) within the Fabaceae family (also known as Leguminosae), the subfamily Faboideae (alternatively referred to as Papilionoideae), and the tribe Fabeae (also recognized as Vicieae) (Smýkal et al., 2011; Schaefer et al., 2012). This genus is organized into 13 sections based on insights from morphological and molecular phylogenetic studies (Kenicer et al., 2005, 2009). However, recent considerations arising from chloroplast DNA and AFLP analyses of certain species have led to proposed reclassifications (Asmussen and Liston, 1998; Badr et al., 2002; Skiba et al., 2007). The genus is broadly distributed across temperate regions of the Northern Hemisphere, with a presence in Europe (52 species), North America (30 species), Asia (78 species), tropical East Africa (24 species), and temperate South America (24 species) (Badr, 2006). In India, Lathyrus species exhibit widespread occurrence, growing at elevations ranging from 1,200 m to 3,000 m (Gautam et al., 1998). The term "Lathyrus" traces its origin to the Greek term "lathyros," reflecting the intriguing connotations associated with the potential aphrodisiac properties of grass pea (Loudon et al., 1855). Recent phylogenetic and historical investigations have pointed to the Eastern Mediterranean or Fertile Crescent as the likely center of origin around 6,000 years before the present (Schaefer et al., 2012). However, this contrasts with Smartt's (1984) hypothesis, which suggested Southwest or Central Asia as the center of origin for Lathyrus. Vavilov (1951) proposed two centers of origin for Lathyrus: Central Asia and the Abyssinian center (Campbell, 1997). Historical evidence highlights the ancient domestication of grass pea, with seeds of Lathyrus species discovered in Turkey and Iraq as collected or cultivated items (Lambein et al., 2019). The *Lathyrus* might have been the first crop domesticated in Europe, and its cultivation was initiated around 6,000 BC. Correspondingly, the seeds of Lathyrus were found in the oldest archeological sites in India and Balkan which belong to 2,500 BC and 8,000 BC, respectively (Kislev, 1989). Moreover, Lathyrus cultivation is mostly distributed in countries like Ethiopia, Nepal, Bangladesh, Pakistan, India, China, and in several countries in the Middle East and Northern Africa and Europe (Dixit et al., 2016; Parihar et al., 2022). Considering taxonomy and morphological attributes, *Lathyrus* species are categorized into five groups, namely, *Lathyrus, Aphaca*, *Nissolia*, *Clymenum*, and *Cicerula* (Kenicer et al., 2005). Among them, *Lathyrus* species are typically perennial type, while the rest of the species are annual type (Asmussen and Liston, 1998; Kupicha, 1983). The *Lathyrus* species, based on cross-compatibility, is further classified into primary, secondary, and tertiary gene pools (Heywood et al., 2007; Kumar et al., 2013). The primary gene pool of *Lathyrus* mainly contains cultivated and wild races of *L. sativus*; however, it is further divided into two gene pools, namely, 1A and 1B. The 1A gene pool contains white-flowered and white-seeded cultivars, while 1B is comprised of blue-flowered and small mottled-seeded cultivars (Heywood et al., 2007; Rizvi et al., 2016). Within the primary gene pool, the availability of cultivars and landraces remains constrained. As a result, advancing conventional breeding efforts hinges on effectively harnessing and utilizing crop wild relatives (Yunus and Jackson, 1991). The secondary gene pool encompasses a total of ten species (Heywood et al., 2007). The remaining species (*Lathyrus* sp.) form the tertiary gene pool, offering potential avenues for genetic enhancement through contemporary biotechnological methods aimed at integrating essential genes from the tertiary gene pool into well-adapted cultivars.

10.3 LATHYRUS GENETIC RESOURCES

The success of any breeding program mainly hinged on the availability of genetic resources and their judicious utilization for the genetic improvement of any cultivated plant species (Yunus and Jackson, 1991). Complete information about crop wild relatives (CWR) is essential to strengthen and speed up any breeding program (Schaefer et al., 2012). The most economical, significant, and widely cultivated *Lathyrus* species are covered under the section *Lathyrus*, which included *L. sativus*, *L. cicera*, and *L. odoratus*. Unfortunately, very limited efforts have been made globally in

FIGURE 10.1 Illustrations to depict the different stages of grass pea: (a) grass pea field, (b) foliage, (c–d) flowers, (e) pods, (f) harvested crop, (g) whole seeds, and (h) broken seeds.

the direction of the genetic enhancement of these species. In the recent past, conservation and maintenance of *Lathyrus* genetic resources captivated extra concern due to the potential role of this climate-resilient species in the present climate-changing scenario (Kumar et al., 2020). At the end of the twentieth century, concerted efforts were initiated globally to collect and conserve the *Lathyrus* crop species by forming a "Lathyrus Genetic Resources Network" (Mathur et al., 1998). The International Treaty on Plant Genetic Resources for Food and Agriculture (ITPGRFA) established a multifaceted platform with the purpose of safeguarding the access and benefit sharing of genetic resources of major food legumes, including Lathyrus (FAO, 2009).

Similarly, the Global Crop Diversity Trust (GCDT) has collaborated with the International Center for Agricultural Research in the Dry Areas (ICARDA) to establish a program dedicated to the long-term conservation of *L. sativus*, *L. cicera*, and *L. ochrus* (GCDT, 2009). At the international scale, numerous institutes and organizations have amassed a substantial collection of cultivated and wild relatives (CWRs), preserving them within their gene banks and natural habitats. The largest compilation of Lathyrus accessions, totaling 4,477, of which 53% are *L. sativus* and 18% are *L. cicera*, is housed at the Conservatoire Botanique National des Pyrenees et de Midi-Pyrenees (CBNPMP) in France. Another significant repository comprising 4,200 accessions of *L. sativus* and *L. cicera* is maintained at the International Center for Agricultural Research in Dry Areas (ICARDA) in Morocco (Sarkar et al., 2019; Parihar et al., 2022). Additionally, the European Cooperative Program on Crop Genetic Resources, based at Bioversity International in Rome, safeguards over 4,000 accessions (Hillocks and Maruthi, 2012). Within India, the ICAR–National Bureau of Plant Genetic Resources (NBPGR) safeguards approximately 2,600 accessions, encompassing 3% wild types and 85% cultivated types. Notably, the Bangladesh Agricultural Research Institute (BARI) houses all cultivated-type accessions (Vaz Patto and Rubiales, 2014). In European countries such as Germany, Hungary, and Spain, collections of 300–500 Lathyrus accessions are conserved. Moreover, a comprehensive set of 2,134 accessions, spanning 44 countries, finds its home at the Svalbard Global Seed Vault (Almeida et al., 2015). Additionally, various other organizations, both on a global and national scale, maintain smaller repositories of grass pea germplasm, as depicted by country-wise illustration

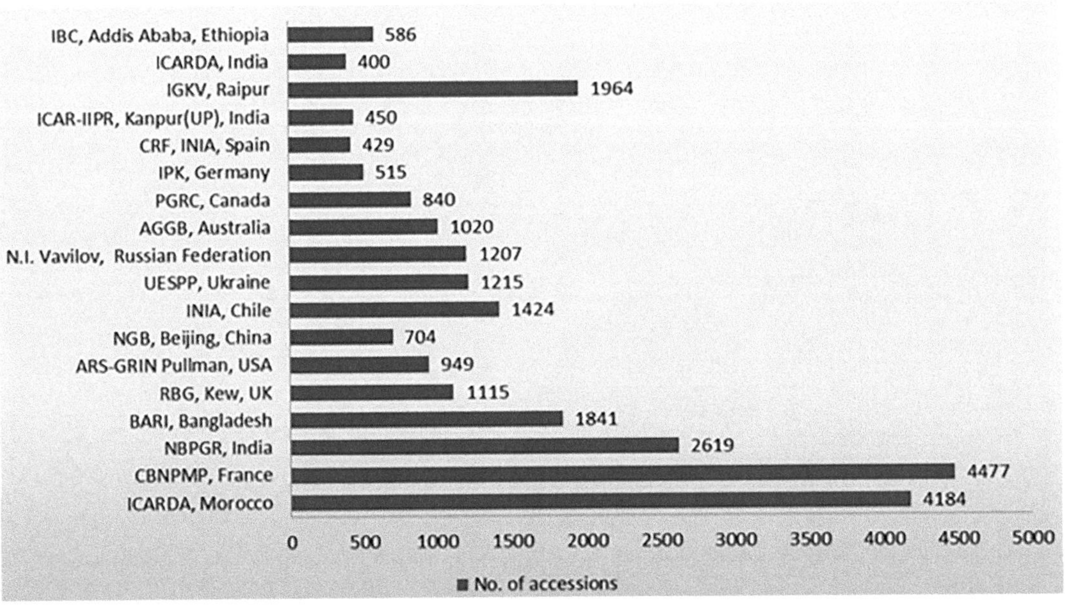

FIGURE 10.2 Lathyrus genetic resources existing at different institutions/organizations worldwide.

in Figure 10.2. In addition to the aforementioned ex situ collections for long-term preservation, in situ conservation methods are also employed (Maxted et al., 1997). However, there has been limited effort to safeguard Lathyrus diversity in its natural habitat, resulting in the vulnerability of native populations to genetic erosion or even extinction (Maxted and Bennett, 2001). For various legume genera, including Lathyrus, a multi-gene pool strategy has been embraced, leading to the collection of 61,081 distinct geo-referenced Lathyrus accessions between 1884 and 2008 (Maxted et al., 2012). These accessions encompass a wide spectrum of genetic diversity (Almeida et al., 2015). Diligent endeavors have been undertaken to explore the genetic diversity of Lathyrus germplasm concerning ODAP content (Fikre et al., 2008; Kumar et al., 2011; Grela et al., 2012; Xiong et al., 2015; Barpete et al., 2021), phenology, and grain yield (Mera, 2010; Grela et al., 2012), resistance to weeds (Fernández-Aparicio et al., 2012), biotic stresses (Vaz Patto and Rubiales, 2009), and other quality traits (Granati et al., 2003; Urga et al., 2005; Tamburino et al., 2012; Sacristán et al., 2015). By implementing accurate and precise phenotyping techniques and harnessing the existing diversity through high-throughput phenotyping and subsequent genotyping, novel alleles can be unveiled. These newly discovered alleles hold the potential to be integrated into regular breeding programs, thereby enhancing the genetic progress of this underutilized crop.

10.4 GENETIC DIVERSITY ANALYSIS

The detailed characterization of the existing genetic diversity is crucial for effective use of germplasm collections in breeding. The genetic diversity is usually performed using morphological, biochemical, and molecular markers. Various genetic diversity studies in Lathyrus using different types of morphological, biochemical, and molecular markers are described in the following.

10.4.1 Genetic Diversity Analysis Using Morphological and Biochemical Markers

Jackson and Yunus (1984) evaluated 49 grass pea accessions originating from 26 countries representing eight different geographical regions using 20 morphological characters. They observed high variations in most of the characters, especially in leaf and leaflet length, flower color (white, blue, and mixed flower color), seed color (white, light brown, or gray), and seed size. They found high correlations between flower and seed colors, for instance, white-flower color accessions produce only white seeds. These accessions exhibited a clustering pattern into two primary groups. One group consisted of blue-flowered accessions originating from regions such as Southwest and South Asia, as well as Ethiopia. The second group comprised white and mixed flower color accessions originating from areas spanning the Canary Isles to the western republics of the Soviet Union. The differentiation between these two groups arises from both their geographical isolation and human-driven selection for grain or forage utilization. Besides, they also studied the interspecific relationship between grass pea and other Lathyrus species of section Lathyrus using morphological variations and showed that grass pea is closely related to *L. cicera* and *L. gorgoni* of the section Lathyrus. They also proposed that blue flowers and smaller seeds in grass pea are more primitive than that of white flower and larger seed types. Przybylska et al. (1998), using electrophoretic analysis of seed albumin for 105 grass pea accession from diverse geographical regions, showed a clear distinction between white-seeded accessions with large seeds and colored-seeded accessions with small seeds. However, there have been no correlation between seed albumin variation and geographical origin of the accessions used in the study. Hanbury et al. (1999) evaluated 407 grass pea accessions from diverse origin using phenology, seed yield, and seed ODAP content traits and found a clear difference between lines of different geographical origin for these traits. Grass pea accessions originating from the Indian subcontinent have been shown to have lower yield, smaller seed size, early phenology, and lower ODAP concentration compared to the accessions originating from Mediterranean/European and Ethiopian origin. The Ethiopian accessions showed higher ODAP concentration and slightly later phenology than the Mediterranean/European lines. Later, genetic diversity studies in grass pea using morphological traits were focused on grass pea germplasm collected from a restricted geographical region. Tay et al. (2000) assessed a population of 74 Chilean grass pea landraces, analyzing both morphological and biochemical attributes. The study revealed significant diversity within the population, encompassing seed size, shape, and color, as well as characteristics such as days to flowering, protein content (ranging from 24% to 32%), and ODAP content (ranging from 0.18% to 0.52%). Similarly, 50 grass pea landraces population of Ethiopian origin were evaluated for morphological characters, ODAP content, and protein contents, which showed significant variations in traits like days to maturity, pods per plant, seed yield per plant, harvest index and dry matter yield, ODAP content (0.149 to 0.916%), and protein content (23.93 to 31.94%) (Tadesse and Bekele, 2003). Morphological traits were assessed in a collection of 18 Indian landrace populations, revealing notable diversity in attributes such as flowering time, podding time, pod count per plant, seed count per pod, and 100-seed weight across the populations. In a separate study, Tavoletti et al. (2005) conducted an evaluation of 16 Italian grass pea accessions examining morphological and agronomic traits as well as ODAP content. The results highlighted substantial genetic variation between populations, effectively distinguishing commercial populations from household populations of grass pea in Italy. Abd El Moneim et al. (2001) studied variation in β-ODAP content in grass pea germplasm originating from diverse geographical region, which indicated that accessions originating from Indian sub-continent and Ethiopia were higher in β-ODAP content (0.02–1.2%) than that from accessions originating from Middle Eastern countries (0.02–1.2%). Genetic variations in ODAP content were also examined by several research groups using grass pea germplasm of restricted origin and found ODAP content in the range of 0.02% to 2.59%, with ODAP content being highest (2.59%) in Indian accession and lowest (0.02%) in ICARDA germplasm lines (Kumar et al., 2013).

10.4.2 Genetic Diversity Analysis Using Molecular Markers

Genetic mapping and QTL analysis enable the utilization of markers to ascertain the count, location, and individual impacts of genes/QTLs governing significant attributes like disease or pest resistance, reduced ODAP concentration, enhanced protein content, and other crucial agronomic traits (Gupta et al., 2018; Arslan et al., 2020). DNA markers are based on differences in nucleotide sequence of DNA. They offer powerful means of analyzing genetic diversity compared to morphological and biochemical markers, as they have greater genome abundance and are not influenced by tissue, developmental stage, and environment. Various molecular markers, such as RAPD (Croft et al., 1999; Chtourou-Ghorbel et al., 2001; Barik et al., 2007; Nosrati et al., 2012), RFLP (Chtourou-Ghorbel et al., 2001), AFLP (Tavoletti and Iommarini, 2007), ISSR (Belaid et al., 2006), and SSR (Lioi et al., 2011; Ponnaiah et al., 2011; Shiferaw et al., 2012; Sun et al., 2012; Yang et al., 2014; Wang et al., 2015; Soren et al., 2015; Gupta et al., 2018; Arslan et al., 2020), have been used for genetic diversity studies in various grass pea germplasm collections. RAPD markers were used for assessing the intra- and interspecific genetic diversity of grass pea accessions of diverse geographical origin and other related species in 1999 (Croft et al., 1999), for analyzing the genetic variability among five selected grass pea genotypes of Indian origin in 2007 (Barik et al., 2007), and for measuring the genetic variation among different accessions of grass pea originating from diverse geographical region worldwide in 2012 (Nosrati et al., 2012). Comparisons of RAPDs and RFLPs markers in determining genetic diversity in the genus Lathyrus, representing five species originating from diverse geographical regions, showed that these marker systems provide same levels of resolution for determining genetic diversity in populations of Lathyrus. Results of RFLP and RAPD analysis indicated that grass pea is closely related to *L. cicera* and *L. sylvestris* and *L. latifolius* (Chtourou-Ghorbel et al., 2001). AFLP marker has clarified that 20 grass pea accessions from Central Italy were divided into the commercial and household populations (Tavoletti and Iommarini, 2007). This information will be suitable to enhance the use of Italian grass pea germplasm in crop improvement programs. Evaluating the genetic diversity across diverse populations originating from a broad geographical spectrum, encompassing three distinct species within the Lathyrus genus (*Lathyrus sativus*, *Lathyrus cicera*, *Lathyrus ochrus*), through the application of ISSR markers, showcased notable levels of variation within and between species. These findings underscore the significance of both wild and cultivated Lathyrus species as a valuable reservoir of genetic diversity. However, there has been no correlation between ISSR polymorphisms and geographical origin of accessions. Lathyrus section–specific molecular marker has also been identified in this study (Belaid et al., 2006). SSR markers were created by Lioi et al. (2011) using cDNA sequences from EMBL database of *L. sativus* and *Lotus japonicus*, aiming to investigate diversity within grass pea accessions from Southern Italy. Ponnaiah et al. (2011), using 11 EST-SSR (7 from grass pea and 4 from *Medicago truncatula*) markers, have identified the moderate level of diversity among the 20 Ethiopian grass pea accessions. However, SSR polymorphism has no correlation with the geographical region of the genotypes, and most of the genetic variations (84%) were due to accessions within regions. Shiferaw et al. (2012) undertook a comparable investigation employing 19 EST-SSR markers, comprising 7 markers specific to grass pea and 12 from *Medicago truncatula*, on grass pea accessions. Their study unveiled a moderate degree of diversity among different accessions, with a majority of the diversity (84%) stemming from variations within accessions. In contrast, the diversity among collection regions accounted for a minimal proportion (1%). They argued that seed exchange among farmers and cross-pollination might have contributed to the existence of comparable levels of diversity among different collection regions of grass pea. Another study using 17 polymorphic grass pea–derived EST-SSR in 176 grass pea accessions collected from different parts of India identified the clustering of these accessions into two major groups with some admixtures, and differentiation of accessions was independent of their geographical region and suggested the possibility of gene flow between the accessions of different geographical origins (Soren et al., 2015). Yang et al. (2014) generated large-scale SSR-enriched sequence data using Roche 454 next-generation sequencing and identified 651,827 genomic SSR loci. After validation of 288 SSR loci, 74 SSRs were found polymorphic. These 74 genomic SSR markers were effective enough to separate *L. sativus* accessions from the *L. cicera* accessions. Wang et al. (2015), by using 30 polymorphic genomic SSR markers for assessing the genetic diversity of 283 accessions of grass pea from diverse geographical region and its wild relatives, revealed that a genomic SSR marker system is able to discriminate Asian accessions from European and African grass pea accessions and from the wild species. They have also indicated the possibility of high gene flow between African and European grass pea accessions, but not in Asian accessions. High genetic diversity of Asian grass pea accessions than European and African accessions suggests that the spread of Asian grass pea accessions occurs in a much broader way. Mekonen et al. (2022) conducted a study examining the genetic diversity of 25 Ethiopian germplasm accessions, employing ten SSR markers. Their findings indicated the presence of gene flow between different geographic regions. Arslan et al. (2020) delved into the genetic diversity of low-β-ODAP accessions (with β-ODAP content ranging from 0.24% to 0.58%), which originated from Bangladesh. They employed 31 EST-SSR markers and identified elevated genetic distances among accessions, particularly GP6/GP11 (β-ODAP content 0.244%/0.319%), GP4/GP11 (β-ODAP content 0.566%/0.319%), and GP5/GP8 (β-ODAP content 0.316%/0.267%). These findings suggest potential utility for grass pea breeding programs. In a study by Rahman et al. (2022), an analysis was conducted on 400 grass pea accessions sourced from the Institute of Crop Science, Chinese Academy of Agricultural Sciences (CAAS), Beijing, China, and the Plant Genetic Resources Center (PGRC), Bangladesh Agricultural Research Institute (BARI), Gazipur, Bangladesh. This investigation involved the utilization of 56 polymorphic SSR markers. The findings revealed the capacity of SSR markers to categorize these accessions into two primary populations, with some instances of admixture. This outcome suggests a noteworthy level of diversification present within the grass pea populations.

10.5 TRADITIONAL BREEDING IN LATHYRUS

Although Lathyrus has served as a pulse crop for over 8,000 years, its genetic enhancement has been minimally pursued due to the neurotoxin (ODAP) content. This is reflected in limited efforts within this crop, as documented by Rahman et al. (1995) and Vaz Patto et al. (2006). Nevertheless, in recent times, several initiatives have been launched to simultaneously address diverse aspects of enhancing grass pea. These endeavors encompass the creation of climate-resilient cultivars with reduced ODAP content and heightened nutrient density. These cultivars are designed to exhibit superior nutrient uptake efficiency while ensuring elevated food, fodder, and forage yields, thereby bolstering food, feed, and nutritional security amid challenging environmental conditions (Hillocks and Maruthi, 2012; Dixit et al., 2016; Parihar et al., 2022). To attain the aforementioned goals, a combination of traditional and innovative breeding approaches has been employed to advance the genetic qualities of grass pea. Interestingly, a noteworthy 50% of low-ODAP varieties have been produced using the direct selection breeding method (Vaz Patto et al., 2006). Traditional breeding methods mainly focused on the hybridization between good agronomic base and trait-specific genotypes and evaluation of their succeeding generations for targeted traits and selection of superior progenies. Earlier, most of the hybridization efforts were made between the high-yielding good agronomic base and low-ODAP genotypes to reduce ODAP content (Campbell, 1997). In most crop improvement programs, high seed yield has been considered as an important factor during selection, while very less effort was made for other yield-attributing traits. In recent decades, the attribute of substantial biomass has garnered notable interest within the context of grass pea, primarily due to its remarkable potential as a feed and fodder source, particularly in the regions of North Africa and South Asia (Campbell, 1997; Abd El Moneim et al., 2001; Vaz Patto et al., 2006). Additionally, prostrate and indeterminate growth habits, coupled with extended duration and pod shattering, have been included as an objective in several breeding programs (Rybinski, 2003). As previously mentioned, a substantial proportion of grass pea varieties was historically derived through the selection process from available germplasm and local landraces. The enhancement efforts for grass pea have been categorized into three distinct phases. The initial phase (1940–1960) focused on generating high-yielding cultivars through the selection of superior single plant progenies from indigenous landraces. This approach led to the development of several high-yielding varieties suitable for regions like Madhya Pradesh and West Bengal. Notable examples include

BR13, LC76, T2–12, and B-19 (Gautam et al., 1998; Dixit et al., 2016). The next phase (1974–1990s) of grass pea improvement was committed to developing low-ODAP (≤0.2%) varieties with high productivity, particularly for upland areas (Gautam et al., 1998). Consequently, a systematic grass pea improvement program was begun in many countries in between 1966 and 1986, for example, India (Lal et al., 1986), Canada (Campbell, 1988), Bangladesh (Kaul and Islam, 1981), and Nepal. The systematic national and international breeding efforts delivered numbers of improved varieties that had less than 0.1% β-ODAP content (Kumar et al., 2011, 2013; Dixit et al., 2016; Lambein et al., 2019). The first variety with low ODAP content (0.2%) was "Pusa-24," which was selected in 1966 from Bihar and successively released for upland areas (Jain et al., 1974; Mehra et al., 1995). Later on, this variety was used as a parent to develop low-ODAP varieties worldwide (Campbell and Briggs, 1987). In subsequent years, several varieties, viz., LSD1, LSD2, SD3, LSD6, Pusa-305, and Selection 1276, which had low ODAP content (0.15–0.20%) and were suitable for different ecologies, were developed (Gautam et al., 1998; Sethi et al., 1981). On a similar note, in 1983, grass pea variety "Quila-blanco" was released for marketable cultivation in Chile, which was synchronous in maturity and white bold-seeded variety coupled with good protein content (Campbell et al., 1994). During the third phase (1990s onward) of grass pea improvement, primary objectives were shifted to low ODAP and high yield coupled with tolerance to powdery mildew and thrips, which were not covered in the previous two phases. In pursuit of achieving their targeted objectives, various countries, including Australia, Bangladesh, Canada, China, Ethiopia, India, Nepal, Syria, Poland, Italy, the USA, and Chile, have initiated conventional breeding programs (Kumar et al., 2020; Das et al., 2021; Parihar et al., 2022). Among these programs, some remain actively ongoing, while most are executed on a smaller scale (Vaz Patto et al., 2011). Germplasm characterized by low ODAP content is associated with undesirable traits, such as delayed flowering, reduced yield, and vulnerability to biotic and abiotic stresses. Consequently, breeding strategies have been devised to combine low ODAP levels with high yield, appropriate phenology, and stress resilience (Hanbury et al., 2000; Crino et al., 2004; Vaz Patto et al., 2006). As a result, a considerable number of high-yielding varieties and genotypes have been developed, showcasing low β-ODAP content (<0.1%) and robust resistance against prevailing biotic and abiotic stressors (Kumar et al., 2013; Dixit et al., 2016). The occurrence of lathyrism in human populations prompted a notable shift in the primary grass pea breeding efforts toward reducing ODAP content. This shift yielded multiple varieties of *L. sativus* or *L. cicera* with diminished ODAP levels (Siddique et al., 2006; Kumar et al., 2011; Almeida et al., 2015). Notably, in India, several high-yielding lines, such as BioR-202, BioL-203, BioL-212, BioR-231, and BioL-208, were developed, showcasing elevated harvest indices. Among these, BioL-212 (Ratan) emerged through somaclonal variation of Pusa-24 and was recommended for commercial cultivation in 1997 (Gautam et al., 1998; Pandey et al., 1998). In succeeding years, other high-yielding varieties, like Prateek and Mahateora, were developed with low ODAP content (<0.1%) using hybridization (Dixit et al., 2016). Drawing parallels, two hybridization-derived products characterized by high yield and low ODAP content (<0.29%) were introduced in Bangladesh during 1995–1996 (Rahman et al., 2001). Similarly, in Australia, hybridization efforts resulted in the development of the "Ceora" variety, exhibiting low ODAP levels (0.04–0.09%) (Siddique et al., 2006). Another notable instance is the "Chalus" variety, which boasts a protein-rich composition (26.5%) and low ODAP content (0.09%), originating from selective breeding of IFLA1279 (Hanbury and Siddique, 2000). In Chile, the "Luanco-INIA" variety, characterized by white seed color and substantial seed size (30–35 g/100-seed weight), was achieved through germplasm selection (Mera et al., 2003). Furthermore, in the realm of grass pea research, Barpete et al. (2020) developed an enhanced flowering protocol, facilitating the attainment of 4.33 generations per year and thereby reducing the time required to obtain homozygous lines. In recent times, the mutagenesis approach has been adopted to induce genetic variability for the development of zero/low-ODAP varieties (Talukdar, 2009b). Notably, two varieties (Poltavskaya, BinaKhesari) have been successfully created using EMS and gamma rays in Russia and Bangladesh, respectively (Kumar et al., 2013). Intriguingly, a large-seeded somaclonal variant (NGOG 5) has emerged, showcasing high seed yield and low neurotoxin content, making it a promising candidate for future breeding endeavors. Additionally, the presence of dwarf mutants characterized by upright and determinate growth traits, accompanied by various leaflet and tendril mutations (Talukdar, 2009a), presents an opportunity to reshape the growth habit of grass pea. Typically, this plant species exhibits low and sprawling growth tendencies, often leading to lodging (Rybinski, 2003). To date, a diverse collection of approximately 37 distinct varieties has been established, with a primary emphasis on the direct selection method, supplemented by a few instances arising from hybridization, mutation, and tissue culture techniques. The substantial enhancement of grass pea plant type can be achieved by adopting traits similar to field peas, including the development of dwarf and leafless plants. This alteration involves the conversion of leaflets into tendrils, enhancing upright stability and rendering them suitable for mechanical harvesting (Ali and Kumar, 2009).

10.6 MOLECULAR MAPPING OF GENES AND QTLS

During the past few years, considerable progress has been made on the generation of genomic resource in grass pea. As a result, several transcriptomes and draft genome sequence of grass pea have now become available (Chapman, 2015; Hao et al., 2017; Yang et al., 2014; Tan et al., 2017; Xu et al., 2018; Rathi et al., 2019, Emmrich et al., 2020). However, construction of linkage maps and mapping and tagging of genes and QTL still remains very slow in grass pea. Until now, only three linkage maps in grass pea have been developed. The first genetic linkage map of grass pea was constructed by Chowdhury and Slinkard (1999) using F2 population (100 individual) and 75 markers (71 RAPD, 3 isozymes, and 1 morphological markers). The map consists of 14 linkage groups, spanning 898 centimorgans (cM), and the average genetic distance between markers was 17·2 cM. The second genetic linkage map was constructed by Skiba et al. (2004) using backcrossed population (92 individual) derived from a cross between Ascochyta blight resistance accessions (ATC80407) and susceptible accessions (ATC 80407). They assigned the 64 markers (47 RAPD, 7 STMS, and 13 STS/CAPS markers) into nine linkage groups, with a map distance of 803.1 cM, and average genetic distance between markers was 15.8 cM (Skiba et al., 2004). Stem resistance to Ascochyta blight during the seedling stage was subjected to QTL mapping. This process encompassed single-marker analysis, as well as both simple and composite interval mapping techniques, leading to the identification of two distinct QTLs associated with Ascochyta blight in grass pea. The third linkage map was constructed in *Lathyrus cicera*, a close relative of grass pea. This map was constructed based on 307 RNA sequencing–derived markers (E-SSR, SNPs, ITAP) in a 103 RIL population derived from a cross between two *Lathyrus cicera* genotypes, BGE008277 (susceptible to rust) and BGE023542 (partially resistant to rust). The resulting map consists of 7 major and 2 minor linkage groups, spanning a cumulative map distance of 724.2 cM, with an average marker density of one marker per 2.4 cM (Santos et al., 2018). The estimated number of linkage groups in these maps was more than the seven haploid chromosome number of grass pea, indicating poor genome coverage by these markers. Now, the WGS (whole-genome sequence) of grass pea is published (Emmrich et al., 2020) and available in public domain (accession no. PRJEB33571), making it possible to develop a large number of molecular markers directly from the grass pea genome and to use them for construction of high-density genetic linkage map. This would facilitate in mapping and tagging of genes and QTL underlying traits of interest in grass pea.

10.7 GENOMIC RESOURCES DEVELOPED IN GRASS PEA

From the last one decade, major progress has been done on generation of genomic resource in grass pea. As of June 1, 2023, in the NCBI GenBank (http://www.ncbi.nlm.nih.gov/GenBank/) 118 transcriptome sequences and two whole-genome assemblies of grass pea are available. These genomic resources are invaluable for the grass pea community. The recently identified SSR and SNP markers derived from these genomic resources hold promise for enhancing grass pea through genetic breeding endeavors.

10.8 TRANSCRIPTOME SEQUENCES

Hao et al. (2017) performed RNA-Seq using the Illumina NextSeqTM 500 platform in two grass pea (*L. sativus*) accessions, viz., RQ23 from Africa and RQ36 from Europe, using RNA from mixtures of root, stem, and leaf in the seedling stage. The RNA sequencing data was assembled into 1,970,104 contigs, resulting in 142,053 transcripts and 27,431 unigenes. Unigenes exhibited an average length of 1,250 bp, with the maximum reaching 48,515 bp. Utilizing these unigenes, researchers detected 3,204 EST-SSR and 146,406 SNP variations. Validation was conducted on a total of 284 SSR markers and 50 SNP loci across 43 grass pea accessions from diverse geographical regions. Among these, 87 polymorphic EST-SSR primers and 40 KASP markers were successfully identified.

10.9 ASSOCIATION MAPPING IN GRASS PEA

Unlike linkage mapping, association mapping relies on the presence of natural variation in the germplasm collection, where several alleles for a particular trait will be segregated. Availability of fully assembled reference genome sequence of grass pea is prerequisite for association mapping study to map the genotypic data along chromosomes and calculate the LD patterns across the genome. Pea is most closely related to grass pea, and with the recent availability of reference genome sequence of pea (Kreplak et al., 2019), few studies have been performed on genome-wide association mapping in grass pea. One recent attempt looked for fusarium wilt resistance using 161 grass pea accessions from worldwide collection and diversity arrays technology sequencing (DArTseq) and Beijing Genomic Institute–based SNP markers. Across the various genotypes, a spectrum of responses to *Fusarium oxysporum f.* sp. *pisi* (Fop) race 2 was observed, ranging from high resistance to partial resistance and susceptibility. This assessment was carried out under controlled conditions, leading to the identification of 17 noteworthy marker–trait associations through GWAS. This implies that the observed resistance to Fop race 2 is likely influenced by multiple genes. In the absence of a grass pea reference genome sequence, the researchers employed the reference genome of peas to map the corresponding SNPs. Among these SNPs, seven were successfully mapped onto pea chromosomes 1, 6, and 7. The suggested candidate genes are associated with secondary metabolism, amino acid metabolism, RNA transcription regulation, transport, and developmental processes (Sampaio et al., 2021). The same grass pea accessions (Sampaio et al., 2021) were also used for screening rust resistance, and sources of partial rust resistance were identified (Martins et al., 2022) among these accessions. A comprehensive study employing genome-wide association analysis was conducted to pinpoint the genetic locus responsible for partial rust resistance in grass pea. This investigation utilized diversity arrays technology sequencing (DArTseq) and SNP markers from the Beijing Genomic Institute. Seven SNP markers exhibited notable associations with rust disease severity, suggesting a multifactorial basis for partial rust resistance. Additionally, the study proposed 19 potential candidate genes linked to rust resistance, encompassing genes encoding leucine-rich repeat and NB-ARC domains, as well as the TGA transcription factor family (Martins et al., 2022).

10.10 ALLELE MINING

Allele mining is a procedure of finding all alleles of a fully characterized gene available in a germplasm collection. Strategies like EcoTILLING and sequence-based allele mining are commonly used for exploring alleles in the germplasm collection that comprises landraces, breeding lines, varieties, and wild relatives. Collections of Lathyrus germplasm have been made worldwide and are being maintained in the gene bank of several countries, such as Lebanon, France, Bangladesh, Chile, Ukraine, Russian Federation, Australia, Canada, USA, Germany, Spain, and India. ICARDA in Lebanon (4,477 accessions) (https://www.genebanks.org/) and Conservatoire Botanique National des Pyre´ne´es et de Midi-Pyre´ne´es (BP 70315) in France (4477 accessions)

(Vaz Patto and Rubiales, 2014) hold the world's largest collection of Lathyrus accessions. These germplasm collections can serve as an important resource for developing EcoTILLING platform of grass pea. EcoTILLING (ecotype targeting–induced local lesions in genome) is similar to TILLING, except that it is used for the identification of naturally occurring polymorphism/ novel alleles in a gene of interest from germplasm collection as opposed to induced mutations as in TILLING. EcoTILLING has been used successfully in plants and animals to discover SNP/ or small InDels. It was used for the first time in 2004 by Comai and his co-workers in the natural population of Arabidopsis. Technically, EcoTILLING involves procurement of geographically, morphologically, and genetically diverse accessions from the global diversity collections center of the crop. These accessions are then grown in the field to collect tissues for DNA isolation and seed from the plant used for DNA isolation to produce single-seed descent lines for further phenotyping. The isolated DNA is arrayed in a 96-well format followed by DNA pooling of reference and target genotypes in a 1:1 ratio. Next is performing PCR using pooled DNA as a template and gene-specific primers, followed by heteroduplex formation using heating and cooling of PCR product. *CelI* digestion is required to cleave heteroduplex mismatches. The cleaved product undergoes analysis within an automated capillary electrophoresis system that employs fluorescence-based technology to detect polymorphic variations. EcoTILLING facilitates rapid screening of extensive germplasm collections to uncover naturally existing SNPs or minor InDels within the target gene. This technique accelerates SNP discovery, resulting in time and cost savings, as it negates the necessity for sequencing all individuals within a population to identify polymorphism. Only a confirmatory sequencing is required in those individual which exhibited polymorphism by EcoTILLING. In a highly heterozygous species, EcoTILLING can also be used to determine the level of heterozygosity within a gene (Barkley and Wang, 2008). Therefore, the EcoTILLING approach is highly suitable in crops like grass pea, where outcrossing by bees can go up to 27.8% (Rahman et al., 1995). Since EcoTILLING is a reverse genetic approach, prior sequence information of the genes and its regulatory region for the trait of interest should be known. The grass pea genome has now been sequenced (Emmrich et al., 2020), and few genes related to ODAP biosynthesis pathway and its regulation have been characterized, viz., Oxalyl-CoA synthetase (Kushwah et al., 2022; Goldsmith et al., 2022b), ODAP synthase (Goldsmith et al., 2022a; Kushwah et al., 2022) in Genbank (GenBank:MZ127288), and beta-cyanoalanine synthase (Xu et al., 2017; Song et al., 2021). This rise in sequence information paved the way for routine integration of TILLING/ EcoTILLING genomic tools for deducing the function of a gene and developing allele-specific markers for introgression of superior allele to suitable genetic background of grass pea. Recently, John Inn Centre, UK, is in the process of developing an EcoTILLING platform of Lathyrus using 435 germplasm lines obtained from Lathyrus global diversity collections ICARDA.

10.11 FUTURE PERSPECTIVES

Over the past few years, substantial endeavors have been dedicated to advancing the creation of genomic assets for grass pea. As a result, a fair amount of transcriptome resources from various tissues and various lines of grass pea has been generated, and assembled genome sequence of grass pea has now been available in the public domain. These genomic resources will serve as important resource for gene annotations, trait discovery, and markers development. Genome sequence of grass pea has also opened the doors for CRISPR/Cas-based genome editing for precisely manipulating the genome of grass pea for the desired trait. CRISPR/Cas technology is particularly important for reducing or eliminating the β-ODAP in the edible portion of the grass pea. Besides generating genomic resources, now focus should be more on the development of mapping populations, mutant population, genetic transformation protocol, and TILLING/EcoTILLING platform for identifying and validating the functions of useful genes (for multiple stress-tolerant, health-promoting compounds, etc.) present in the grass pea for general crop improvement and human health.

REFERENCES

Abd El Moneim AM, Van Dorrestein B, Baum M et al. (2001) Role of ICARDA in improving the nutritional quality and yield potential of grass pea (*Lathyrus sativus* L.), for subsistence farmers in dry areas. *Lath Lath Newsl* 2:55–58.

Ali M, Kumar S (2009) Major technological advances in pulses: Indian scenario. In: Ali M, Kumar S (eds) *Milestones in Food Legumes Research*. Indian Institute of Pulses Research, Kanpur, India, pp. 1–20.

Almeida NF, Krezdorn N, Rotter B et al. (2015) Lathyrus sativus transcriptome resistance response to Ascochyta lathyri investigated by deepSuperSAGE analysis. *Front Plant Sci* 6:178.

Arslan M, Basak M, Aksu E, Uzun B, Yol E (2020) Genotyping of Low β-ODAP grass pea (Lathyrus sativus L.) germplasm with EST-SSR markers. *Braz Arch Biol Technol* 63:e20190150.

Asmussen CB, Liston A (1998) Chloroplast DNA characters, phylogeny, and classification of Lathyrus (Fabaceae). *Am J Bot* 85:387–401.

Barpete S, Gupta P, Khawar KM, Ozcan S, Kumar S (2020) In vitro approaches for shortening generation cycles and faster breeding of low N-oxalyl-L-diaminopropionic acid content of grass pea (*Lathyrus sativus* L.). *Fresenius Environ Bull* 2698–2706.

Badr A, ElShazly H, ElRabey H et al. (2002) Systematic relationships in *Lathyrus* sect. Lathyrus (Fabaceae) based on amplified fragment length polymorphism (AFLP) data. *Can J Bot* 80:962–969.

Badr SF (2006) Karyotype analysis and chromosome evolution in species of *Lathyrus* (Fabaceae). *Cytologia* 71:447–455.

Barik DP, Acharya L, Mukherjee AK, Chand PK (2007) Analysis of genetic diversity among selected grass pea (*Lathyrus sativus* L.) genotypes using RAPD markers. *Z Naturforsch C J Biosci* 62:869–874.

Barkley NA, Wang ML (2008) Application of TILLING and EcoTILLING as reverse genetic approaches to elucidate the function of genes in plants and animals. *Curr Genom* 9(4):212–226.

Barpete S, Gupta P, Khawar KM et al. (2021) Effect of cooking methods on protein content and neurotoxin (β-ODAP) concentration in grass pea (*Lathyrus sativus* L.) grains. *CyTA – J Food* 19(1):448–456.

Belaid Y, Chtourou-Ghorbel N, Marrakchi M, Trifi-Farah, N (2006) Genetic diversity within and between populations of the Lathyrus genus (Fabaceae) revealed by ISSR markers. *Genet Resour Crop Evol* 53:1413–1418.

Campbell CG (1988) Improvement of Lathyrus sativus L: In: Spencer PS (ed) *The Grasspea, Threat and promise. Proc. Intern. Net Work Workshop for the improvement of Latbyrus sativus and Eradication of Lathyrism, London, 5–8 May, 1988*. Third World Medical Research Foundation, New York, pp. 139–145.

Campbell CG (1997) *Grass Pea (Lathyrus sativus L.), Promoting the Conservation and Use of Underutilized and Neglected Crops*. Institute of Plant Genetic and Crop Plant Research. Gatersleben/International Plant Genetic Resources Institute, Rome, Italy, vol. 18.

Campbell CG, Briggs CJ (1987) Registration of low neurotoxin content Lathyrus germplasm LS 8246. *Crop Sci* 27:821.

Campbell CG, Mehra RB, Agrawal SK et al. (1994) Current status and future research strategy in breeding grasspea (*Lathyrus sativus*). *Euphytica* 73:167–175.

Chapman MA (2015) Transcriptome sequencing and marker development for four underutilized legumes. *Appl Plant Sci* 3:1400111.

Chowdhury M, Slinkard A (1999) Linkage of random amplified polymorphic DNA, isozyme and morphological markers in grasspea (*Lathyrus sativus*). *The J Agric Sci* 133(4):389–395.

Chtourou-Ghorbel N, Lauga B, Combes D, Marrakchi, M (2001) Comparative genetic diversity studies in the genus Lathyrus using RFLP and RAPD markers. *Lathyrus Lathyrism Newsl* 2:62–68.

Crino P, Polignano GB, Tavoletti S (2004) Grass pea–a potentially important crop in Mediterranean agriculture. *Grain Legumes* 40:6–7.

Croft A, Pang E, Taylor P (1999) Molecular analysis of *Lathyrus sativus* L. (grasspea) and related Lathyrus species. *Euphytica* 107:167–176.

Das A, Parihar, AK, Barpete S, Kumar S. Gupta S (2021) Current perspectives on reducing the β-ODAP content and improving potential agronomic traits in grass pea (*Lathyrus sativus* L.). *Front Plant Sci* 12:703275.

Dixit GP, Parihar AK, Bohra A et al. (2016) Achievements and prospects of grass pea (*Lathyrus sativus* L.) improvement for sustainable food production. *The Crop Journal* 4:407–416. https://doi.org/10.1016/j.cj.2016.06.008

Emmrich PMF, Sarkar A, Njaci I, Sarkar A et al. (2020) A draft genome of grass pea (Lathyrus sativus), a resilient diploid legume. *BioRvix* [preprint], 2020–04. https://doi.org/10.1101/2020.04.24.058164

FAO (2009) *International Treaty on Plant Genetic Resources for Food and Agriculture*. Food and Agriculture Organization.

Fernández-Aparicio M, Flores F, Rubiales D (2012) Escape and true resistance to crenate broomrape (*Orobanche crenata* Forsk.) in grass pea (*Lathyrus sativus* L.) germplasm. *Field Crops Res* 125:92–97.

Fikre A, Korbu L, Kuo YH et al. (2008) The contents of the neuro-excitatory amino acid beta-ODAP (beta-N-oxalyl-L-alpha, beta-diaminopropionic acid), and other free and protein amino acids in the seeds of different genotypes of grass pea (*Lathyrus sativus* L.). *Food Chem* 110:422–427.

Gautam PL, Singh IP, Karihaloo JL (1998) Need for a crop network on Lathyrus genetic resources for conservation and use. In: Mathur PN, Ramanatha Rao V, Arora RK (eds) *Lathyrus Genetic Resources Network (Proceedings of the IPGRI-ICARDA-ICAR Regional Working Group Meeting, 8–10 December 1997, New Delhi, India)*. International Plant Genetic Resources Institute-Office for South Asia, New Delhi, pp. 15–21.

GCDT (2009) *Strategy for the Ex Situ Conservation of Lathyrus, with Special Reference to Lathyrus Sativus, L. cicera, and L. ochrus*. Global Crop Diversity Trust (GCDT), Rome, Genetics, IARI, New Delhi, India, pp. 1–7.

Goldsmith M, Barad S, Knafo M, Savidor A, Ben-Dor S et al. (2022a) Identification and characterization of the key enzyme in the biosynthesis of the neurotoxin β-ODAP in grass pea. *J Biol Chem* 298(5):101806.

Goldsmith M, Barad S, Peleg Y, Albeck S, Dym O, Brandis A, Mehlman T, Reich Z (2022b) The identification and characterization of an oxalyl-CoA synthetase from grass pea (*Lathyrus sativus* L.). *RSC Chem Biol* 3(3):320–333.

Gonçalves L, Rubiales D, Bronze MR, Vaz Patto MC (2022) Grass pea (*Lathyrus sativus L.*) – a sustainable and resilient answer to climate challenges. *Agronomy* 12(6):1324.

Granati E, Bisignano V, Chiaretti D et al. (2003) Characterization of Italian and exotic *Lathyrus* germplasm for quality traits. *Genet Resour Crop Evol* 50:273–280.

Grela E, Rybiński W, Matras J et al. (2012) Variability of phenotypic and morphological characteristics of some *Lathyrus sativus* L. and Lathyrus cicera L. accessions and nutritional traits of their seeds. *Genet Resour Crop Evol* 59:1687–1703.

Gupta P, Udupa SM, Gupta DS, Kumar J, Kumar S (2018) Population structure analysis and determination of neurotoxin content in a set of grass pea (*Lathyrus sativus* L.) accessions of Bangladesh origin. *Crop J* 6:435–442.

Hanbury CD, Siddique KHM (2000) Registration of 'Chalus' *Lathyrus cicera* L. *Crop Sci* 40:1199.

Hanbury CD, Siddique KHM, Galwey NW, Cocks PS (1999) Genotype-environment interaction for seed yield and ODAP concentration of *Lathyrus sativus* L. and *L. cicera* L. in Mediterranean-type environments. *Euphytica* 110:45–60.

Hanbury CD, White CL, Mullan BP et al. (2000) A review of the potential of Lathyrus sativus L. and L. cicera L. grain for use as animal feed. *Anim Feed Sci Technol* 87:1–27.

Hao X, Yang T, Liu R, Hu J, Yao Y, Burlyaeva M et al. (2017) An RNA sequencing transcriptome analysis of grass pea (Lathyrus sativus L.) and development of SSR and KASP markers. *Front Plant Sci* 8:1873.

Heywood V, Casas A, Ford-Lloyd B et al. (2007) Conservation and sustainable use of crop wild relatives. *Agric Ecosyst Environ* 121:245–255.

Hillocks R, Maruthi M (2012) Grass pea (*Lathyrus sativus*): Is there a case for further crop improvement? *Euphytica* 186:647–654.

Jackson MT, Yunus AG (1984) Variation in the grass pea (*Lathyrus sativus* L.) and wild species. *Euphytica* 33:549–559.

Jain HK, Somayajulu PLN, GK Barat (1974) *Final Technical Report of PL 480 Scheme on Investigations on Lathyrus sativus L.* Division of Genetics, IARI, New Delhi.

Joshi M (1998) Status of grass pea (Lathyrus sativus L.) genetic resources in Nepal. In: Mathur PN, Rao VR, Arora RK (eds) *Lathyrus Genetic Resources Network: Proceedings of a IPGRI-ICARDA-ICAR Regional Working Group Meeting, December 8–10, 1997*. National Bureau of Plant Genetic Resources, New Delhi 1998, pp. 22–29 (IPGRI Office for South Asia, New Delhi, India).

Kaul AK, Islam MQ, Hamid A (1986) Screening of Lathyrus germplasm of Bangladesh for BOAA content and some agronomic characters. In: Kaul AK, Combes D (eds) *Lathyrus and Lathyrism*. Third World Medical Research Foundation, New York, pp. 130–141.

Kaul AK, Islam Q (1981) Khesari cultivation in Bangladesh and Lathyrism. *Proceed Nat Workshop on Pulses* 18–19.

Kenicer G, Kajita T, Pennington R et al. (2005) Systematics and biogeography of *Lathyrus* (Leguminosae) based on internal transcribed spacer and cpDNA sequence data. *Am J Bot* 92:1199–1209.

Kenicer G, Nieto-Blásquez EM, Mikić A et al. (2009) *Lathyrus* – diversity and phylogeny in the genus. *Grain Legumes* 54:16–18.

Kislev ME (1989) Origins of the cultivation of *Lathyrus sativus* and *L. cicera* (Fabaceae). *Econ Bot* 43:262–270.

Kreplak J, Madoui MA, Cápal P et al. (2019) A reference genome for pea provides insight into legume genome evolution. *Nat Genet* 51:1411–1422.

Kumar S, Gupta P, Barpete S, Sarker A, Amri A et al. (2013) Grass pea. In: Singh M Upadhyaya HD Bisht IS (eds) *Genetic and Genomic Resources of Grain Legumes Improvement Newnes*. Elsevier, Amsterdam, The Netherlands, pp. 269–293.

Kumar S, Bejiga G, Ahmed S, Nakkoul H, Sarker A (2011) Genetic improvement of grass pea for low neurotoxin (ODAP) content. *Food Chem Toxicol* 49:589–600.

Kumar S, Gupta P, Barpete S et al. (2020) Grass pea. In: Pratap A, Gupta S (eds) *The Beans and the Peas, from Orphan to Mainstream Crops*. Elsevier, Amsterdam, pp. 273–287. eBook ISBN: 9780128214442

Kupicha FK (1983) The infrageneric structure of Lathyrus. *Notes RBG Edinburgh* 41(2):209–244.

Kushwah NS, Shanmugavadievel PS, Das A, Rathore M, Singh A et al. (2022) Identification and characterization of oxalyl-CoA synthetase gene in grasspea (*Lathyurs sativus* L.). *J Food Legumes* 35(1):27–40.

Lal MS, Agrawal I, Chitale MW (1986) Genetic improvement of chickling vetch (*Lathyrus Sativus* L.) in Madhya Pradesh, India. In: Kaul AK, Combes D (eds) *Lathyrus and Lathyrism*. Third World Medical Research Foundation, New York, pp. 146–160.

Lambein F, Travella S, Kuo YH et al. (2019) Grass pea (*Lathyrus sativus* L.): Orphan crop, nutraceutical or just plain food? *Planta* 250(3):821–838.

Lioi L, Sparvoli F, Sonnante G, Laghetti G, Lupo F et al. (2011) Characterization of Italian grass pea (*Lathyrus sativus* L.) germplasm using agronomic traits, biochemical and molecular markers. *Genet Resour Crop Evol* 58:425–437.

Loudon JC, Don G, Wooster D (1855) *Loudon's encyclopædia of plants*. Longman, Brown, Green and Longmans, London.

Martins DC, Rubiales D, Vaz Patto MC (2022) Association mapping of Lathyrus sativus L. disease response to Uromyces PISI reveals novel loci underlying partial resistance. *Front Plant Sci* 13:842545.

Mathur PN, Ramanatha Rao V, Arora RK (1998) *Lathyrus Genetic Resources Network: Proceeding of a IPGRI-ICARDA-ICAR Regional Working Group Meeting*, 8–10 December 1997. NBPGR, IPGRI Office for South Asia, New Delhi, India.

Maxted N, Bennett S (2001) *Plant Genetic Resources of Legumes in the Mediterranean*. Kluwer Academic Publishers, Dordrecht, The Netherlands.

Maxted N, Ford-Lloyd BV, Hawkes JG (1997) Complementary conservation strategies. In: Maxted N, Ford-Lloyd BV, Hawkes JG (eds) *Plant Genetic Conservation: The in situ Approach*. Chapman and Hall, London, pp. 20–55.

Maxted N, Hargreaves S, Kell SP et al. (2012) Temperate forage and pulse legume genetic gap analysis. *Bocconea* 24:115–146.

Mehra RB, Raju DB, Himabindu K (1995) Status paper on *Lathyrus* Research. Division of Genetics, IARI, New Delhi, India, pp. 1–7.

Mekonen DA, Abraham A, Oselebe H et al. (2022) Genetic diversity and population structure analysis of Grass pea (Lathyrus sativus L.) accessions collected from North-Western Ethiopia using SSR markers. *Genet Resour Crop Evol* 69:1247–1260.

Mera M (2010) Inheritance of seed weight in large-seed grass pea *Lathyrus sativus* L. *Chil J Agr Res* 70:357–364.

Mera M, Tay J, France A et al. (2003) Luanco-INIA, a large-seeded cultivar of *Lathyrus sativus* released in Chile. *Lathyrus Lathyrism Newsl* 3:26.

Mikic´ A, Mihailovic VC´ upina, Ethuric´B, Krstic´ D, Vasic´ M et al. (2011) Towards the re-introduction of grass pea (*Lathyrus sativus*) in the West Balkan Countries: The case of Serbia and Srpska (Bosnia and Herzegovina). *Food Chem Toxicol* 49:650–654.

Nosrati H, Hosseinpour-Feizi MA, Nikniazi, M, Razban-Haghighi A (2012) Genetic variation among different accessions of Lathyrus sativus (Fabaceae) revealed by RAPDs. *Botanica Serbica* 36(1):41–47.

Pandey RL, Sharma RN, Chitale MW (1998) Status of *Lathyrus* genetic resources in India. In: Mathur PN, Rao VR, Arora RK (eds) *Lathyrus* Genetic Resources Network. Proc IPGRI-ICARDA-ICAR Regional Working Group Meeting, ICARDA, New Delhi, India, pp. 7–14.

Parihar AK, Barpete S, Das A, Lamichaney A, Gupta S (2022) Lathyrus Breeding. In *Fundamentals of Field Crop Breeding*. Springer Nature, Singapore, pp. 1323–1389.

Polignano GB, Bisignano V, Tmaselli V, Uggenti P, Alba V et al. (2009) Genotype environment interaction in grass pea (*Lathyrus sativus* L.) lines. *Int J Agron* 2009:1–7.

Ponnaiah M, Shiferaw E, Pe ME, Porceddu E (2011) Development and application of EST-SSRs for diversity analysis in Ethiopian grass pea. *Plant Genet Resour* 9(2):276–280.

Przybylska J, Zimniak-Przybylska Z, Krajewski P (1998) Diversity of seed albumins in the grasspea (*Lathyrus sativus* L.): An electrophoretic study. *Genet Resour Crop Evol* 45:423–431.

Rahman MA, Rahman MM, AkhtaruzzamanM (2001) Progress in isolation and purification of Lathyrus sativus breeding lines. *Lathyrus Lathyrism Newsl* 2:39–40.

Rahman MM, Kumar J, Rahaman MA et al. (1995) Natural outcrossing in *Lathyrus sativus* L. *Indian J Genet* 55:204–207.

Rahman MM, Quddus MR, Ali MO, Liu R, Li M et al. (2022) Genetic diversity of Lathyrus sp collected from different geographical regions. *Mol Biol Rep* 49:519–529.

Rathi D, Gayali S, Pareek A, Chakraborty S, Chakraborty N (2019) Transcriptome profiling illustrates expression signatures of dehydration tolerance in developing grasspea seedlings. *Planta* 250:839–855.

Rizvi AH, Sarker A, Dogra A (2016) Enhancing grass pea (*Lathyrus sativus* L.) production in problematic soils of South Asia for nutritional security. *Indian J Genet* 76:583–592.

Rubiales D, Emeran AA, Flores F (2020) Adaptation of grass pea (*Lathyrus sativus*) to Mediterranean environments. *Agronomy* 10:1295.

Rybinski W (2003) Mutagenesis as a tool for improvement of trait in grass pea (*Lathyrus sativus* L.). *Lathyrus Lathyrism Newsl* 3:27–31.

Sacristán M, Varela A, Pedrosa MM, Burbano C, Cuadrado C, Legaz ME, Muzquiz M (2015) Determination of β-N-oxalyl-l-α, β-diaminopropionic acid and homoarginine in Lathyrus sativus and Lathyrus cicera by capillary zone electrophoresis. *J Sci Food Agric* 95(7):1414–1420.

Sampaio A, Margarida A, Mara L, Pereira, P et al. (2021) Grass pea natural variation reveals oligogenic resistance to *Fusarium oxysporum* f. sp. *pisi*. *Plant Genome* 14:20154.

Santos C, Almeida NF, Alves ML, Horres R, Krezdorn N, Leitão ST et al. (2018) First genetic linkage map of *Lathyrus cicera* based on RNA sequencing-derived markers: Key tool for genetic mapping of disease resistance. *Hort Res* 5:1–4.

Sarkar A, Emmrich PM, Sarker A et al. (2019) Grass Pea: Remodeling an ancient insurance crop for climate resilience. In: Kole C (ed) *Genomic Designing of Climate-Smart Pulse Crops*. Springer, Cham, pp. 425–469.

Schaefer H, Hechenleitner P, Santos-Guerra A et al. (2012) Systematics, biogeography, and character evolution of the legume tribe Fabeae with special focus on the middle-Atlantic island lineages. *BMC Evol Biol* 12(1):1–19.

Sethi KL, Mehra RB, Lal BM (1981) Studies on variability and correlation between toxin content and seed weight in *Lathyrus sativus*. L. *Sci Cult* 47(7):263–264.

Shiferaw E, Pe ME, Porceddu E, Ponnaiah M (2012) Exploring the genetic diversity of Ethiopian grass pea (*Lathyrus sativus* L.) using EST-SSR markers. *Mol Breed* 30:789–797.

Siddique KHM, Hanbury CL, Sarker A (2006) Registration of 'Ceora' grass pea. *Crop Sci* 46:986.

Skiba B, Ford R, Pang EC (2004) Construction of a linkage map based on a *Lathyrus sativus* backcross population and preliminary investigation of QTLs associated with resistance to ascochyta blight. *Theor Appl Genet* 109(8):1726–1735.

Skiba B, Gurung AM, Pang EC (2007) Genome mapping and molecular breeding in Lathyrus. In: Kole C (ed) *Pulses, Sugar and Tuber Crops*. Springer Science & Business Media, Berlin, Heide, pp. 123–132.

Smartt J (1984) Evolution of grain legumes. 1. Mediterranean pulses. *Exp Agric* 20:275–296.

Smýkal P, Kenicer G, Flavell AJ et al. (2011) Phylogeny, phylogeography and genetic diversity of the *Pisum* genus. *Plant Genet Resour* 9:4–18.

Song Y, Wang L, Liu F, Jiao C, Nan H, Shen X, Chen H, Li Y, Lei B, Jiang J, Chen P, Xu Q. (2021) β-Cyanoalanine Synthase regulates the accumulation of β-ODAP via interaction with serine acetyltransferase in *Lathyrus sativus*. *J Agric Food Chem* 69(6):1953–1962.

Soren KR, Yadav A, Pandey G, Gangwar P, Parihar AK, Bohra A, Dixit GP, Datta S, Singh NP (2015) EST-SSR analysis provides insights about genetic relatedness, population structure and gene flow in grass pea (*Lathyrus sativus*). *Plant Breed* 134:338–344.

Sun XL, Yang T, Guan JP, Ma Y, Jiang JY, Cao R, Burlyaeva M, Vishnyakova M, Semenova E, Bulyntsev S, Zong XX (2012) Development of 161 novel EST-SSR markers from Lathyrus sativus (Fabaceae). *Am J Bot* 99(10):e379–e390.

Tadesse W, Bekele E (2003) Variation and association of morphological and biochemical characters in grass pea (*Lathyrus sativus* L.). *Euphytica* 130:315–324.

Talukdar D (2009a) Recent progress on genetic analysis of novel mutants and aneuploid research in grass pea (*Lathyrus sativus* L.). *Afr J Agric Res* 4(11):1549–1559.

Talukdar D (2009b) Dwarf mutations in grass pea (*Lathyrus sativus* L.): Origin, morphology, inheritance and linkage studies. *J Genet* 88:165–175.

Tamburino R, Guida V, Pacifico S, Rocco M, Zarelli A, Parente A, Di Maro A (2012) Nutritional values and radical scavenging capacities of grass pea (*'Lathyrus sativus'* L.) seeds in Valle Agricola district, Italy. *Aust J Crop Sci* 6(1):149–156.

Tan RY, Xing GY, Zhou GM, Li FM, Hu WT, Lambein F et al. (2017) Plant toxin β-ODAP activates integrin β1 and focal adhesion: A critical pathway to cause neurolathyrism. *Sci Rep* 7:40677.

Tavoletti S, Iommarini L (2007) Molecular marker analysis of genetic variation characterizing a grass pea (*Lathyrus sativus*) collection from central Italy. *Plant Breed* 126:607–611.

Tavoletti S, Iommarini L, Crinó P, Granati E (2005) Collection and evaluation of grasspea (*Lathyrus sativus* L.) germplasm of central Italy. *Plant Breed* 124:388–391.

Tay J, Valenzuela A, Venegas F (2000) Collecting and evaluating Chilean germplasm of grasspea (*Lathyrus sativus* L.). *Lathyrus Lathyrism Newsl* 1:21.

Urga K, Fufa H, Biratu E, Husain A (2005) Evaluation of Lathyrus sativus cultivated in Ethiopia for proximate composition, minerals, β-ODAP and anti-nutritional components. *Afr J Food Agric Nutr Dev* 5(1).

Vavilov NI (1951) The origin, variation, immunity and breeding of cultivated plants (Translated by S. K. Chestitee). *Chron Bot* 13:1–366.

Vaz Patto MC, Hanbury CD, Moorhem MV et al. (2011) Grass pea. In: de la Vega MP, Torres AM, Cubero JI, Kole C (eds) *Genetics, Genomics and Breeding of Cool Season Grain Legumes*. Science, Enfield, pp. 151–204.

Vaz Patto MC, Rubiales D (2009) Identification and characterization of partial resistance to rust in a germplasm collection of *Lathyrus sativus* L. *Plant Breed* 128:495–500.

Vaz Patto MC, Rubiales D (2014) Lathyrus diversity: Available resources with relevance to crop improvement–*L. sativus* and *L. cicera* as case studies. *Ann Bot* 113(6):895–908.

Vaz Patto MC, Skiba B, Pang ECK et al. (2006) Lathyrus improvement for resistance against biotic and abiotic stresses: From classical breeding to marker assisted selection. *Euphytica* 147:133–147.

Wang F, Yang T, Burlyaeva M, Li L, Jiang J, Fang L et al. (2015) Genetic diversity of grasspea and its relative species revealed by SSR markers. *PLoS One* 10(3):e0118542.

Xiong JL, Xiong YC, Bai X et al. (2015) Genotypic variation in the concentration of β-N-Oxalyl-L-α, β-diaminopropionic acid (β-ODAP) in Grass Pea (*Lathyrus sativus* L.) seeds is associated with an accumulation of leaf and pod β-ODAP during vegetative and reproductive stages at three levels of water stress. *J Agric Food Chem* 63(27):6133–6141.

Xu Q, Liu F, Chen P, Jez JM, Krishnan HB (2017) β-N-Oxalyl-l-α, β-diaminopropionic Acid (β-ODAP) Content in Lathyrus sativus: The integration of nitrogen and sulfur metabolism through β-cyanoalanine synthase. *Int J Mol Sci* 18:526.

Xu Q, Liu F, Qu R, Gillman JD, Bi C, Hu X et al. (2018) Transcriptomic profiling of Lathyrus sativus L. metabolism of β-ODAP, a neuroexcitatory amino acid associated with neurodegenerative lower limb paralysis. *Plant Mol Biol Rep* 36:832–843.

Yang T, Jiang J, Burlyaeva M, Hu J, Coyne CJ, Kumar S et al. (2014) Large-scale microsatellite development in grasspea (*Lathyrus sativus* L.), an orphan legume of the arid areas. *BMC Plant Biol* 14:65.

Yunus AG, Jackson MT (1991) The gene pool of the grass pea (*Lathyrus sativus* L). *Plant Breed* 106:319–328.